Statistical Thermodynamics for Beginners

Statistical Thermodynamics for Beginners

Howard D Stidham

University of Massachusetts Amherst, USA

World Scientific

NEW JERSEY · LONDON · SINGAPORE · BEIJING · SHANGHAI · HONG KONG · TAIPEI · CHENNAI · TOKYO

Published by

World Scientific Publishing Co. Pte. Ltd.

5 Toh Tuck Link, Singapore 596224

USA office: 27 Warren Street, Suite 401-402, Hackensack, NJ 07601

UK office: 57 Shelton Street, Covent Garden, London WC2H 9HE

Library of Congress Cataloging-in-Publication Data

Names: Stidham, Howard D., author.

Title: Statistical thermodynamics for beginners / Howard D. Stidham
   (University of Massachusetts Amherst, USA).

Description: Hackensack, NJ : World Scientific, [2016] | Includes bibliographical references.

Identifiers: LCCN 2016040516| ISBN 9789813149922 (hardcover ; alk. paper) |
   ISBN 9813149922 (hardcover ; alk. paper) | ISBN 9789813149939 (pbk. ; alk. paper) |
   ISBN 9813149930 (pbk. ; alk. paper)

Subjects: LCSH: Statistical thermodynamics.

Classification: LCC QC311.5 .S485 2016 | DDC 536/.7015195--dc23

LC record available at https://lccn.loc.gov/2016040516

British Library Cataloguing-in-Publication Data

A catalogue record for this book is available from the British Library.

Typeset by Stallion Press

Email: enquiries@stallionpress.com

Preface

This book is intended to serve as a relatively thorough introduction to statistical mechanics for interested students who can regard themselves as beginners in the area. It is assumed that the reader has an abiding interest in mathematics and physics, and has had introductory courses in advanced calculus, thermodynamics, and quantum mechanics. The book is based on a series of notes developed over approximately 50 years of teaching the subject to first and second year graduate students and to some fortunate undergraduates. There is much more material than can be reasonably presented in one semester, developed bit by bit over the years by the artifice of reserving the latter part of the course for coverage of material not included in previous years.

The book is organized by chapter only, but has an unstated structure. The first 5 chapters are concerned with the fundamentals of statistical mechanics. The next 4 chapters are concerned with calculation of thermodynamic properties from spectroscopic data. This is followed by three chapters on polymers of increasing difficulty. The chapter on excluded volume is presented near the research level. The remaining four chapters cover quantum statistics, adsorption, heat capacity of crystals, and imperfect gases and liquids, not in this order. Although the book is intended as a textbook, there are relatively few problems at the ends of the chapters. Some chapters have no problems at all. The reason for this is that graduate students are expected to do research, generally working in another field to which statistical mechanics is not directly applicable, and do not have much time to solve routine problems. In addition, there are many very good problems in other books on statistical thermodynamics. McQuarrie's book is an especially rich source of endless problems, should an instructor feel motivated to augment the text. The problems given are largely intended to emphasize some point in the text.

A close reading of the book will reveal the great debt owed T. L. Hill, W. H. Stockmayer, Marshall Fixman and E. A. DiMarzio. In particular, the lectures given by Marshall at Yale in 1965 resulted in my notes on which I based several developments in the text, notably the Gaussian probability for the infinite polymer chain, the transfer method in the helix-coil transition, and the argument for ensemble equivalence. It is probably impossible to surpass Hill's *Introduction to Statistical Thermodynamics*, and this book does not try. But Hill's book is out of print, and is no longer useful as a textbook since it is of undependable availability. There are a number of textbooks presently available, each with its own strengths and weaknesses. This book is a unique addition, containing some material not easily available elsewhere.

Any serious student of statistical mechanics will sooner or later run into a need to consult some of the classic works in the area. A brief list of some of the prominent books follows. In particular, the book by Mayer and Mayer has some surprising inclusions that are not often seen in modern applications, but deserve inspection. In his lectures on statistical mechanics at Yale in the 1960's, Lars Onsager always came to lecture carrying a copy of Mayer and Mayer, which he never consulted, but which always occupied a prominent place on the lecture table. It is especially recommended.

The list of books that should be available to a practicing statistical mechanician includes several of a mathematical nature. Those from my personal collection are:

M. Abramowitz and I. A. Stegun, Nat'l Bur. Stds. AMS 55, 1964, 5$^\text{th}$ Printing.

J. H. Hildebrand, "Advanced Calculus for Engineers," Prectice-Hall, New York, 1949.

L. B. W. Jolley, "Summation of Series," Dover reprint.

E. T. Whittaker, "Analytical Dynamics,: Dover reprint.

E. T. Whittaker and G. N. Watson, "A Course of Modern Analysis," 4th Ed., Cambridge, 1927.

In addition, there are a number of books on statistical mechanics that are useful. These include, but are not limited to, the following:

W. Band, "Introduction to Quantum Statistics," Van Nostrand, New York, 1955.

M. Born and K. Huang, "Dynamical Theory of Crystal Lattices," Oxford, 1954.

D. Chandler, "Introduction to Modern Statistical Mechanics," Oxford, 1987.

N. Davidson, "Statistical Mechanics," McGraw-Hill, New York, 1962.

P. J. Flory, "Principles of Polymer Chemistry," Cornell, Ithaca NY, 1953.

R. H. Fowler and E. A. Guggenheim, "Statistical Thermodynamics," Cambridge, 1939.

J. W. Gibbs, "The Collected Works of J. Willard Gibbs," Vols. I and II, Yale, 1948.

D. ter Haar, "Elements of Statistical Mechanics," Constable, London, 1954.

D. ter Haar, "Elements of Thermostatistics," 2nd Ed., Holt, Rinehart and Winston, New York, 1966.

T. L. Hill, "Statistical Mechanics," McGraw-Hill, New York, 1956.

T. L. Hill, "Thermodynamics of Small Systems," Benjamin, New York, 1963.

T. L. Hill, "Introduction to Statistical Thermodynamics," Addison-Wesley, Reading, MA, 1960.

T. L. Hill, "Cooperativity Theory in Biochemistry," Springer Verlag, New York, 1985.

J. O. Hirschfelder, C. F. Curtiss and R. B. Bird, "Molecular Theory of Gases and Liquids," Wiley, New York, 1954.

L. D. Landau and E. M. Lifshitz, "Statistical Physics," Pergamon, London, 1958.

J. E. Mayer and M. G. Mayer, "Statistical Mechanics," Wiley, New York, 1940.

D. A. McQuarrie, "Statistical Mechanics," Harper and Row, New York, 1976.

D. A. McQuarrie, "Statistical Thermodynamics," Harper and Row, New York, 1973.

L. K. Nash, "Elements of Statistical Thermodynamics," Addison-Wesley, Reading MA, 1968.

G. S. Rushbrooke, "Introduction to Statistical Mechanics," Oxford, 1949.

E. Schrodinger, "Statistical Mechanics,"Cambridge, 1960.

J. C. Slater, "Introduction to Chemical Physics," McGraw-Hill, New York, 1939.

M. Toda, R. Kubo, N. Saito, "Statistical Physics," 2nd Ed., Springer Verlag, Berlin, 1983, 1992.

R. C. Tolman, "Principles of Statistical Mechanics," Oxford, 1938.

G. H. Wannier, "Elements of Solid State Theory," Cambridge, 1959.

One important area not covered in the first edition of this book is phase transitions. The interested reader may find it useful to consult good textbooks of physical chemistry for the groundwork and then graduate to such books as the following:

H. E. Stanley, "Introduction to Phase Transitions and Critical Phenomena,"Oxford, 1971.

J. M. Yoemans, "Statistical Mechanics of Phase Transitions," Clarendon Press, Oxford, 1992.

"New Kinds of Phase Transitions," V. V. Brazhkin, S. V. Buldyrev, V. N. Ryzhov, H. E. Stanley, Eds., NATO Science Series, Kluwer, Dordrecht/Boston, London, 2002.

Amherst, Massachusetts

May 2016

Contents

Chapter 1

Review of Classical Mechanics and Thermodynamics

1.1. Introduction

Spectroscopy and statistical mechanics developed nearly simultaneously. Max Planck developed the theory of the black body radiator, introducing for the first time the idea of quantized oscillators. A little later, Albert Einstein took the idea of quantized oscillation and developed an essentially correct theory of the heat capacity of substances. Independently of the European intellectual ferment, J. Willard Gibbs in the USA introduced the ensemble and investigated its properties. By 1906, all of these theoretical constructs were in the literature, waiting for applications. Classical mechanics was fully developed, but quantum mechanics took another 20 years to appear. Thermodynamics had been long since summarized, and produced many exact equivalences. For example, the constant pressure and constant volume heat capacities of substances are related by a well-known formula involving commonly measured quantities such as the bulk modulus of elasticity (the reciprocal of the isothermal compressibility) and the coefficient of thermal expansion. However, thermodynamics provided no insight into the reasons for the behavior of bulk matter, only relationships between various coefficients. Quantum mechanics provided a basis for understanding spectroscopy, but the understanding was, and is, model dependent. Thermodynamics is model independent. The object of statistical mechanics is to provide a model independent bridge between the behavior of bulk matter and that of the manifold quantum systems of which bulk matter is composed. In order to do this, it is necessary to assume that the ensemble average is equivalent to the long time average of a measurement, something called the ergodic hypothesis. This is a reasonable assumption, but it is one that is unprovable. Ultimately, the justification for the use of the ergodic hypothesis is that the statistical mechanical formulas to which it leads actually do produce quantitative agreement with experiment. It is interesting that the quantum behavior of isolated molecules is completely suppressed in bulk matter. The reason for this is partly that individual quantum systems fluctuate about a central value when in bulk, and that it takes an astronomical number of molecules to make up one mole of bulk matter. The fact that this number is so large is fortunate, as it leads to statistically valid averaging procedures that solve the problem of calculating the properties of bulk matter from those of the quantum systems on which these are based.

This book assumes a reasonable introduction to thermodynamics, classical and quantum mechanics on the part of the reader. The book starts with a brief review of classical mechanics, then of thermodynamics. The next several chapters are concerned with the postulates of statistical mechanics. Applications begin with gases, heteroatomic diatomic molecules, homonuclear diatomic molecules and polyatomic molecules, in which the quantum mechanics and spectroscopy are regarded as completed studies. Three chapters of graded difficulty follow on polymers, then further investigations of the structure of statistical mechanics with the introduction of phase space and quantum statistics. Imperfect gases and liquids are briefly considered, and applications to chemical equilibrium, adsorption, and solid state finish the book.

1.2. Review of Classical Mechanics

The classical equations of motion are familiar in Cartesian coordinates, but usually less so in either LaGrange's or Hamilton's formulation. A brief study of the relationship between these different ways of essentially saying the same thing is often profitable.

In Cartesian coordinates, the kinetic energy and any potential energy that may be acting in the system are expressed in either coordinates x_i or velocities v_i. It is convenient to express the velocities formally as the time derivatives of the corresponding coordinate, that is, as $\dot{x}_i$, where the dot denotes time differentiation. The kinetic energy is

$$T = \frac{1}{2} \sum_i m_i \dot{x}_i^2 \tag{1.1}$$

Differentiation with respect to the velocity produces the linear momentum

$$p_i = \frac{\partial T}{\partial \dot{x}_i} = m_i \dot{x}_i \tag{1.2}$$

where m_i is the ith mass. Newton defined force as the time rate of change of the momentum. Time differentiation yields the force on the particle of mass m_i

$$F = \frac{\partial}{\partial t} \frac{\partial T}{\partial \dot{x}_i} = m_i \ddot{x}_i \tag{1.3}$$

If a potential $V(\mathbf{x})$ acts, the force on the particle is given by the negative gradient of the potential, and the equations of motion are obtained by setting the two forces equal.

$$m_i \ddot{x}_i = -\frac{\partial V(\mathbf{x})}{\partial x_i} \tag{1.4}$$

provided no forces not derivable from a potential (such as those of a frictional nature) act. Here, the bold-faced x represents the entire set of Cartesian coordinates x_i. If a frictional force F' acts and needs to be taken into account, the equation becomes

$$m_i \ddot{x}_i + \frac{\partial V(\mathbf{x})}{\partial x_i} = F_i' \tag{1.5}$$

Note that the momentum is defined in this approach by the velocity derivative of the kinetic energy, and the energy is the basic quantity from which all else flows. The same is true for both LaGrange's and Hamilton's formulation. If generalized coordinates g_i are introduced, and each Cartesian coordinate is expressed as a function of these generalized coordinates alone, then the equations of motion are more complex, and become

$$\frac{\partial}{\partial t}\frac{\partial T}{\partial \dot{g}_i} = -\frac{\partial V(\boldsymbol{g})}{\partial g_i} + \frac{\partial T}{\partial g_i} \qquad (1.6)$$

If forces not derivable from a potential are involved, these must be added to the right side of the equation, after conversion into generalized coordinates. These are LaGrange's equations of motion, and the relation is proven in the Appendix.

In LaGrange's approach, both the kinetic and the potential energy are combined in a single function, the Lagrangian, L. The definition is sublimely simple, but renders treatment of some curvilinear coordinate systems relatively simple, rather than the algebraic nightmare that often ensues in the Cartesian system. The definition is

$$L(\dot{g}_i, g_i) = T(\dot{g}_i, g_i) - V(g_i) \qquad (1.7)$$

Both the kinetic energy and the potential energy must be expressed as functions of the set of basis coordinates chosen. Thus L must be expressed as a function of the generalized coordinates and their velocities. When this is done, the generalized momenta are defined as

$$p_i = \frac{\partial L}{\partial \dot{g}_i} \qquad (1.8)$$

Including the forces not derivable from a potential F', the equations of motion may be written

$$\frac{\partial}{\partial t}\frac{\partial L}{\partial \dot{g}_i} - \frac{\partial L}{\partial g_i} = F' \qquad (1.9)$$

If there are no forces not derivable from a potential, then set F' equal to zero.

In terms of these momenta and the corresponding coordinates q, the Hamiltonian is defined as

$$H(\boldsymbol{p}, \boldsymbol{q}) = \sum_s p_s q_s(\boldsymbol{p}, \boldsymbol{q}) - L(\boldsymbol{p}, \boldsymbol{q}) \tag{1.10}$$

Note that the Hamiltonian has to have the generalized coordinates expressed as functions of both the momenta and the coordinates. That is, the Lagrangian in Equation (1.10) has to be re-expressed in terms of p and q, not q and $\dot{q}$. The momentum is still the velocity derivative of the Lagrangian function. The Hamiltonian defined in this way proves to be equal to the energy,

$$H = T + V \tag{1.11}$$

Velocity and force are defined in Hamilton's formulation by

$$\dot{q}_i = \frac{\partial H}{\partial p_i} \quad \text{velocity} \tag{1.12}$$

$$-\dot{p}_i = \frac{\partial H}{\partial q_i} \quad \text{force} \tag{1.13}$$

If forces F' not derivable from a potential are active, the force equation becomes

$$-\dot{p}_i = \frac{\partial H}{\partial q_i} + F' \tag{1.14}$$

When the forces not derivable from a potential are acting, the rate of dissipation of energy from the system is given by

$$\frac{\partial H}{\partial t} = \sum_i F'_i \dot{q}_i \tag{1.15}$$

1.3. Review of Essential Features of Thermodynamics

Thermodynamics has three experimentally derived Laws at its base. The first law is variously stated, but basically means that energy does not get lost in any change in state. It may be dissipated through forces of a frictional nature, for example, but the totality of energy remains unchanged. Mathematical statements of the first law are

$$dE = dQ + dW \quad \text{definition of energy function}$$

$$\Delta E = E_{final} - E_{initial}$$

$$= Q + W = \int_{E_{initial}}^{E_{final}} dE = \int_{E_{initial}}^{E_{final}} (dQ + dW)$$

$$\oint_{cycles} dE = 0 \quad \text{for all cycles} \tag{1.16}$$

In these equations, dE is an exact differential of the energy, dQ is a very small quantity of heat and dW is a very small quantity of work, neither of which is, in general, an exact differential. There is no reversibility implied or needed in these first law relations. Note, in particular, that the defining relation for the energy function is a differential equation. This means that the energy itself must always contain an arbitrary assignable additive constant of integration, that may be chosen once, but only once, in any problem.

The second law is also variously stated, but the meaning of the second law is that heat may not be completely converted to work. The mathematical statement of the second law is

$$dS = \frac{dQ_{reversible}}{T} \quad \text{definition of entropy function}$$

$$\Delta S = S_{final} - S_{initial} = \int_{S_{initial}}^{S_{final}} dS = \int_{S_{initial}}^{S_{final}} \frac{dQ_{reversible}}{T}$$

$$\oint_{cycles} dS = 0 \quad \text{for all cycles} \tag{1.17}$$

In these relations, dS is an exact differential and S is the entropy. $dQ_{reversible}$ is a very small amount of heat that crosses the boundary between the system and the surroundings when the process generating it is reversible.

The equations for the Carnot cycle show that if it were possible to attach the Carnot engine to a heat reservoir maintained at the absolute zero and to another maintained at some higher temperature, it would be possible to convert heat rejected by the high temperature heat reservoir completely into work. But the sense of the second law is that this is impossible. Hence, the third law is simply that the absolute zero of temperature is inaccessible by any process in a finite number of steps. The mathematical statement is

$$T \neq 0 \qquad (1.18)$$

for all finite processes. The first and second laws can be combined in mathematical statement. It is convenient to define the enthalpy $H = E + pV$, the Helmholtz Free Energy $A = E - TS$ and the Gibbs Free Energy $G = H - TS = E + pV - TS$. In terms of these, the combined first and second law equations for open systems in which work W_x other than compression work (pdV) may be done are as follows.

$$dE = TdS - pdV + \sum_i \mu_i dn_i + dW_x$$

$$dH = TdS + Vdp + \sum_i \mu_i dn_i + dW_x$$

$$dA = -SdT - pdV + \sum_i \mu_i dn_i + dW_x$$

$$dG = -SdT + Vdp + \sum_i \mu_i dn_i + dW_x \qquad (1.19)$$

These equations may be integrated only along reversible paths. In these equations, μ_i is the chemical potential, a true potential energy per unit measure of the mass of the ith species or component, and n_i represents a measure of the mass of the ith species or component. Note that the chemical potential may be variously calculated.

$$\mu_i = \left(\frac{\partial E}{\partial n_i}\right)_{S,V,n_j,x} = \left(\frac{\partial H}{\partial n_i}\right)_{S,p,n_j,x}$$

$$= \left(\frac{\partial A}{\partial n_i}\right)_{T,V,n_j,x} = \left(\frac{\partial G}{\partial n_i}\right)_{p,T,n_j,x} \qquad (1.20)$$

Many authors take the last equality as the definition of chemical potential. In fact, any one will do. The symbol W_x stands for any reversible work other than the compression work pdV. Note that

$$T = \left(\frac{\partial E}{\partial S}\right)_{V,n,x} = \left(\frac{\partial H}{\partial S}\right)_{p.n.x} \tag{1.21}$$

In addition to the enthalpy H and the two free energies A and G, Mathieu defined a function J as

$$J = S - \frac{E}{T} \tag{1.22}$$

but this function serves essentially the same purposes as A, and so is not often used. Planck defined a function Y as

$$Y = S - \frac{E}{T} - \frac{pV}{T} \tag{1.23}$$

but this function serves essentially the same purposes as G, and again is not often used.

The combined first and second law equations allow relations among the coefficients based on the fact that in partial differentiation, the order of differentiation is immaterial. These relations have often been called the Maxwell relations, and are such differential relations as

$$\left(\frac{\partial T}{\partial V}\right)_{S,n,x} = -\left(\frac{\partial p}{\partial S}\right)_{V,n,x}$$

$$\left(\frac{\partial T}{\partial p}\right)_{S,n,x} = \left(\frac{\partial V}{\partial S}\right)_{p,n,x}$$

$$-\left(\frac{\partial S}{\partial V}\right)_{T,n,x} = -\left(\frac{\partial p}{\partial T}\right)_{V,n,x}$$

$$-\left(\frac{\partial S}{\partial p}\right)_{T,n,x} = \left(\frac{\partial V}{\partial T}\right)_{p,n,x} \tag{1.24}$$

In addition, there are similar relations among the chemical potentials. For example, from the variation in the Gibbs free energy written for a multicomponent mixture,

$$dG = -SdT + Vdp + \sum \mu_i dn_i + dW_x \tag{1.25}$$

there follows

$$\left(\frac{\partial \mu_i}{\partial n_j}\right)_{p,T,n_k} = \left(\frac{\partial \mu_j}{\partial n_i}\right)_{p,T,n_l} \tag{1.26}$$

where the index k does not include j and the index l does not include i.

All of the extensive thermodynamic state functions (S, H, A, E, G, V, etc.) are homogeneous functions of order one in the mole numbers n_i, and thus obey Euler's theorem on homogeneous functions. Specifically, if $\bar{V}_i$ is the partial molal volume $(\frac{\partial V}{\partial n_i})_{p,T,n_j,x}$ then the theorem states that

$$V = \sum_i \bar{V}_i n_i$$

$$G = \sum_i \mu_i n_i \tag{1.27}$$

and so on for all of the other extensive state functions. Consider a system in which the W_x term does not occur. A variation in the Gibbs free energy is

$$dG = \sum_i \mu_i dn_i + \sum_i n_i d\mu_i = -SdT + Vdp + \sum_i \mu_i dn_i \tag{1.28}$$

The terms $\sum_i \mu_i dn_i$ cancel from the expression and there remains

$$\sum_{i=1}^{c} n_i d\mu_i = -SdT + Vdp \tag{1.29}$$

where c is the number of components. The chemical potential of a substance considered as a species is the same as the chemical potential of the substance considered as a component. In a multi-phased system at equilibrium, this equation may be written once for each phase, and all must hold simultaneously. At equilibrium, the temperatures are equal, the pressures are equal and the chemical potentials are either equal, if the corresponding component can exchange throughout the phases, or the chemical potential is greater than or equal to zero in phases in which it occurs. Since this equation may be written once for each of the P phases, and there are

C components, and temperature and pressure are independently variable, there remain $C - P + 2$ independent variables, which is the phase rule. The proof that the temperatures, pressures and chemical potentials are equal in a multi-phased system at equilibrium proceeds from the energy variation equation, which may be written once for each phase. The total energy variation for the system as a whole is the sum of these variations. Similarly, the sum of the variations in the entropy of each phase makes the total variation in entropy for the whole system. Likewise, the sum of the volume variations must yield the variation in the volume of the whole system, and the sum of all the variations in the numbers of moles of each of the components comprises the change in the number of moles of each in the system. But each of these changes is zero, since no heat flows into or out of the equilibrium system, the volume of the whole system does not change, and no amount of any component is varied, the system remaining at equilibrium. Thus we have, for the system at equilibrium,

$$dS = 0 = \sum_i dS_i$$

$$dV = 0 = \sum_i dV_i$$

$$dn_j = 0 = \sum_i dn_{ij}$$

$$dE \geq 0 \tag{1.30}$$

where the subscript i refers to the phase, while the subscript j refers to individual components Not all of these variations are independently variable and one of them may be eliminated by solving the restrictive condition for one of them. For example,

$$dn_{11} = -dn_{12} - dn_{13} - \cdots \tag{1.31}$$

If all of the dependent variables are eliminated from the energy variation, the total energy variation becomes

$$dE = (T_2 - T_1)dS_2 + (T_3 - T_1)dS_3 + \cdots - (p_2 - p_1)dV_2$$
$$- (p_3 - p_1)dV_3 - \cdots + \cdots \tag{1.32}$$

Now set all the independent variables equal to zero except dS_2. There results

$$(T_2 - T_1)dS_2 \geq 0 \qquad (1.33)$$

There are three possibilities. Either dS_2 is greater than zero, zero, or less than zero. If dS_2 is greater than zero, $(T_2 - T_1)$ must be either greater than zero, or zero. If the difference is greater than zero, then $T_2 > T_1$. If dS_2 is zero, the inequality is satisfied and no statement about $T_2 - T_1$ can be made. However, if dS_2 is less than zero, then $(T_2 - T_1)$ must also be less than zero to satisfy the inequality. In that case, $T_1 > T_2$. In the one case, T_2 is greater than T_1 and in the other, T_1 is greater than T_2. Since neither can be greater, the temperatures must be equal. Similar logic shows equality of all the temperatures in all the phases, as well as all the pressures in all the phases and all the chemical potentials in all of the phases. That is, at equilibrium,

$$T_1 = T_2 = T_3 = T_4 = \cdots$$

$$p_1 = p_2 = p_3 = p_4 = \cdots$$

$$\mu_{11} = \mu_{12} = \mu_{13} = \mu_{14} = \cdots$$

$$\mu_{21} = \mu_{22} = \mu_{23} = \mu_{24} = \cdots$$

$$\cdots \qquad (1.34)$$

In the case of chemical equilibrium conducted at constant temperature and pressure, the stoichiometric equation for the equilibrium may be written in general as

$$\sum_i \nu_i S_i = 0 \qquad (1.35)$$

where S_i represents one of the species involved and ν_i its signed stoichiometric coefficient. For example, the Haber process for conversion of nitrogen and hydrogen into ammonia is

$$N_2 + 3H_2 \rightarrow 2NH_3$$

and ν_i for nitrogen is -1, for hydrogen -3 and for ammonia $+2$. These coefficients are useful in defining a progress variable ξ by writing

$$d\xi = \frac{dn_i}{\nu_i} \qquad (1.36)$$

The progress variables for nitrogen, hydrogen and ammonia are respectively $-d\xi$, $-3d\xi$ and $+2d\xi$. Since dT and dp both vanish (T and p are held constant), there results

$$\sum_i n_i d\mu_i = 0 \qquad (1.37)$$

Then a variation in the Gibbs free energy (in general, $dG = \Sigma n_i d\mu_i + \Sigma \mu_i dn_i$, since $G = \Sigma \mu_i n_i$) becomes

$$dG = \sum_i \mu_i dn_i$$

$$= \sum_i \mu_i \nu_i d\xi \qquad (1.38)$$

Plotting G against the progress variable ξ must show a minimum for equilibrium to occur. At the minimum, $dG/d\xi = 0$. Hence,

$$\sum \nu_i \mu_i = 0 \qquad (1.39)$$

as was first shown by Gibbs. That is, replacing the substances S_i with the corresponding chemical potentials produces a relation among the chemical potentials, the system remaining at equilibrium.

Thermodynamics is a much more highly developed branch of science than this brief review of essential relations can address. However, these relations will be needed in the part of statistical mechanics that is concerned with calculating thermodynamic functions from spectroscopic data.

Appendix

Often it is difficult to solve a problem stated in rectangular, or Cartesian, coordinates x_i but if the problem is recast in more general coordinates g_j the burden of solution is eased considerably. The functional dependences of coordinates, velocities and energies are important to the transformation. In general, if n Cartesian coordinates are required to describe a problem, these define an n-dimensional space which the coordinates are said to span. The generalized coordinates must in general span the same space, and there will be at most n of them. It is sometimes possible for the generalized coordinates to span a sub-space. For example, in the Wilson GF matrix method of treating molecular vibrations, the translations and rotations of the rigid molecular frame are removed by the use of internal coordinates as the generalized coordinates. In general, though, the same space is spanned in both Cartesian and generalized coordinates, and the functional dependences of the coordinates are

$$x_i = x_i\left(g_1, g_2, g_3, \ldots g_n\right) \quad i = 1, 2, 3, \ldots n$$

$$g_j = g_j\left(x_1, x_2, x_3, \ldots x_n\right) \quad j = 1, 2, 3, \ldots n \qquad (1.40)$$

and those for the velocities (the dot denotes time differentiation)

$$\dot{x}_i = \dot{x}_i(g_1, g_2, g_3, \ldots g_n, \dot{g}_1, \dot{g}_2, \dot{g}_3, \ldots \dot{g}_n) \quad i = 1, 2, 3, \ldots n$$

$$\dot{g}_j = \dot{g}_j(x_1, x_2, x_3, \ldots x_n, \dot{x}_1, \dot{x}_2, \dot{x}_3, \ldots \dot{x}_n) \quad j = 1, 2, 3, \ldots n$$

$$(1.41)$$

The kinetic energy T is defined in terms of the Cartesian velocities and the masses m_i as

$$T = \frac{1}{2} \sum_i m_i \dot{x}_i^2 \qquad (1.42)$$

Thus, T may be regarded as a function of the generalized coordinates and their velocities. The potential energy V is a function of coordinates alone, and forces F_i derivable from a potential are given by the negative gradient

$$F_i = -\frac{\partial V}{\partial x_i} \qquad (1.43)$$

The equations of motion are

$$F_i = m_i \ddot{x}_i = -\frac{\partial V}{\partial x_i} \quad i = 1, 2, 3, \ldots n \tag{1.44}$$

The problem is to convert the equations of motion into generalized coordinates. First, establish a surprising relation between coordinate and velocity derivatives. Due to the functional relation between coordinates

$$dx_i = \sum_{j=1}^{n} \frac{\partial x_i}{\partial g_j} dg_j \tag{1.45}$$

Then the time derivative is given by

$$\dot{x}_i = \sum_{j=1}^{n} \frac{\partial x_i}{\partial g_j} \dot{g}_j \tag{1.46}$$

The partial derivative of the x velocity with respect to the generalized velocity is given by

$$\frac{\partial \dot{x}_i}{\partial \dot{g}_j} = \frac{\partial x_i}{\partial g_j} \tag{1.47}$$

which is startling. Since it is so, evidently

$$\ddot{x}_i \frac{\partial \dot{x}_i}{\partial \dot{g}_j} = \ddot{x}_i \frac{\partial x_i}{\partial g_j}$$

$$\ddot{x}_i \frac{\partial x_i}{\partial g_j} = \frac{d}{dt}\left(\dot{x}_i \frac{\partial \dot{x}_i}{\partial \dot{g}_j}\right) - \dot{x}_i \frac{d}{dt}\left(\frac{\partial x_i}{\partial g_j}\right) \tag{1.48}$$

Now

$$\frac{\partial x_i}{\partial g_k} = \frac{\partial x_i}{\partial g_k}(g_1, g_2, g_3, \ldots g_n) \tag{1.49}$$

so

$$d\left(\frac{\partial x_i}{\partial g_k}\right) = \sum_j \frac{\partial^2 x_i}{\partial g_j \partial g_k} dg_j \tag{1.50}$$

and then

$$\frac{d}{dt}\left(\frac{\partial x_i}{\partial g_k}\right) = \sum_j \frac{\partial^2 x_i}{\partial g_j \partial g_k}\dot{g}_j = \frac{\partial}{\partial g_k}\left(\sum_j \frac{\partial x_i}{\partial g_j}\dot{g}_j\right) = \frac{\partial \dot{x}_i}{\partial g_k} \quad (1.51)$$

Then

$$\ddot{x}_i \frac{\partial x_i}{\partial g_k} = \frac{d}{dt}\left(\dot{x}_i \frac{\partial \dot{x}_i}{\partial \dot{g}_k}\right) - \dot{x}_i \frac{\partial \dot{x}_i}{\partial g_k} \quad (1.52)$$

$$\ddot{x}_i \frac{\partial x_i}{\partial g_k} = \frac{d}{dt}\frac{\partial}{\partial \dot{g}_k}\left(\frac{1}{2}\dot{x}_i^2\right) - \frac{\partial}{\partial g_k}\left(\frac{1}{2}\dot{x}_i^2\right) \quad (1.53)$$

The Cartesian equations of motion when all forces are derivable from a potential are

$$m_i \ddot{x}_i = -\frac{\partial V}{\partial x_i}$$

therefore

$$m_i \ddot{x}_i \frac{\partial x_i}{\partial g_k} = -\frac{\partial V}{\partial x_i}\frac{\partial x_i}{\partial g_k} = \frac{d}{dt}\frac{\partial}{\partial \dot{g}_k}\left(\frac{1}{2}m_i \dot{x}_i^2\right) - \frac{\partial}{\partial g_k}\left(\frac{1}{2}m_i \dot{x}_i^2\right)$$

Summing over i produces

$$\sum_i m_i \ddot{x}_i \frac{\partial x_i}{\partial g_k} = -\sum_i \frac{\partial V}{\partial x_i}\frac{\partial x_i}{\partial g_k} = \frac{d}{dt}\frac{\partial}{\partial \dot{g}_k}\left(\sum_i \frac{1}{2}m_i \dot{x}_i^2\right)$$

$$-\frac{\partial}{\partial g_k}\left(\sum_i \frac{1}{2}m_i \dot{x}_i^2\right)$$

Noting that $\frac{\partial V}{\partial g_k} = \sum_i \frac{\partial V}{\partial x_i}\frac{\partial x_i}{\partial g_k}$ and that $T = \frac{1}{2}\sum_i m_i \dot{x}_i^2$ this equation amounts to

$$-\frac{\partial V}{\partial g_k} = \frac{d}{dt}\frac{\partial T}{\partial \dot{g}_k} - \frac{\partial T}{\partial g_k}$$

This equation is sometimes called LaGrange's equation of motion. The term $\frac{\partial T}{\partial g_k}$ vanishes identically when g_k is taken as a Cartesian coordinate, since in Cartesian coordinates T is a function of the velocities alone and is independent of any coordinate. For example, for curvilinear coordinates, such as cylindrical, this is obviously not the case. If $x = r \cos \Theta$, then the time derivative $\dot{x} = \dot{r} \cos \Theta - r \sin \Theta \dot{\Theta}$, and although $x = x(r, \Theta)$, one has $\dot{x} = \dot{x}(r, \Theta, \dot{r}, \dot{\Theta})$. It is then a short step to the definition of the LaGrangian function, $L = T - V$.

Problems

1. Consider a system of two point particles with masses m_1 and m_2 moving in two dimensions. This means one point has mass m_1 and the other m_2. The potential energy of interaction U is a function of Cartesian difference coordinates $x_1 - x_2$ and $y_1 - y_2$. The total energy is

$$E = \frac{m_1}{2}(\dot{x}_1^2 + \dot{y}_1^2) + \frac{m_2}{2}(\dot{x}_2^2 + \dot{y}_2^2) + U(x_1 - x_2, y_1 - y_2)$$

where the dot denotes time differentiation. Introduce four new variables

$$X = \frac{m_1 x_1 + m_2 x_2}{m_1 + m_2}$$

$$Y = \frac{m_1 y_1 + m_2 y_2}{m_1 + m_2}$$

$$x_{12} = x_1 - x_2$$

$$y_{12} = y_1 - y_2$$

Show that this two-body problem can be reduced to two one-body problems, one involving the center of mass of the system and the other involving the relative motion of the two points. Give a physical interpretation of the quantity $m_1 m_2/(m_1 + m_2)$ that arises naturally in the relative motion. What is this quantity named?

2. Consider the rotation of a diatomic molecule with fixed internuclear separation l (the letter l, not the numeral 1). Let the masses be m_1 and m_2. Use Cartesian relative coordinates x_{12}, y_{12}, z_{12} to

describe the motion of the reduced mass μ and Cartesian coordinates X, Y, Z for the motion of the mass M at the center of mass. Show that the rotational kinetic energy can be written in spherical coordinates as

$$\frac{1}{2}I(\dot{\vartheta}^2 + \dot{\varphi}^2 \sin^2 \vartheta)$$

where $I = \mu l^2$ is the moment of inertia of the diatomic combination. Now show that the Hamiltonian for rotation can be written as

$$H_{rot} = \frac{1}{2I}\left(p_\vartheta^2 + \frac{p_\varphi^2}{\sin^2 \vartheta}\right)$$

3. Derive LaGrange's equations for a particle moving in two dimensions under the action of a central potential $u(r)$. Which of these equations shows that angular momentum is conserved? If the potential $u(r, \theta)$ depends on both r and θ, is angular momentum still conserved?

4. Start with LaGrange's equations in Cartesian coordinates

$$\frac{\partial}{\partial t}\left(\frac{\partial L}{\partial \dot{x}}\right) - \frac{\partial L}{\partial x} = 0$$

and similarly for y and z. Now introduce three generalized coordinates q_1, q_2, q_3 which are related to the Cartesian coordinates by $x = x(q_1, q_2, q_3)$, and similarly for y and z. Show that by transforming LaGrange's equations from Cartesian coordinates $x, \dot{x}, y, \dot{y}, z, \dot{z}$ to the generalized coordinates and their velocities $q_1, \dot{q}_1, q_2, \dot{q}_2, q_3, \dot{q}_3$ the form of LaGrange's equations stands unmodified, i.e., that

$$\frac{\partial}{\partial q_1}\left(\frac{\partial L}{\partial \dot{q}_1}\right) - \frac{\partial L}{\partial q_1} = 0$$

5. By considering the entropy of a pure substance (a single component) as a function of p and $T, S = S(p, T)$, show that the constant volume and constant pressure heat capacities C_p and C_v

are related to the bulk modulus of elasticity ε and the thermal coefficient of expansion α as

$$C_p - C_v = TV\alpha^2\varepsilon$$

Useful definitions are as follows.

$$\alpha = \frac{1}{V}\left(\frac{\partial V}{\partial T}\right)_p \qquad \text{thermal expansion coefficient}$$

$$\kappa = -\frac{1}{V}\left(\frac{\partial V}{\partial p}\right)_T \qquad \text{isothermal compressibility}$$

$$\varepsilon = \frac{1}{\kappa} \qquad \text{isothermal bulk modulus of elasticity}$$

Chapter 2

The Basis of Statistical Mechanics

2.1. Averaging

The first postulate on which statistical mechanics is based is hardly a postulate at all. It is the atomic constitution of matter. Of course, once atomic scale events are involved, quantum effects dominate. The atoms and molecules of which bulk matter is composed are all regarded as existing in stationary states of definite energy. To specify a state, the quantum numbers associated with the atom or molecule must be given, and when a specific energy is assigned to a collection of identical atoms or molecules, the probability of occurrence of any state consistent with the given energy must be the same as that of any other state consistent with that energy. Finally, the behavior of bulk matter originates in averages of the desired property over all the states the atoms or molecules are capable of occupying that are consistent with the energy specified. Averaging is accomplished in the ordinary way. That is, if there are Ω ways of specifying all of the quantum numbers of all of the atoms or molecules in a system consistent with a specified energy, and if the value of the observable property being averaged is f_i in the ith state, then the average of the observable f_{obs} is

$$f_{obs} = \frac{\sum_{i=1}^{\Omega} f_i}{\Omega} \qquad (2.1)$$

A typical problem in teaching is calculating the average grade on an examination. The individual grades G_i are summed up over the entire class and divided by the number N of students in the class. That is, if $\langle G \rangle$ is the average grade,

$$\langle G \rangle = \frac{\sum_{i=1}^{N} G_i}{N} \tag{2.2}$$

This is exactly the same relation as the one above that defines the average value of an observable in an atomic or molecular system . . . only the variable and parameter names have been changed. Suppose the number of students n_j with a particular grade G_j is known for all grade categories. Then if there are n such grade categories, the total number of students in the class must be

$$N = \sum_{j=1}^{n} n_j \tag{2.3}$$

To get the average grade, instead of summing over students, the sum may be carried out over grade categories, and the average becomes

$$\langle G \rangle = \frac{\sum_{j=1}^{n} n_j G_j}{N} \tag{2.4}$$

Equation (2.4) may also be written as

$$\langle G \rangle = \frac{\sum_{j=1}^{n} n_j G_j}{\sum_{j=1}^{n} n_j} \tag{2.5}$$

The probability P_j of occurence of grade category j is

$$P_j = \frac{n_j}{\sum_{n_j=1}^{n} n_j} \tag{2.6}$$

Note that the probability may also be written simply as $P_j = n_j/N$. This allows the fundamental averaging procedure to be cast in terms of the probability as

$$\langle G \rangle = \sum_{i=1}^{n} P_i G_i \tag{2.7}$$

The same definition of probability and of averaging will occur frequently throughout this book.

Specification of the state of a system is critical, and is different in thermodynamics, classical mechanics and quantum mechanics. The differences are emphasized in Table 2.1.

In thermodynamics, one hopes that the average value of the observable is given by the short time average shown in the table rather than by the indefinitely long time average $\langle f_{obs}^{LT} \rangle$

$$\langle f_{obs}^{LT} \rangle = \lim_{\Delta t \to \infty} \frac{1}{\Delta t} \int_t^{t+\Delta t} f(t) dt \tag{2.8}$$

Similarly, in classical mechanics, $P(j)$ is regarded as invariant with time:

$$P(j) = \lim_{\Delta t \to \infty} \frac{1}{\Delta t} \int_t^{t+\Delta t} P(j, t) dt \tag{2.9}$$

2.2. Ensembles

With these preliminaries out of the way, it is now time to consider the definition of an ensemble and to investigate its properties. Gibbs (Collected Works, Vol. II, page 5) is almost casual in his first definition. He just writes, "Let us imagine a great number of independent systems, identical in nature, but differing in phase, that is, in their condition with respect to configuration and velocity." This is already almost an ensemble. As an ensemble is understood today, it consists of an indefinitely large collection of identical systems. The distinguishing feature that differentiates one type of ensemble from another is the communication allowed between the systems. The microcanonical ensemble consists of identical systems, each with the same energy E, volume V, and number of atoms or molecules N, and communication between the systems is not allowed. Thus, the energy of a system of this ensemble remains the same forever, as does its volume and numbers of atoms or molecules. Nothing is exchanged with the other systems in the microcanonical ensemble.

The canonical ensemble is the same as the microcanonical with one exception: its systems are allowed to exchange energy in the form of heat. This exchange constitutes the communication among the systems of this ensemble.

Table 2.1.   State specification.

	Thermodynamics	Classical Mechanics	Quantum Mechanics
Composition	1 system	N systems, each with 3 degrees of freedom	N systems, each with at least 3 degrees of freedom, perhaps more (e.g., spin)
State	2 intensive variables plus the kind, quantity and phase of the matter	$3N$ coordinates q and $3N$ momenta p	At least $3N$ quantum numbers, plus any more needed for such degrees of freedom as spin
Averages of Observables	$f_{obs} = \dfrac{1}{\Delta t}\displaystyle\int_t^{t+\Delta t} f(t)dt$ t is time, Δt is a time interval and $f(t)$ is the value of f at time t	$f_{obs} = \displaystyle\int_{all\,j} f(j)P(j)dp_j dq_j$ $P(j)$ is the probability for one specification of p and q, and $f(j)$ the value of f for that specification.	$f_{obs} = \dfrac{\sum\limits_{i=1}^{\Omega} f_i}{\Omega}$ f_i is the value of f in the state i, and there are Ω different states

The grand canonical ensemble is the same as the microcanonical but now with two exceptions: not only are its systems allowed to exchange energy in the form of heat, its systems may also exchange atoms or molecules. Communication may take place in two ways in this ensemble.

Other ensembles may be constructed by limiting the exchange to volume or to molecules or atoms alone, or any combination of the three exchangeable variables, heat, volume or numbers of atoms or molecules. The different ensembles have different applications, but, as will be shown, all are equivalent, and the choice of ensemble to treat a particular problem is made at the convenience of the investigator.

Consider the canonical ensemble, in which heat may be exchanged amongst the systems of the ensemble, but not volume and not numbers of atoms or molecules. Surround the ensemble with a rigid adiabatic wall, and attach it to a large heat reservoir until the temperatures equilibrate. Then detach the ensemble from the heat reservoir and thereafter let it evolve in time, exchanging heat among its systems but maintaining temperature equality throughout the interior of the ensemble. It may help to think of the ensemble as a block of lead which is indefinitely large but which has been cut up into an extremely large number of tiny but identical cubes of lead, all in thermal contact with their neighbors. Each system of the ensemble is characterized by the three variables temperature T, volume V and number of atoms or molecules N. Any given state of the ensemble is specified completely by giving all of the quantum numbers. The total energy E_t of the ensemble is constant, but may be specified by giving all of the quantum numbers in the ensemble. There are a very large number of really different ways Ω_t of specifying quantum numbers that are consistent with the energy E_t, which means that E_t is Ω_t fold degenerate.

In the ensemble, there are N_t systems, each characterized by the variables T, V and N. The problem is to calculate Ω_t, for, given Ω_t, average values of observables may be calculated using the fundamental definition of average given in equation (2.1).

To get Ω_t, let there be n_1 systems in state 1 of energy E_1, n_2 systems in state 2 of energy E_2, and so on until the N_t systems of the ensemble are exhausted. Note that E_1 may equal E_2, or it may be different. The specification of all of the n_i is called a distribution, since it tells how the quantum number specifications are distributed over the systems of the ensemble. The number of ways of doing this for one distribution is

$$\Omega_t(n) = \frac{N_t!}{\prod_i n_i!} \tag{2.10}$$

There are a very large number of ways of specifying quantum numbers that are consistent with the total energy E_t and number of systems N_t. Summing over all possible distributions (n) gives Ω_t. That is,

$$\Omega_t = \sum_{(n)} \Omega_t(n) \tag{2.11}$$

$$\Omega_t = \sum_{(n)} \frac{N_t!}{\prod_i n_i!} \tag{2.12}$$

The sum is subject to the restrictive conditions

$$N_t = \sum_i n_i \tag{2.13}$$

$$E_t = \sum_i E_i n_i \tag{2.14}$$

In all the distributions over which the sum for Ω_t is performed, there is one, designated n_i^*, which is overwhelmingly more probable than all of the others, and which contributes the dominant majority to the sum. Then, to a very good approximation

$$\Omega_t \approx \frac{N_t!}{\prod_i n_i^*!} \tag{2.15}$$

subject to

$$N_t = \sum_i n_i^* \tag{2.16}$$

$$E_t = \sum_i n_i^* E_i \tag{2.17}$$

To determine n_i^*, the typical term in Ω_t has to be maximized subject to the restrictive conditions. It proves to be easier to work with the logarithm of $\Omega_t(n)$ than it is to work with $\Omega_t(n)$ itself. This can be done without causing unanticipated mathematical difficulties since the logarithm is a monotonically increasing function of its argument for positive arguments, and the nature of the problem requires positive arguments. Thus,

$$\ln \Omega_t(n) = \ln N_t! - \sum_i \ln n_i! \qquad (2.18)$$

To proceed, assume that Stirling's approximation applies in this case. The approximation is that $\ln x! \approx x \ln x - x$. The differential of $\ln x!$ is then approximately

$$d \ln x! = \ln x \, dx \qquad (2.19)$$

and the derivatives of $\ln \Omega_t(n)$, N_t and E_t are

$$d \ln \Omega_t(n) = \sum_i - \ln n_i dn_i \qquad (2.20)$$

$$0 = \sum_i dn_i \qquad (2.21)$$

$$0 = \sum_i E_i dn_i \qquad (2.22)$$

Here introduce LaGrange multipliers α and β. Multiply the derivative of the first restrictive condition by α and that of the second by β, then add both to the derivative of $\ln \Omega_t(n)$. The result is still $d \ln \Omega_t(n)$, since zero has been added. The condition for a maximum is $d \ln \Omega_t(n) = 0$, and the result of these manipulations is

$$0 = \sum_i (- \ln n_i + \alpha + \beta E_i) dn_i \qquad (2.23)$$

Not all of the dn_i are independently variable. However, α and β are as yet undefined, and in order to define them, two equations are needed.

Let those be

$$-\ln n_1 + \alpha + \beta E_1 = 0 \tag{2.24}$$

$$-\ln n_2 + \alpha + \beta E_2 = 0 \tag{2.25}$$

These two definitions eliminate dn_1 and dn_2 from the equation, and there remains

$$0 = \sum_{i=3}(-\ln n_i + \alpha + \beta E_i)dn_i \tag{2.26}$$

All of the remaining dn_i are independently variable, and may assume positive, negative or zero values at pleasure. Then the only way to satisfy the equality is to require

$$-\ln n_i + \alpha + \beta E_i = 0 \quad \text{for all } i \tag{2.27}$$

This is the condition that defines the most probable distribution, and n_i is really n_i^*. Solving for n_i^*

$$n_i^* = e^\alpha e^{\beta E_i} \tag{2.28}$$

This allows several results to be written in terms of the undetermined multipliers α and β

$$N_t = e^\alpha \sum_i e^{\beta E_i} \tag{2.29}$$

$$E = e^\alpha \sum_i E_i e^{\beta E_i} \tag{2.30}$$

$$\frac{n_i^*}{n_j^*} = \frac{e^{\beta E_i}}{e^{\beta E_j}} \tag{2.31}$$

$$P_i = \frac{n_i^*}{N} = \frac{e^{\beta E_i}}{\sum_i e^{\beta E_i}} \tag{2.32}$$

$$E = \frac{E_t}{N_t} = \frac{\sum_i E_i e^{\beta E_i}}{\sum_i e^{\beta E_i}} = \sum_i E_i P_i \tag{2.33}$$

Note especially the last equation, $E = \sum_i E_i P_i$. This is the ordinary average of the energy, taken in the ordinary way. As these results stand, they are not of very much use, since the parameter β occurs in all of them and is as yet completely undefined. The next section addresses this issue, and provides a definition of β that at once renders these results useful and defines the size of the degree.

2.3. Determination of β

In order to get at β, consider the function $F = N_t \ln \sum_i e^{\beta E_i}$. This function is a function of the total number of systems in the ensemble, N_t, of the parameter β, and of all the energy levels E_i. The volume dependence of a system of the ensemble is incorporated implicitly in the energy levels E_i. A variation in the function F is given by

$$dF = \frac{\partial F}{\partial N_t} dN_t + \frac{\partial F}{\partial \beta} d\beta + \sum_i \frac{\partial F}{\partial E_i} dE_i \qquad (2.34)$$

where

$$\frac{\partial F}{\partial N_t} = \ln \left(\sum_i e^{\beta E_i} \right) \qquad (2.35)$$

$$\frac{\partial F}{\partial \beta} = N_t \langle E \rangle = E_t \qquad (2.36)$$

$$\frac{\partial F}{\partial E_i} = N_t \beta P_i = \beta n_i^* \qquad (2.37)$$

These are not quite the most convenient variables, and it proves fruitful to define a new function $F - \beta E_t$. Then a variation in this function may be written

$$d(F - \beta E_t) = dF - \beta dE_t - E_t d\beta$$

$$= \ln \left(\sum_i e^{\beta E_i} \right) dN_t + E_t d\beta$$

$$+ \sum_i \beta n_i^* dE_i - E_t d\beta - \beta dE_t$$

$$= \ln \left(\sum_i e^{\beta E_i} \right) dN_t + N_t \sum_i \beta P_i dE_i - \beta dE_t$$

$$(2.38)$$

Now, close the ensemble. This sets dN_t equal to zero. Further, connect the ensemble to a gigantic heat bath and allow heat dQ_{rev} to flow reversibly into the ensemble, thus changing the populations in all the energy levels but not the levels themselves. Changing n_i^* reversibly changes P_i reversibly. Now put the ensemble into a pressure chamber and increase the pressure, changing the volumes of the systems. Doing so also changes the energy levels E_i. The work of compression is

$$dW_c = \sum_i n_i^* dE_i \qquad (2.39)$$

The total energy change for the ensemble is

$$dE_t = dQ_{rev} + dW_c$$

$$= dQ_{rev} + \sum_i n_i^* dE_i$$

$$= dQ_{rev} + N_t \sum_i P_i dE_i \qquad (2.40)$$

But from the above work $d(F - \beta E_t)$ is

$$d(F - \beta E_t) = -\beta dE_t + \beta N_t \sum_i P_i dE_i$$

$$= -\beta(dQ_{rev} + dW_c) + \beta N_t \sum_i P_i dE_i$$

$$= -\beta \left(dQ_{rev} + N_t \sum_i P_i dE_i \right) + \beta N_t \sum_i P_i dE_i$$

$$= -\beta dQ_{rev} \qquad (2.41)$$

Now, $d(F - \beta E_t)$ is an exact differential, Hence, $-\beta dQ_{rev}$ must also be an exact differential, even though dQ_{rev} is not, showing that $-\beta$ is an integrating factor for dQ_{rev}.

Quite generally, an integrating factor is an algebraic construct that is common to all the partial differential coefficients in a total differential expression, That is, if the independent variables are $x_1, x_2, \ldots$, and if the total differential of a function $g(x_1, x_2, \ldots)$ is given by

$$dg = \alpha(x_1, x_2, \ldots)A(x_1, x_2, \ldots)dx_1$$
$$+ \alpha(x_1, x_2, \ldots)B(x_1, x_2, \ldots)dx_2 + \cdots \qquad (2.42)$$

and this pattern is repeated for all of the coefficients of all of the independent variables, then α is an integrating factor for the quantities $A, B, \ldots$

In this case, there is only one coefficient and one quantity Adx_1 in the expression, namely dQ_{rev}. Clearly, β is a function of volume, V, and temperature, T, both of which were held constant in the argument, but not of N_t, since the extensive nature of the relation is embedded in the heat term, dQ_{rev}. Thus, we have

$$\beta = \beta(T, V) \qquad (2.43)$$

The next argument shows β to be a function of T alone. Note that an integrating factor for dQ_{rev} must produce an entropy difference, allowing the proportionality

$$dS(T, V) \propto -\beta dQ_{rev} \qquad (2.44)$$

If dQ_{rev} is zero, the path for a finite change in state is adiabatic and $\Delta S = 0$. Then $S(T, V)$ is constant along that path. If dQ_{rev} is not zero, the reversible path is some more general one, dS is not zero and $S(T, V)$ is not constant along such paths. Since $S(T, V)$ describes the unique reversible adiabatic path through T, V space, it must be a function of the state of the system. Then $dS(T, V)$ is independent of the path by which the change with dQ_{rev} not zero is effected. The function $\beta(T, V)$ is in this case an integrating factor for dQ_{rev}.

Now consider a system consisting of two chambers of different volumes V_1 and V_2 with a common rigid diathermal wall, but with both surrounded by a rigid adiabatic wall. After some time, the temperatures in the two compartments will become equal, but the volumes will remain different since the walls are rigid. Now let heat flow reversibly from one of the compartments to the other through the diathermal wall. If heat dq_1 leaves the compartment with volume V_1, and heat dq_2 enters the compartment of volume V_2, then we can write

$$-dq_{1,rev} = dq_{2,rev} \qquad (2.45)$$

In this change in state, the change in entropy of the entire system must vanish since the system is surrounded by adiabatic walls through which no heat may flow. Thus

$$dS = dS_1 + dS_2 = 0 \qquad (2.46)$$

Therefore

$$\beta(T, V_1)dq_{1,rev} + \beta(T, V_2)dq_{2,rev} = 0 \qquad (2.47)$$

Substituting $-dq_{1,rev}$ for $dq_{2,rev}$ from the work above, there results

$$dq_{1,rev}(\beta(T, V_1) - \beta(T, V_2)) = 0 \qquad (2.48)$$

By hypothesis, $dq_{1,rev}$ can not be zero, and that must mean that the two integrating factors $\beta(T, V_1)$ and $\beta(T, V_2)$ must be equal. In short, the integrating factor β is a function of T alone, and is independent of the volume.

$$\beta = \beta(T) \qquad (2.49)$$

Now, the definition of the entropy function is $dS = \frac{dQ_{rev}}{T}$, and the proportionality is that

$$dS \propto d(F - \beta E_t) = -\beta dQ_{rev} \qquad (2.50)$$

Hence, $-\beta T = $ constant, the constant of proportionality between dS and $d(F - \beta E_t)$. The constant effectively determines the size of the degree, and is normally assigned to Boltzmann's constant k_B. Then

$$\beta = -\frac{1}{k_B T} \tag{2.51}$$

All the results stated above in terms of β may now be stated in terms of ordinary thermodynamic temperature as defined by the second law. The sum over states is called "*Zustandsumme*" in German, but in English it is usually referred to as the partition function, since its terms tell how energy is partitioned among the states of the ensemble. The previous results for the population ratios, average energy and occupation probability become

$$\frac{n_i^*}{n_j^*} = \frac{e^{-E_i/k_B T}}{e^{-E_j/k_B T}} \tag{2.52}$$

$$\langle E \rangle = \frac{\sum_i E_i e^{-E_i/k_B T}}{\sum_i e^{-E_i/k_B T}} \tag{2.53}$$

$$P_i = \frac{e^{-E_i/k_B T}}{\sum_i e^{-E_i/k_B T}} \tag{2.54}$$

and of course the average energy is given in terms of the probabilities of occupation as

$$E = \sum_i P_i E_i \tag{2.55}$$

It is now necessary to identify the function F. Note that the differential $d(F - \beta E_t)$ is proportional to dS_t, the entropy of the entire ensemble, and in fact

$$d(F - \beta E_t) = -\beta dQ_{rev} = \frac{dQ_{rev}}{k_B T} = \frac{dS}{k_B} \tag{2.56}$$

This implies that to within a constant (the constant of integration)

$$F - \beta E_t = \frac{S_t}{k_B} = F + \frac{E_t}{k_B T} \tag{2.57}$$

This may be rearranged to

$$k_B T F = -E_t + T S_t = -A_t \tag{2.58}$$

where A_t is the Helmholz Free Energy for the ensemble. Referring to the definition of F, this allows writing the Helmholz Free Energy as

$$A_t = -N_t k_B T \ln \left(\sum_i e^{-E_i/k_B T} \right) \qquad (2.59)$$

which gives $A_t = A_r(T, V, N_t)$, since each E_i is a function of V. This is a fundamental equation in the sense of Gibbs, and a general variation is

$$dA_t = -S_t dT - p dV_t + \mu dN_t \qquad (2.60)$$

Immediately,

$$-S_t = \left(\frac{\partial A_t}{\partial T} \right)_{V_t, N_t} \qquad (2.61)$$

and similar partial differential coefficients for the pressure and chemical potential. Performing the differentiation yields a formal expression for the entropy

$$S_t = N_t k_B \ln \left(\sum_i e^{-E_i/k_B T} \right) + \frac{N_t \langle E \rangle}{T} \qquad (2.62)$$

This is the total ensemble entropy. The entropy of one system, S, is S_t/N_t, or

$$S = k_B \ln \left(\sum_i e^{-E_i/k_B T} \right) + \frac{\langle E \rangle}{T} \qquad (2.63)$$

The sum recurs so frequently in these considerations that it is useful to give it a special symbol. Most people choose Q, making the equation

$$S = k_B \ln Q + \frac{\langle E \rangle}{T} \qquad (2.64)$$

It is important to keep in mind that in all these expressions, the sums run over states, not levels, and that each energy level E_i is Ω_t fold degenerate.

Now consider the nature of the probability, P_i. It is

$$P_i = \frac{e^{-E_i/k_B T}}{Q} \qquad (2.65)$$

where the partition function Q has been explicitly introduced. The logarithm is

$$\ln P_i = -\frac{E_i}{k_B T} - \ln Q \qquad (2.66)$$

Multiplying by P_i and summing over states yields

$$-\sum_i P_i \ln P_i = \ln Q \sum_i P_i + \frac{\sum_i P_i E_i}{k_B T} \qquad (2.67)$$

But $\sum_i P_i = 1$, leaving the leading term just $\ln Q$. The second is by definition $\langle E \rangle / k_B T$, and cross multiplying by k_B gives the entropy

$$S = -k_B \sum_i P_i \ln P_i = k_B \ln Q + \frac{\langle E \rangle}{T} \qquad (2.68)$$

This allows writing the entropy of any system of the ensemble as

$$S = -k_B \sum_i P_i \ln P_i \qquad (2.69)$$

Some authors take this relation as the fundamental definition of the entropy. It is important to keep in mind that the sum is over states, and not over energy levels.

So far, the discussion has centered on the canonical ensemble, the distinguishing feature of which is the communication among systems afforded by heat exchange. If this communication is shut off, and each system of the ensemble is given exactly the same energy, volume and number of molecules or atoms, the microcanonical ensemble is formed. In this case, the probability of occupancy of each state of the system is exactly the same as that of any other. For a given energy, E, volume, V, and number of atoms or molecules, N, the degeneracy of the energy is $\Omega(E, V, N)$. This is the number of states that correspond to E, V and N. Then the probability of occupancy P_i is, by definition of probability, the reciprocal of the degeneracy, $1/\Omega(E, V, N)$. The entropy of a system of the microcanonical ensemble is then

$$S = -k_B \sum_{i=1}^{\Omega} \frac{1}{\Omega} \ln \frac{1}{\Omega} = k_B \ln \Omega \qquad (2.70)$$

since $\ln(1/\Omega) = -\ln \Omega$ and the sum adds Ω identical terms, each of which is $1/\Omega$ times the same quantity, $-\ln \Omega$. Note that with this definition of the entropy, whenever the energy is non-degenerate and $\Omega = 1$, the entropy vanishes. Boltzmann arrived at the same result first, but by a more arduous route.

2.4. Comparison of Ensembles

Now compare the several ensembles Gibbs investigated. These are the Canonical, the Grand Canonical and the Microcanonical Ensembles. Each has natural thermodynamic variables, displayed in a fundamental equation. Each has its characteristic function, which is a function of the natural variables. Each has its own expression of the occupancy probability, and all suggest that there are yet other ensembles waiting to be found.

2.4.1. *The microcanonical ensemble*

The systems of this ensemble are closed, adiabatic and rigid, and there is correspondingly no communication among the systems of this ensemble. The systems are isolated. The characteristic thermodynamic equation that specifies the natural variables for this ensemble is derived directly from the combined first and second law variation in the energy, and is

$$dS = \frac{dE}{T} + \frac{p}{T} dV - \frac{\mu}{T} dN \qquad (2.71)$$

This equation shows that if the entropy can be captured as a function of E, V and N, then all the thermodynamic functions can be derived from that expression by algebra or at most differentiation. For this reason, the variables E, V and N are called the natural variables. Since the degeneracy of the energy level is such a function, the entropy expressed as a function of the degeneracy is the desired characteristic function. That is,

$$S(E, V, N) = k_B \ln \Omega(E, V, N) \tag{2.72}$$

The probability of occupation of a state of a system of the ensemble P_i is, of course,

$$P_i(E, V, N) = \frac{1}{\Omega(E, V, N)} \tag{2.73}$$

Once again, there is no communication among the systems of the ensemble, each of which is an isolated system equipped with rigid, adiabatic and impermeable walls. None of the variables fluctuates.

2.4.2. *The canonical ensemble*

In the canonical ensemble, the walls of the systems of the ensemble are diathermal, rigid and impermeable to matter. The appropriate combined first and second law variation is

$$dA = -SdT - pdV + \mu dN \tag{2.74}$$

establishing the natural variables as T, V and N. The probability of occupation P_i is

$$P_i(T, V, N) = \frac{e^{-E_i(V,N)/k_B T}}{\sum_i e^{-E_i(V,N)/k_B T}} \tag{2.75}$$

in terms of which the average energy of a system of the ensemble is

$$\langle E \rangle = \sum_i E_i P_i \tag{2.76}$$

As noted above, the sum over states is called the partition function, Q, and since each energy level is Ω-fold degenerate, the sum may be performed over levels rather than states by writing

$$Q(T, V, N) = \sum_E \Omega(E, V, N) e^{-E(V,N)/k_B T} \tag{2.77}$$

Here, Q is a function of T, V and N rather than of E, V and N since the E dependence has been summed out. The Helmholz Free Energy characterizes the Canonical Ensemble, and may be written in either of two ways

$$A(T, V, N) = -k_B T \ln \sum_i e^{-E_i(V,N)/k_B T} \qquad \text{sum over states}$$

(2.78)

$$A(T, V, N) = -k_B T \ln \sum_E \Omega(E, V, N) e^{-E/k_B T} \qquad \text{sum over levels}$$

(2.79)

In either case, $A = -k_B T \ln Q$. The entropy is no longer the characteristic function for the ensemble, but may be calculated for a system of the ensemble by

$$S = -k_B \sum_i P_i \ln P_i \qquad (2.80)$$

Note that the energy of a system of the ensemble is no longer necessarily equal to that of another, but fluctuates from system to system. In effect, summing over a variable is equivalent to fluctuation in the variable, a theme that is often repeated in statistical thermodynamics.

2.4.3. *The grand canonical ensemble*

The Grand Canonical Ensemble is formed by making the walls of the systems of the ensemble rigid, diathermal and permeable to matter. Communication among the systems of the Grand Canonical Ensemble is not only through heat exchange but also through matter exchange. This changes the characteristic combined first and second law equation to

$$dpV = SdT + pdV + Nd\mu \qquad (2.81)$$

establishing the natural variables for the thermodynamic function pV as T, V and μ. An analysis similar to that for the Canonical Ensemble leads to a partition function involving a double sum, one over energy and one over the numbers of atoms or molecules in a system of the ensemble. The result is

$$pV = k_B T \ln \left(\sum_N e^{\mu N / k_B T} \sum_E \Omega(E, V, N) e^{-E(V,N)/k_B T} \right)$$

$$(2.82)$$

The N dependence is made explicit in the second sum. It is, of course, the partition function in the Canonical Ensemble, $Q(T, V, N)$. The probability now includes information pertaining to the number of molecules or atoms, and may be written

$$P(E, V, N) = \frac{\Omega(E, V, N) e^{\mu N / k_B T} e^{-E/k_B T}}{\sum_N \sum_E \Omega(E, V, N) e^{\mu N / k_B T} e^{-E/k_B T}} \qquad (2.83)$$

Then the entropy, the average number of molecules or atoms in a system, and the average energy of a system may be written

$$S = -k_B \sum_N \sum_E P(E, V, N) \ln P(E, V, N) \qquad (2.84)$$

$$\langle E \rangle = \sum_N \sum_E E P(E, V, N) \qquad (2.85)$$

$$\langle N \rangle = \sum_N \sum_E N P(E, V, N) \qquad (2.86)$$

The partition function for the Grand Canonical Ensemble is usually named $\Xi(T, V, \mu)$. This is given in general by either of the relations

$$\Xi = \sum_N \sum_i e^{\mu N / k_B T} e^{-E_i / k_B T} \qquad (2.87)$$

$$\Xi = \sum_N \sum_E \Omega(E, V, N) e^{\mu N / k_B T} e^{-E/k_B T} \qquad (2.88)$$

Thus, the characteristic function for this ensemble may be expressed

$$pV = k_B T \ln \Xi \qquad (2.89)$$

In this ensemble, both E and N fluctuate, but V is fixed.

Problems

1. The energy levels of the simple harmonic oscillator are $(v+1/2)h\nu$, where ν is the classical oscillational frequency in Hz and v $=$ $0, 1, 2, 3, \ldots \infty$, that is, v is a non-negative integer. Consider a microcanonical ensemble whose systems are each 4 such oscillators, weakly coupled to allow energy exchange amongst the 4 oscillators but not with adjacent systems. The total energy of a system of the ensemble is

$$E = (v_1 + 1/2)h\nu + (v_2 + 1/2)h\nu + (v_3 + 1/2)h\nu + (v_4 + 1/2)h\nu$$

in which the subscript designates the oscillator. The oscillators are distinguished by their position in space, but are otherwise indistinguishable. Let 6 vibrational quanta be distributed over the 4 oscillators (this specifies the energy). Discuss this ensemble. This direction means at least the following:

a. Since the zero point energies sum to $2h\nu$, the total energy is $8h\nu$. How many different states are there for this energy? Answer: There are exactly 84 states possible. You can show this by making a table listing the possible values of the quantum numbers that could be assigned to give a total energy of $8h\nu$.

b. A distribution is defined by stating the number n_v of oscillators with each value of the quantum number v. Give the values of n_v for each of the nine possible distributions. Verify by direct arithmetic calculation that the number of states that correspond to each distribution is correctly calculated by $N!/\Pi_i n_i!$. Which distribution is the most probable? What would you expect to find for the most probable values of the occupation numbers n_v in a system of Avogadro's number of oscillators of this type?

c. Show that the total number of states (84) is given by the coefficient of z^6 in the polynomial expansion of the function

$$f(z) = \left(\sum_{n=0}^{\infty} z^n \right)^4$$

You may find it convenient to recall the geometric series, $\frac{1}{1-z} = \sum_{n=0}^{\infty} z^n$ provided $z^2 < 1$.

d. How many states would there be for 5 quanta shared amongst 4 oscillators? How many for 4 quanta shared amongst 4 oscillators? For 3 quanta shared amongst 4 oscillators? For 2 shared amongst 4 oscillators? For 1? For none? How many states must there be for no quanta shared amongst Avogadro's number of such oscillators?

e. Refer to the table prepared in part a. Find the number of states for which the quantum number v assigned to oscillator 1 (v_1) has the value 0. Repeat the count for $v = 1, 2, 3, 4, 5, 6$. Assign equal weight to each state and calculate the probability of finding oscillator 1 with each of the seven possible values of v_1. The average number of oscillators (n_v) in the vth state is obtained by multiplying the probability by N, taken to be 4 in this problem.

f. The mean occupation numbers can also be calculated from the fundamental postulate for averaging,

$$n_{\text{v}} = \frac{\sum_i n_{v_i}}{\Omega}$$

where the sum runs over all the Ω states. In this example, $\Omega = 84$. The symbol n_{v_i} means the number of times the quantum number v occurs in the ith state. Calculate the seven mean occupation numbers in this way and compare the results with those obtained in part e.

g. The total energy of a system of the ensemble is given by

$$E = N \frac{\sum_i E_i e^{-E_i/k_B T}}{\sum_i e^{-E_i/k_B T}}$$

where E_i is the energy of the ith oscillator. Show that

$$E = 2h\nu + \frac{4h\nu}{e^{h\nu/k_B T} - 1}$$

and use this relation to define a temperature for the system.
Result: $T = \frac{h\nu}{k_B \ln(5/3)}$

h. Find the most probable distribution numbers n_v^* for $v = 0$, 1, 2, 3, 4, 5, 6, 7. Use the most probable distribution number relation for large ensembles:

$$n_v^* = N \frac{e^{-E_v/k_B T}}{\sum_v e^{-E_v/k_B T}}$$

By summing the series and using the temperature $T = \frac{h\nu}{k_B \ln(5/3)}$, show that

$$n_v^* = \frac{8}{5}(3/5)^v$$

You may find it useful to recall that $x = e^{\ln x}$. Check your result by summing all the n_v^* (the sum has to be N, which is 4 in this example).

i. Make a plot of $\langle n_v \rangle$ versus v, and, on the same graph and to the same scale, plot n_v^* versus v.

j. In part h, a value for n_7^* was calculated. What value must $\langle n_7 \rangle$ assume? Why is n_7^* different from $\langle n_7 \rangle$?

2. Assume for simplicity that the eigenstates of a nuclear spin Hamiltonian consist of n different energy levels W_m equally spaced W joules apart. Take the zero of energy at the center of gravity of the energy levels. Then the mth energy level $W_m = mW$, where m is an integer or half integer that lies between the extreme values $-(n-1)/2$ and $+(n-1)/2$. All states are accessible. For example, if there are only 4 spin states, $n = 4$ and the energy level diagram sketched below is appropriate.

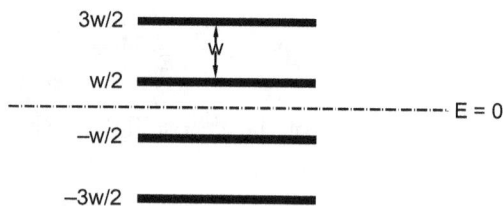

Such spin states can arise in a number of different ways. One common example is in the interaction of an externally applied large magnetic field with a nuclear magnetic dipole moment, leading to precisely defined orientations of the nuclear spin with the field direction. The nuclear spin I may assume half integer values, and, depending on the nucleus, may be 0, 1/2, 1, 3/2, 2, 5/2 ... There are then $2I + 1$ distinct orientational states. The levels are not degenerate, and there is one state for each level. The diagram above applies to a nucleus of spin 3/2 in a large magnetic field. In any spin orientation problem, however generated, the energy levels possess the important property of being truncated not only from below, as are the harmonic oscillator energy levels, but also from above, which is very different from the unlimited number of energy levels of the harmonic oscillator.

a. The properties considered below require the following mathematical preparation. If you are not familiar with the properties of the hyperbolic functions, look them up. The hyperbolic sine, for example, is defined as

$$\sinh(x) = \frac{e^x - e^{-x}}{2}$$

Make a table of the functions, their definitions and their derivatives. It is useful in what comes to have memorized these.

b. Use synthetic division to show that $\frac{1-z^n}{1-z} = \sum_{k=0}^{n-1} z^k$.

c. Write the partition function $Z = \sum_i e^{-E_i/k_B T}$, where the sum runs over all accessible states, for the case of nuclei of spin 3/2 oriented in a magnetic field. Factor $e^{3w/2k_B T}$ from the expression. Note that there are only 4 non-degenerate levels, and that because of this there are only four terms in the sum that represents the partition function.

d. Use the result of part b to show that the partition function may be written

$$Z = e^{3w/2k_B T} \frac{1 - e^{-4w/k_b T}}{1 - e^{-w/k_B T}}$$

e. Show that

$$Z = \frac{\sinh(2w/k_B T)}{\sinh(w/2k_B T)}$$

f. Write the Helmholtz Free Energy A for N spins (N is a number of the order of Avogadro's number).

g. Set $\beta = 1/k_B T$. Show that

$$E = Nk_B T^2 \left(\frac{\partial \ln Z}{\partial T}\right)_{N,V} = \left(\frac{\partial \beta A}{\partial \beta}\right)_{N,V}$$

$$S = -\left(\frac{\partial A}{\partial T}\right)_{N,V} = k_B \beta^2 \left(\frac{\partial A}{\partial \beta}\right)_{N,V}$$

$$C_v = Nk_B \left(\frac{\partial}{\partial T} T^2 \frac{\partial \ln Z}{\partial T}\right)_{N,V} = -k_B \beta^2 \left(\frac{\partial^2 \beta A}{\partial \beta^2}\right)_{N,V}$$

where the first written derivatives (those in terms of T) may be taken as given, and the second (those in terms of β) are to be shown.

h. Show that for the four spin state problem above,

$$E = -\frac{Nw}{2}(4\coth(2\beta w)) - \coth(\beta w/2)$$

$$S = Nk_B \ln\left(\frac{\sinh(2\beta w)}{\sinh(\beta w/2)}\right)$$

$$-\frac{Nk_B \beta w}{2}(4\coth(2\beta w) - \coth(\beta w/2))$$

$$C_v = Nk_B (\beta w/2)^2 (\operatorname{csch}^2(\beta w/2) - 16\,\operatorname{csch}^2(2\beta w))$$

i. In problem 1, specification of the energy E of a system of the ensemble led to a definition of the temperature T. In this case, also, specification of E will lead to definition of a temperature for the systems of the ensemble. This in turn determines β, upon which S entirely depends once w is given (w is a property of the spin system). Thus, S may be regarded as a function of E through T. Make a schematic plot of S/Nk_B versus E/Nw. Be sure to note that E can be negative (corresponds to putting an excess of the spins in negative energy states).

j. Let $dE = TdS - \mathbf{M} \bullet d\mathbf{H}$, where $\mathbf{H}$ is the magnetic field and $\mathbf{M}$ is the bulk magnetization for the collection of spins. How may temperature be defined from this relation?

k. Interpret the slope of your E versus S plot obtained in part i of the above at various points. Is there a region of the plot that corresponds to positive temperatures? Is there a region that corresponds to negative temperatures? Where is the temperature infinite, and where is it zero? To what physical situation would negative temperatures correspond? What would be the likely consequence of allowing electromagnetic radiation of a frequency ω/h to be incident on a system in a state of negative temperature in a waveguide with perfectly reflecting ends?

3. Stirling's approximate formula is (Abramowitz and Stegun, NBS 55)

$$\Gamma(n+1) = n! = \sqrt{2\pi} n^{n+\frac{1}{2}} e^{-n+\frac{\Theta}{12n}}$$

where $0 < \Theta < 1$ and n is positive.

a. Show that the common statement of Stirling's approximation ($\ln n! = n \ln n - n$) omits several terms.
b. Prepare a table with columns headed $n, n!, n \ln n!, n \ln n - n$, and $\ln(\sqrt{2\pi} n^{n+\frac{1}{2}} e^{-n+\frac{\Theta}{12n}})$, where $\theta = 1/2$. Set the n values to integers $1, 2, 3, \ldots, 10$. Compare $\ln n!$ with $n \ln n - n$ in a separate column (use the difference divided by $\ln n!$).

4. The slope of a monotonically increasing function of argument x can be approximated by

$$\frac{df\left(x+\frac{1}{2}\right)}{dx} \cong f(x+1) - f(x)$$

Let the function be $\ln x!$. Using the fact that $\int \ln x \, dx = x \ln x - x + \text{constant}$, show that

$$\ln x! \cong x \ln x - x + \frac{1}{2} \ln x + \frac{3}{2} - \frac{3}{2} \ln \frac{3}{2} + \frac{1}{4x}$$

Make a column in the table prepared in problem 3 above and enter appropriate values for $\ln n!$ calculated from the formula just derived. How do these values compare with $\ln n!$?

5. The nuclear spin of ^{203}Tl is 1/2, and the energy levels of this isotope are split by 0.278 cm^{-1}. The partition function for nuclear spin in ^{203}Tl is

$$q_{NS} = 1 + e^{-\Theta_{NS}/T}$$

where $\Theta = \frac{hc\varepsilon}{k_B}$ and $\varepsilon = 0.278$ cm^{-1} while h is Planck's constant and c is the velocity of light in cm s^{-1} $(2.99792458 \times 10^{10}$ cm s$^{-1})$.

a. The dimensionless heat capacity contribution due to nuclear spin in ^{203}Tl is

$$\frac{C_{NS}}{R} = \frac{\partial}{\partial T} T^2 \frac{\partial \ln q_{NS}}{\partial T}$$

Plot this function from -1 K to 1 K.

b. The dimensionless entropy contribution due to nuclear spin is

$$\frac{S_{NS}}{R} = \frac{\partial}{\partial T} T \ln q_{NS}$$

Plot this function from -1 K to 1 K.

c. What is the value of C_{NS} and S_{NS} at 10, 100, 1000 and 10000 K?

6. Nuclei with spin 1 or greater have an electric quadrupole moment that interacts with an electrostatic field gradient at the quadrupolar nucleus to provide orientational states of several energies, just as the nuclear magnetic moment interacts with an externally applied magnetic field to produce different orientational states of various energies. All ^{14}N nuclei have spin 1, and in crystals where the electrostatic field gradient is constant from nucleus to nucleus due to crystal symmetry, these orientational states may be interrogated by radio frequency waves and the energy differences measured. The measurement yields a product of the quadrupole moment eQ and the field gradient eq, where e is the electronic charge, and is measured in MHz units. To avoid confusion with the partition functions q and Q, in this problem the quadrupole coupling constant $(e^2 Qq)/h = \chi$ will be used.

a. The three orientational states of nitrogen nuclei have energies

$$\varepsilon_0 = 0, \quad \varepsilon_1 = \frac{3\chi}{4}\left(1 - \frac{\eta}{3}\right), \quad \varepsilon_2 = \frac{3\chi}{4}\left(1 + \frac{\eta}{3}\right)$$

where η is a dimensionless variable called the asymmetry parameter. It varies between 0 and 1 and measures the departure of the electrostatic field gradient from cylindrical symmetry. When $\eta = 0$, the gradient is cylindrically symmetric. For example, in crystalline thiourea, $\eta = 0.3954$ and $\chi = 3.1216$ MHz. These parameters are temperature independent. Calculate a characteristic temperature from the quadrupole coupling constant of crystalline thiourea. Answer: 0.0001124 K.

b. Show that the partition function q may be written

$$q = 1 + 2\cosh\left(\frac{\Theta\eta}{3T}\right)$$

c. Give an expression for the average quadrupolar energy in a canonical ensemble of such spins.

d. Show that the quadrupolar heat capacity in crystalline $^{14}N_2$, in which the asymmetry parameter is zero, is

$$C_v/R = \frac{2x^2 e^x}{(2 + e^x)^2}$$

where $x = \frac{\Theta}{T}$.

e. Plot C_v/R as a function of T from $-\infty$ to $+\infty$. Locate the maxima and explain why quadrupolar states do not contribute to the specific heat at chemically useful temperatures.

6. The partition function Φ is given by the sum

$$\Phi = \sum_{G_1}\sum_{G_2}(G_1, G_2, G_3)e^{g_1 G_1}e^{g_2 G_2}$$

The characteristic thermodynamic function is

$$F(g_1, g_2, G_3) = \ln \Phi$$

where g_1 and g_2 are intensive and G_3 is extensive. Show that the ensemble averages of G_1 and G_2 are given by the derivatives

$$\langle G_1 \rangle = \left(\frac{\partial F}{\partial g_1} \right)_{g_2 G_3}$$

$$\langle G_2 \rangle = \left(\frac{\partial F}{\partial g_2} \right)_{g_1 G_3}$$

What is the result of differentiating twice?

7. Use the fact that the specific heat must be always positive to show that Boltzmann's constant k_B is always positive, even in a region of negative temperatures. Keep in mind that the average energy in the canonical ensemble is

$$\langle E \rangle = \frac{\sum_i E_i e^{-\frac{E_i}{k_B T}}}{\sum_i e^{-\frac{E_i}{k_B T}}}$$

where the sums are over the states of the ensemble.

Chapter 3

Steepest Descent

In the last chapter, the equations that characterize the canonical ensemble were worked out, based on the use of the Stirling approximation, which depends for accuracy on the size of the argument. If $\ln n! \cong n \ln n - n$ is to be approximately correct, n must be large. However, in the most probable distribution, the n_i are all either 0 or 1, and the approximation is at its worst for these numbers. At least for this reason the results obtained must be suspect, and require substantiation by more rigorous means. There is another, more subtle, error made in the analysis that proves fortuitously just to compensate for the error made by using Stirling's approximation, and the results obtained are in fact correct results, just obtained by an incorrect method. That can be proved by using a mathematical technique called "steepest descent", and the purpose of this chapter is to proceed through the steepest descent method and show that exactly the same results are obtained. The original analysis was done by Darwin and Fowler and is described in Fowler's book.[1] The presentation here follows suggestions made by Schrodinger in a seminar later published by Cambridge in a paperback edition.[2]

First, however, the reader will recall that in the last chapter the degeneracy Ω was calculated by summing up all the states that were compatible with the total energy and number of systems. If, instead, it is desired to sum over the energy levels rather than the states, then the degeneracy of each level has to be taken into account. The degeneracy Ω_i is just the number of states with exactly the same energy E_i. Thus, the partition function $Q(T, V, N)$ can be written in either of two completely equivalent ways:

$$Q = \sum_i e^{-E_i/k_B T} \quad \text{sum over states} \tag{3.1}$$

$$Q = \sum_E \Omega(E, V, N) e^{-E/k_B T} \quad \text{sum over levels} \tag{3.2}$$

When the degeneracy $\Omega(E, V, N)$ is itself evaluated, the sum that evaluates Ω is over all possible distributions compatible with the restrictive conditions on E and N

$$\Omega = \sum_{(n)} \frac{N!}{\prod_i n_i!} \quad \text{sum over all distributions subject to}$$

the restrictive conditions

$$E = \sum_i E_i n_i$$

$$N = \sum_i n_i \tag{3.3}$$

The Darwin–Fowler method of steepest descents evaluates Ω directly, without approximation. Along the way, it affords a deep insight into the nature of temperature, and thus is worth pursuing in detail. It starts with a theorem from algebra intimately related to the famous binomial theorem, called the multinomial theorem. Given the binomial theorem, the multinomial theorem is easily induced once the binomial theorem is written in a suitable form. One of the usual expressions of the binomial theorem is

$$(a + b)^n = \sum_{k=0}^{n} \frac{n!}{(n-k)!k!} a^{n-k} b^k \tag{3.4}$$

This may be rewritten as

$$(a + b)^n = \sum_{(n)} \frac{n!}{n_1! n_2!} a^{n_1} b^{n_2} \quad \text{subject to th econdition } n = n_1 + n_2 \tag{3.5}$$

where the notation (n) means to sum over all possible values of n_1 and n_2 consistent with the restrictive condition $n_1 + n_2 = n$. If a third character is added to a and b, the theorem becomes

$$(a + b + c)^n = \sum_{(n)} \frac{n!}{n_1!n_2!n_3!} a^{n_1} b^{n_2} c^{n_3}$$

$$n = n_1 + n_2 + n_3 \tag{3.6}$$

or in general the multinomial theorem results in the form

$$\left(\sum_i a_i \right)^n = \sum_{(n)} \frac{n!}{\prod_i n_i!} \prod_i a_i^{n_i}$$

$$n = \sum_i n_i \tag{3.7}$$

If a_i is now set equal to $\omega_i z^{\varepsilon_i}$, where z is a complex number, and n is set equal to N, there results a complex number closely related to the argument in the partition function sum

$$\left(\sum_i \omega_i z^{\varepsilon_i} \right)^N = \sum_{(n)} \frac{N!}{\prod_i n_i!} \prod_i (\omega_i z^{\varepsilon_i})^{n_i}$$

$$N = \sum_i n_i \tag{3.8}$$

Now $(\omega_i z^{\varepsilon_i})^{n_i} = \omega_i^{n_i} z^{\varepsilon_i n_i}$ and therefore the sum can be rewritten as

$$\left(\sum_i \omega_i z^{\varepsilon_i} \right)^N = \sum_{(n_i)} \frac{N!}{\prod_i n_i!} z^{\sum_i \varepsilon_i n_i} \prod_i \omega_i^{n_i}$$

$$N = \sum_i n_i \tag{3.9}$$

Upon identifying E with $\sum_i n_i \varepsilon_i$ it is apparent that the sum on the right hand side of the equation is more general than Ω itself, for the sum contains every possible energy, and Ω for a particular energy E occurs as the coefficient of z^E. All that is necessary is to extract it. Fortunately, there is a theorem in functions of a complex variable, due to Cauchy, that allows doing exactly this. The theorem states that if $G(z) = \sum_{n=-\infty}^{n=\infty} C_n z^n$, where z is a complex variable and $G(z)$ is analytic, then

$$C_{-1} = \frac{1}{2\pi i} \oint G(z) dz \qquad (3.10)$$

Then any coefficient can be extracted by forming

$$C_n = \frac{1}{2\pi i} \oint \frac{G(z)}{z^{n+1}} dz \qquad (3.11)$$

Then for $f(z) = \sum_i w_i z^{\varepsilon_i}$ the analytic function $G(z) = (f(z))^N = (\sum_i w_i z^{\varepsilon_i})^N$. For a specific energy E, the degeneracy Ω is

$$\Omega(E) = C_E = \frac{1}{2\pi i} \oint \frac{G(z)}{z^{E+1}} dz \qquad (3.12)$$

The evaluation of Ω reduces to a consideration of the properties of the integrand. For this purpose, it is convenient to introduce two new variables, $\varphi(z) = G(z)/z^{E+1}$ and $g(z) = \ln \varphi(z)$. The first and second derivatives of $g(z)$ are represented by $g'(z)$ and $g''(z)$. These functions are

$$g(z) = \ln \varphi(z) = -(E+1)\ln z + N \ln f(z) \qquad (3.13)$$

$$g'(z) = \frac{1}{\varphi}\frac{d\varphi}{dz} = -\frac{E+1}{z} + \frac{N}{f}\frac{df}{dz} \qquad (3.14)$$

$$g''(z) = \frac{E+1}{z^2} + N\left[\frac{f''(z)}{f(z)} - \left(\frac{f'(z)}{f(z)}\right)^2\right] \qquad (3.15)$$

Consider the behavior of $\varphi(z)$ along the real axis ($z = x + iy$) in the interval $0 \le x \le 1$. The behavior of $g'(z)$ is determined by two terms, $-\frac{E+1}{z}$ and $\frac{N}{f}\frac{df}{dz}$. The first term is negatively infinite at the origin and increases monotonically up to $-(E+1)$ at $z = 1$. Thus, $-\frac{E+1}{z}$ is always negative in the interval $0 \le x \le 1$ along the real axis. Now suppose $\frac{N}{f}\frac{df}{dz}$ always increases with increasing z in the interval $0 \le x \le 1$. Then $\frac{d}{dz}(\frac{N}{f}\frac{df}{dz}) > 0$ would always hold in the interval. Finally, $z\frac{N}{f}\frac{df}{dz}$ would always increase with increasing x

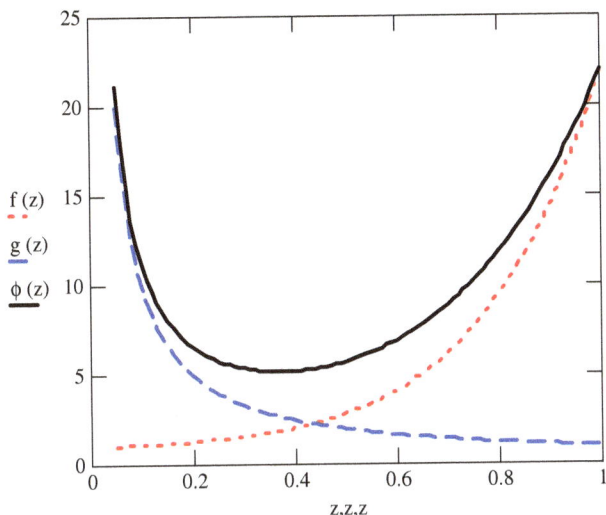

Fig. 3.1. Solid line: typical behavior of $\varphi(x)$. Dotted line: typical behavior of $f(x)$. Dashed line: behavior of $1/x^{E+1}$. All are along the real axis in the interval $0 \leq x \leq 1$.

in the interval. Conversely, if $z\frac{N}{f}\frac{df}{dz}$ increases, then $\frac{N}{f}\frac{df}{dz}$ must do so also. That is, in terms of the slopes,

$$\frac{d}{dz}\left(\frac{N}{f}\frac{df}{dz}\right) > 0 \leftrightarrow \frac{d}{dz}z\left(\frac{N}{f}\frac{df}{dz}\right) > 0 \tag{3.16}$$

The double headed arrow means each implies the other. If the latter term is again multiplied by z, the result will still be positive. Then along the real axis in the interval of interest the chain of logic is

$$\frac{N}{f}\frac{df}{dz} > 0 \leftrightarrow z\frac{N}{f}\frac{df}{dz} > 0 \leftrightarrow \frac{d}{dz}z\left(\frac{N}{f}\frac{df}{dz}\right) > 0 \leftrightarrow z\frac{d}{dz}z\left(\frac{N}{f}\frac{df}{dz}\right) > 0 \tag{3.17}$$

and all that must be shown along the real axis is that everywhere in the interval of interest

$$z\frac{d}{dz}z\left(\frac{N}{f}\frac{df}{dz}\right) > 0 \tag{3.18}$$

Equation (3.18) completely determines the behavior of $\frac{N}{f}\frac{df}{dz}$ in the interval. This may be done by considering the nature of $f(z)$. By definition,

$$f(z) = \sum_i \omega_i z^{\varepsilon_i} \tag{3.19}$$

Differentiating produces

$$z\frac{df}{dz} = \sum_i \varepsilon_i \omega_i z^{\varepsilon_i} \tag{3.20}$$

hence

$$z\frac{N}{f}\frac{df}{dz} = N\frac{\sum_i \varepsilon_i \omega_i z^{\varepsilon_i}}{\sum_i \omega_i z^{\varepsilon_i}} \tag{3.21}$$

Differentiating once again,

$$z\frac{d}{dz}z\left(\frac{N}{f}\frac{df}{dz}\right) = N\frac{\left(\sum_i \omega_i z^{\varepsilon_i}\right)\left(\sum_i \varepsilon_i^2 \omega_i z^{\varepsilon_i}\right) - \left(\sum_i \varepsilon_i \omega_i z^{\varepsilon_i}\right)^2}{\left(\sum_i \omega_i z^{\varepsilon_i}\right)^2}$$
$$\tag{3.22}$$

This expression is greater than zero in the region $0 < x < 1$ along the real axis, since the expression on the right hand side of the equation can be factored as

$$z\frac{d}{dz}z\left(\frac{N}{f}\frac{df}{dz}\right) = N\frac{\left(\sum_k \varepsilon_k \omega_k z^{\varepsilon_k}\right)\left(\varepsilon_k\sqrt{\sum_i \omega_i z^{\varepsilon_i}} - \frac{\sum_i \varepsilon_i \omega_i z^{\varepsilon_i}}{\sqrt{\sum_i \omega_i z^{\varepsilon_i}}}\right)^2}{\left(\sum_i \omega_i z^{\varepsilon_i}\right)^2}$$
$$\tag{3.23}$$

In equation (3.23), the positive branches of the square roots are intended. It was established above that $f(z) = \sum_i \omega_i z^{\varepsilon_i}$ increases along the real axis from 1 at $z = 0$ to some greater positive finite value at $z = 1$, and is always positive in the range $0 < x < 1$. The other quantities in brackets are squared, and are therefore necessarily positive everywhere along the real axis. This proves that $z\frac{d}{dz}z(\frac{N}{f}\frac{df}{dz}) > 0$, and by the logic train given above $\frac{N}{f}\frac{df}{dz} > 0$ along the real axis in the range from 0 to 1. Hence $\varphi(z)$ exhibits only one minimum along the real axis in the interval $0 \le x \le 1$.

Next consider the behavior of $f(z)$ along a circle of radius η in the complex plane. It is convenient to consider the behavior of the logarithm of η instead, which was introduced above as $g(z)$. The condition for a minimum in the slope of $g(z)$ is that $g'(z) = 0$. This leads to a value of z at the minimum in g that is designated η. The condition that results is

$$\frac{E+1}{\eta} = \frac{N}{f(\eta)} \frac{df(\eta)}{dz} \tag{3.24}$$

where the symbol $\frac{df(\eta)}{dz}$ means the derivative of $f(z)$ evaluated at $z = \eta$. To illustrate, assume $f(z) = 1 + z + z^2 + z^3 + z^4 + \ldots$ Now let $z = re^{i\Theta}$. The function f becomes

$$f\left(re^{i\Theta}\right) = 1 + re^{i\Theta} + r^2 e^{2i\Theta} + r^3 e^{3i\Theta} + r^4 e^{4i\Theta} + \cdots \tag{3.25}$$

that is

$$f(re^{i\Theta}) = 1 + r(\cos\Theta + i\sin\Theta) + r^2(\cos 2\Theta + i\sin 2\Theta)$$
$$+ r^3(\cos 3\Theta + i\sin 3\Theta) + \cdots \tag{3.26}$$

and at the radius η, the function becomes

$$f(\eta) = 1 + \eta\cos\Theta + \eta^2\cos 2\Theta + \eta^3\cos 3\Theta + \cdots$$
$$+ i(\eta\sin\Theta + \eta^2\sin 2\Theta + \eta^3\sin 3\Theta + \cdots) \tag{3.27}$$

For any value of Θ other than $2k\pi$, where k is an integer, $|f(z)|^2 < |f(\eta)|^2$ since the trigonometric functions have values between 1 and -1, and the maximum real value is only attained when the cosine is 1. For example, if $\Theta = \pi$, $\cos\pi = -1$, $\cos 2\pi = 1$, $\cos 3\pi = -1$, and so on, while the sine functions all vanish. Then

$$f(\eta)_{\Theta=\pi} = 1 - \eta + \eta^2 - \eta^3 + \cdots < f(\eta)_{\Theta=0} = 1 + \eta + \eta^2 + \eta^3 + \eta^4 + \cdots \tag{3.28}$$

even though the same radius is used. These considerations show that the maximum value of $f(z)$ above the complex plane occurs above the real axis along the circumference of a circle of radius η. The same argument applies to $f(z)/z^{E+1} = \varphi(z)$.

Thus, $\varphi(z)$ has a minimum in the real plane, and a maximum in the plane orthogonal to the real plane at the value of $z = \eta$. Such a point is called a saddle point, for obvious reasons. At the saddle point $z = \eta$, and this is determined by the conditions on $g(z) = \ln \varphi(z)$, namely

$$g'(z) = -\frac{E+1}{z} + \frac{N}{f}\frac{df}{dz} = 0 \qquad (3.29)$$

$$g''(z) = \frac{E+1}{z^2} + N\left[\frac{f''(z)}{f(z)} - \left(\frac{f'(z)}{f(z)}\right)^2\right] \gg 0 \quad \text{at } z = \eta \quad (3.30)$$

since both E and N are very much greater than zero and both terms of $g''(\eta)$ are positive. Now expand $g(z)$ in Taylor's series near η.

$$g(z) = g(\eta) + g'(\eta)(z - \eta) + \frac{1}{2}g''(\eta)(z - \eta)^2 + \cdots \qquad (3.31)$$

But $g'(\eta) = 0$ at the saddle point. In order to investigate the behavior of $\varphi(z)$ in the imaginary direction at η, let $z = \eta + iy$. This produces

$$g(\eta + iy) = g(\eta) - \frac{1}{2}g''(\eta)y^2 + \cdots \qquad (3.32)$$

Since $g''(\eta) \gg 0$, $y \ll 1$ suffices to produce the dominant behavior. Since $g(z) = \ln \varphi(z)$,

$$\varphi(z) = e^{g(z)} = e^{g(\eta)}e^{-\frac{1}{2}g''(\eta)y^2} = \frac{f(z)^N}{\eta^{E+1}}e^{-\frac{1}{2}g''(\eta)y^2} \qquad (3.33)$$

Owing to the very large size of $g''(\eta)$, this is a needle sharp Gaussian. All sensible contributions to the integral $\oint \varphi(z)dz$ around a circle of radius η are made for extremely small values of y. This allows writing Ω very nearly as

$$\Omega = \frac{1}{2\pi i} \int_{-a}^{a} \frac{f(\eta)^N}{\eta^{E+1}}e^{-\frac{1}{2}g''(\eta)y^2} i\,dy \qquad (3.34)$$

where a is a very small quantity in the imaginary direction. Trivially small contributions are made to the integral for values of y greater than a, and nothing prevents replacing the limits with infinity. Then Ω becomes

$$\Omega = \frac{f(\eta)^N}{2\pi\eta^{E+1}} \int_{-\infty}^{\infty} e^{-\frac{1}{2}g''(\eta)y^2} dy \qquad (3.35)$$

The integral is a standard one, of the form

$$\int_{-\infty}^{\infty} e^{-a^2 y^2} dy = \frac{\sqrt{\pi}}{\alpha} \qquad (3.36)$$

Setting $\alpha^2 = \frac{1}{2}g''(\eta)$ evaluates Ω as

$$\Omega = \frac{f(\eta)^N}{\eta^{E+1}\sqrt{2\pi g''(\eta)}} \qquad (3.37)$$

This result embodies all the essential physical results of statistical theory, results that are identical with those obtained by the approximate Boltzmann or maximum term approach illustrated in Chapter 2. To show this, start with the logarithm of Ω.

$$\ln \Omega = N \ln f(\eta) - (E+1) \ln \eta - \frac{1}{2}\ln[2\pi g''(\eta)] \qquad (3.38)$$

Now $g''(\eta)$ is of the order of either E or N, depending on which is larger, and $\ln g''(\eta)$ is very much smaller. For example, if $g''(\eta) = 10^{23}$, $\ln g''(\eta) \approx 53$. In this expression, the term in $g''(\eta)$ is entirely negligible in comparison with the terms multiplied by E or N. Then, recalling the definition of $f(\eta)$, $\ln \Omega$ becomes very nearly

$$\ln \Omega = N \ln \left(\sum_i w_i \eta^{\varepsilon_i} \right) - (E+1) \ln \eta \qquad (3.39)$$

From the condition $g'(\eta) = 0$, there obtains

$$\frac{E+1}{\eta} = N\frac{f'(\eta)}{f(\eta)} = N\frac{\sum_i \varepsilon_i w_i \eta^{\varepsilon_i - 1}}{\sum_i w_i \eta^{\varepsilon_i}} \qquad (3.40)$$

Neglecting 1 in comparison to the very large E, then

$$\frac{E}{N} = \frac{\sum_i \varepsilon_i w_i \eta^{\varepsilon_i}}{\sum_i w_i \eta^{\varepsilon_i}} \qquad (3.41)$$

To get agreement with the Boltzmann results, it is only necessary to identify the radius η with the exponential inverse temperature

$$\eta = e^\beta \qquad (3.42)$$

where ultimately β will be identified as $-1/k_B T$. To do this requires the second law, which guarantees the existence of a temperature scale that is independent of the thermometric substance. Given this, the identification above is particularly pleasing, showing the statistical nature of the temperature, which can be only as sharply defined as the saddle point, and that depends on the magnitudes of E and N. When E and N decline in value, the saddle point is less and less well defined, until when $N = 1$, it is undefined. The saddle point is flat.

Rewriting the principal results above in terms of e^β summarizes the main results of the Darwin–Fowler steepest descent approach.

$$\ln \Omega = N \ln \sum_i \omega_i e^{\beta \varepsilon_i} - \beta E \qquad (3.43)$$

$$\frac{E}{N} = \frac{\sum_i \varepsilon_i \omega_i e^{\beta \varepsilon_i}}{\sum_i \omega_i e^{\beta \varepsilon_i}} \qquad (3.44)$$

To complete results, the mean distribution numbers are required. Start with the definition of average.

$$\langle f_{obs} \rangle = \frac{\sum_{i=1}^{\Omega} f_i}{\Omega} \qquad (3.45)$$

and regard n_j as the quantity to be averaged over all the distributions possible. In one distribution, the total number of states is $N!/\prod_i n_i!$, in which each n_j will have one specific value. The total contribution to the sum is then n_j times the number of states. To get all the contributions, it is necessary to sum over all possible distributions consistent with the restrictive conditions on E and N. If all the energy levels are non-degenerate and the sum in the partition function is over states, the average distribution number would be

$$\langle n_j \rangle = \frac{\sum_{(n)} n_j \frac{N!}{\prod_i n_i!}}{\Omega} \qquad (3.46)$$

and if the sum is over levels, each of which is ω_i-fold degenerate, the expression becomes

$$\langle n_j \rangle = \frac{\sum_{(n)} n_j \frac{N!}{\prod_i n_i!} \prod_i \omega_i^{n_i}}{\Omega} \tag{3.47}$$

The average may be formally calculated by writing

$$\langle n_j \rangle = \frac{\partial \ln \Omega}{\partial \ln \omega_j} \tag{3.48}$$

Since Ω is given by

$$\Omega = \sum_{(n)} \frac{N!}{\prod_i n_i!} \prod_i \omega_i^{n_i} \tag{3.49}$$

subject to $E = \sum_i \varepsilon_i n_i$ and $N = \sum_i n_i$. Differentiating,

$$\frac{\partial \ln \Omega}{\partial \omega_j} = \sum_{(n)} \frac{N!}{\prod_i n_i!} n_j \omega_j^{n_j-1} \prod_{i \neq j} \omega_i^{n_i} \tag{3.50}$$

and on multiplying both sides by ω_j the desired average results. From equation (3.38)

$$\ln \Omega = N \ln \sum_i \omega_i \eta^{\varepsilon_i} - (E+1) \ln \eta - \frac{1}{2} \ln 2\pi g''(\eta) \tag{3.51}$$

Since the term in $g''(\eta)$ is negligible, the logarithmic derivative is essentially

$$n_j = \frac{\partial \ln \Omega}{\partial \ln \omega_j} = N \frac{\omega_j \eta^{\varepsilon_j}}{\sum_i \omega_i \eta^{\varepsilon_i}} \tag{3.52}$$

and on setting $\eta = e^\beta$, the Boltzmann relation results.

$$\langle n_j \rangle = N \frac{\omega_j e^{\beta \varepsilon_j}}{\sum_i \omega_i e^{\beta \varepsilon_i}} \tag{3.53}$$

Several remarks are now in order. First, no real thermodynamic system is ever completely isolated. None has exactly an energy E and a number of molecules N. Instead, the best that can be said is that the energy and number of molecules fluctuates between limits, such that the energy is approximately E_0, where $E_0 - dE \leq E \leq E_0 + dE$, and the number of molecules is approximately N_0, where $N_0 - dN \leq N \leq N_0 + dN$. Second, since $\eta = e^\beta$ and $\beta = -\frac{1}{k_B T}$ there follows $T = \frac{-1}{k_B \ln \eta}$. A determination of η is a determination of T. But η marks the position of the col in $\varphi(z)$, and the exact value of η depends on the values assigned E and N, and these fluctuate. Hence, the temperature fluctuates as it follows the values of E and N. The temperature T is statistically defined, and the col is needle sharp in systems with large E and N. However, the smaller the thermodynamic system, the smaller N and E become, and the less well defined T becomes as the col flattens and fluctuations in E and N become relatively large. In the limit of a single molecule, T is undefined and only the energy is a meaningful variable.

In this chapter it has been shown that the results of the Darwin–Fowler steepest descent method coincide exactly with the results of the less rigorous but mathematically simpler Boltzmann approach in the limit of sufficiently large E and N. Darwin–Fowler is exact, and does not depend on Stirling's approximation for the common cases $n_i = 0$ or 1. That the results are identical is a justification for the use of the Boltzmann method, even to treat problems that are difficult to treat by the Darwin–Fowler approach. The results presented are ultimately justified by the ability of these results to describe accurately physical reality, within the approximations needed to progress, and of course within the limits of unavoidable experimental error.

There remains one job to be done to establish firmly the connection between the mathematical approach of the Darwin–Fowler method and the more intuitive Boltzmann method, namely, the identification of the parameter β as an inverse temperature. This may be done as follows. Start with the entropy in the microcanonical ensemble. Doing so requires identifying ε_i as an energy level of one of the N identical molecules in a system of the ensemble, and ω_i as the degeneracy of the ith energy level. Then start from the combined first and second law equations for the variation of the energy and identify the reciprocal temperature, keeping in mind that the entropy is given by

$$S = k_B \ln \Omega \tag{3.54}$$

$$dE = TdS - pdV + \mu dN \tag{3.55}$$

$$\frac{1}{T} = \left(\frac{\partial S}{\partial E}\right)_{V,N} \tag{3.56}$$

The energy is given in terms of the energy levels and degeneracies as

$$E = N\frac{\sum_i \varepsilon_i \omega_i e^{\beta \varepsilon_i}}{\sum_i \omega_i e^{\beta \varepsilon_i}} \tag{3.57}$$

Showing that β is a function of E.

$$\beta = \beta(E) \tag{3.58}$$

One of the Darwin–Fowler results was

$$\ln \Omega = N \ln(f(\eta)) - E \ln \eta \tag{3.59}$$

And since $\eta = e^{\beta}$ the entropy can be written as

$$S = Nk_B \ln f\left(e^{\beta}\right) - \beta E k_B \tag{3.60}$$

Substituting the expression for $f(\eta)$ and differentiating,

$$S = Nk_B \ln \sum_i \omega_i e^{\beta \varepsilon_i} - \beta E k_B \tag{3.61}$$

$$\left(\frac{\partial S}{\partial E}\right)_{V,N} = Nk_B \frac{\sum_i \varepsilon_i \omega_i e^{\beta \varepsilon_i}}{\sum_i \omega_i e^{\beta \varepsilon_i}} \left(\frac{\partial \beta}{\partial E}\right)_{V,N} - Ek_B \left(\frac{\partial \beta}{\partial E}\right)_{V,N} - k_B \beta \tag{3.62}$$

$$\frac{1}{T} = k_B E \left(\frac{\partial \beta}{\partial E}\right)_{V,N} - k_B E \left(\frac{\partial \beta}{\partial E}\right)_{V,N} - k_B \beta \tag{3.63}$$

Finally, upon setting $E = \langle E \rangle$, there results

$$\beta = -\frac{1}{k_B T} \tag{3.64}$$

This establishes the nature of β as a reciprocal temperature.

Reference

1. E. Schrodinger, "Statistical Thermodynamics", Cambridge University Press, Cambridge, England, 1960.
2. R. H. Fowler and E. A. Guggen Lein, "Statistical Thermodynamics", Cambridge, 1939.

Chapter 4

Gaussian Distribution, Ensemble Equivalence and Fluctuations

4.1. The Effect of the Density of States on the Probability

As noted in Chapter 2, the probability P_i that a system in the canonical ensemble is in state i with energy E_i is

$$P_i(E_i) = \frac{e^{\beta E_i}}{\sum_i e^{\beta E_i}} \qquad (4.1)$$

the sum running over the states of the ensemble. The sum is so frequently encountered that it is given a special name, Q, and called the partition function, since it shows how energy is partitioned among the systems of the ensemble. The partition function may also be written by summing over the levels E, each of degeneracy $\Omega(E, V, N)$. The probability $P(E)$ that a system of the canonical ensemble be found in a state with energy E is

$$P(E) = \frac{\Omega(E)e^{\beta E}}{\sum_E \Omega(E)e^{\beta E}} \qquad (4.2)$$

In this and the following equations, it will be understood that the V and N dependence of the degeneracy are undisturbed, but only the E dependence is explicitly shown. Whichever way is chosen for the expression of the partition function, the energy dependence is summed out and only the temperature dependence remains. Thus,

$Q = Q(T, V, N)$, the natural variables of the Helmholtz Free Energy. Of course, $\beta = -1/k_BT$.

Some systems have sufficiently fine grained energy level structure that a continuous approach is justified. In such cases, the density of states may need to be taken into account. To this end, let $f(E)dE$ be the probability that the energy of a system of the canonical ensemble lie in the range of energies between E and $E + dE$. Now ask, what is the relationship of P to f? To answer this question, let $d\nu(E)$ be the number of energy levels in the range between E and $E + dE$. Then the number of states, $dG(E)$, that lies in this range is given by

$$dG(E) = \Omega(E)d\nu(E) \tag{4.3}$$

and

$$f(E)dE = P(E)d\nu(E) \tag{4.4}$$

since $P(E)d\nu(E)$ is the probability that the system has an energy somewhere in the range between E and $E + dE$. Then $f(E)$ and $P(E)$ are related by

$$f(E)dE = P(E)\frac{d\nu(E)}{dE}dE \tag{4.5}$$

It is convenient to define a special symbol ν' for the gradient of states $d\nu(E)/dE$. Substituting the expression for the probability $P(E)$ yields

$$f(E) = \nu'\frac{\Omega(E)e^{\beta E}}{Q} \tag{4.6}$$

This leads to five different ways of expressing the canonical ensemble partition function Q, collected in Table 4.1.

In Table 4.1, Q is exactly the same partition function in each case, assuming the energy levels are sufficiently close together (the density of states is sufficiently large) that the Heisenberg uncertainty renders individual resolution impossible.

Table 4.1. Probabilities and the partition function Q.

Sum over	Probability	Partition Function
states	$P_i(E_i) = e^{\beta E_i}/Q$	$Q = \sum_i e^{\beta E_i}$
levels	$P(E) = \Omega(E, V, N)e^{\beta E}/Q$	$Q = \sum_E \Omega(E, V, N)e^{\beta E}$
levels	$f(E) = \nu'\Omega(E, V, N)e^{\beta E}dE/Q$	$Q = \int \nu'\Omega(E, V, N)e^{\beta E}dE$
levels	$f(E)dE = \Omega(E, V, N)e^{\beta E}d\nu(E)/Q$	$Q = \int \Omega(E, V, N)e^{\beta E}d\nu(E)$
states	$f(E)dE = e^{\beta E}dG(E)/Q$	$Q = \int e^{\beta E}dG(E)$

4.2. Gaussian Nature of $f(E)$ or $P(E)$

On qualitative grounds, a maximum in $f(E)$ and $P(E)$ is expected. As E increases, $\Omega(E)$ does so also, and $e^{-E/kT}$ decreases exponentially rapidly. The product must display a maximum, and since both changes are rapid, the maximum must be sharp. Now make these qualitative considerations more quantitative.

Consider how the properties of a system of the canonical ensemble depart from the average, specifically, the average energy $\langle E \rangle$. Further suppose that one system of the canonical ensemble has the average energy and that average energy is equal to the energy E_m assigned to a system of a microcanonical ensemble with the same V and N. The other properties of a system of the microcanonical ensemble are designated with the subscript m, such as T_m for the temperature or S_m for the entropy. Finally, let E represent the energy of any system of the canonical ensemble. In order to find the maximum in $f(E)$, write the probability function as

$$f(E) = \frac{\nu'(E)e^{\frac{S_m(E)}{k_B}}e^{-\frac{E}{k_B T}}}{Q(T, V, N)} \tag{4.7}$$

where use has been made of the characteristic equation of the micro-canonical ensemble,

$$S_m(E) = k_B \ln \Omega(E, V, N) \qquad (4.8)$$

It is easier to maximize the logarithm than it is the function itself. Taking the logarithm, then,

$$\ln f(E) = \ln \nu'(E) + \frac{S_m(E)}{k_B} - \frac{E}{k_B T} - \ln Q(T, V, N) \qquad (4.9)$$

Differentiating with respect to E produces

$$\frac{d \ln f(E)}{dE} = \frac{1}{k_B T} \left(T \frac{dS_m(E)}{dE} - 1 \right) \qquad (4.10)$$

Since Q is independent of E and ν' is nearly so. In the microcanonical ensemble, $\frac{dS_m}{dE} = \frac{1}{T_m}$, and on setting the derivative equal to zero to find the condition for a maximum, there results

$$\frac{T}{T_m} - 1 = 0 \qquad (4.11)$$

or $T = T_m$. This implicitly establishes the value of the average energy in the canonical ensemble, since in the microcanonical ensemble the energy is specified and T_m is therefore a function of it, as $T_m = T_m(E)$. Now expand the Helmholtz Free Energy A in the canonical ensemble about the average energy $\langle E \rangle$

$$A(E) = A_m(\langle E \rangle) + \frac{\partial A_m}{\partial E}(E - \langle E \rangle) + \frac{1}{2}\frac{\partial^2 A_m}{\partial E^2}(E - \langle E \rangle)^2 + \cdots \qquad (4.12)$$

The derivatives are to be evaluated at $E = \langle E \rangle$. To do this, set $A(E) = E - T S_m(E)$. Then

$$\frac{\partial A_m(E)}{\partial E} = 1 - T \frac{\partial S_m(E)}{\partial E} \qquad (4.13)$$

But $\frac{\partial S_m(E)}{\partial E} = \frac{1}{T_m(E)}$ and as $T = T_m$, when $E = \langle E \rangle$, the first derivative vanishes. The second derivative is

$$\frac{\partial^2 A_m(E)}{\partial E^2} = 0 - T \frac{\partial T_m(E)^{-1}}{\partial E} \qquad (4.14)$$

Completing the derivative

$$\frac{\partial^2 A_m(E)}{\partial E^2} = -T(-T_m^{-2})\frac{\partial T_m}{\partial E} \qquad (4.15)$$

Since the derivative is the reciprocal of the constant volume heat capacity in the microcanonical ensemble, and $T = T_m$ at $E = \langle E \rangle$, the second derivative reduces to $1/T_m C_v$. Now $A(E)$ may be restated as

$$A(E) \cong A_m(E) + \frac{1}{2T_m C_v}(E - \langle E \rangle)^2 \qquad (4.16)$$

Recall the expression for $f(E)$ given in Table 4.1. Since $\Omega(E) = e^{S_m(E)/k_B}$, and $\beta = -\frac{1}{k_B T}$, $f(E)$ becomes

$$f(E) = \nu'(E)e^{\beta(E-TS_m(E))}/Q \qquad (4.17)$$

Or, more compactly,

$$f(E) = \nu'(E)e^{-A(E)/k_B T}/Q \qquad (4.18)$$

Substituting the expansion of $A(E)$ obtained above

$$f(E) = \nu'(E)\frac{e^{-A_m(E)/k_B T}e^{-(E-\langle E \rangle)^2/2k_B T^2 C_v}}{Q} \qquad (4.19)$$

This equation establishes the Gaussian character of the probability function $f(E)$. The argument for the probability $P(E)$ is identical, with omission of the rate of change of the density of states with energy $\nu'(E)$.

4.3. Ensemble Equivalence

There remains normalization of the Gaussian probability. In doing this, the equivalence of the canonical and microcanonical ensembles becomes apparent. First consider that in the canonical ensemble, the characteristic equation involving the natural variables is

$$A(T, V, N) = -k_B T \ln Q(T, V, N) \qquad (4.20)$$

This may be expressed alternatively as

$$Q = e^{\beta A} \qquad (4.21)$$

But Q is the sum over all energy space of the numerator in the expression for $f(E)$, and is therefore given by

$$Q = \int_{-\infty}^{\infty} \nu' e^{-A_m(\langle E \rangle)/k_B T} e^{-(E - \langle E \rangle)^2 / 2k_B T^2 C_v} dE \qquad (4.22)$$

Generally, ν' is a slowly varying function of E over the width of the Gaussian, and the product $\nu' e^{-A_m(\langle E \rangle)/k_B T}$ is essentially constant over the region near the average energy The integral then becomes a standard one, encountered in the last chapter, and the result of the integration is

$$Q = \nu' e^{\beta A_m(\langle E \rangle)} \sqrt{2\pi k_B T^2 C_v} \qquad (4.23)$$

But this is also $e^{\beta A}$. Taking the logarithm of both sides of the equation,

$$\beta A = \beta A_m(\langle E \rangle) + \ln \nu' + \frac{1}{2} \ln(2\pi k_B T^2 C_v) \qquad (4.24)$$

This equation shows how the Helmholz Free Energy of the canonical ensemble differs from that of the microcanonical ensemble. The difference can be made more quantitative by observing that ν', C_v, A and A_m are extensive, and therefore of order N. Then the difference between A and A_m relative to A_m is of the order of 10^{-21} or so for systems containing molar amounts of material. For such systems

$$\frac{A - A_m}{A_m} = O\left(\frac{\ln N}{N}\right) \qquad (4.25)$$

If N is of the order of 10^{23}, $\ln N$ is of the order 50, and $50/10^{23}$ is of the order of 10^{-21} or so. Thus, no matter which ensemble is used, essentially the same Helmholtz Free Energy results, and the two ensembles are in this sense equivalent.

The argument for $P(E)$ in the canonical ensemble is identical to the above with omission of the effect of the density of states, and the result may be expressed in terms of the standard deviation for the dispersion of the energy

$$P(E) = \frac{1}{\sqrt{2\pi}\sigma_E} e^{-\frac{(E-\langle E\rangle)^2}{2\sigma_E^2}} \tag{4.26}$$

where the squared standard deviation is $\sigma_E^2 = k_B T^2 C_v$.

Similar considerations apply to other ensembles that may be constructed by initiating or relaxing the communication between the systems of the ensembles. There are three ensembles in which the communication is by exchanges of E (the canonical), V or N, each with its own characteristic function. There are three more in which two of the three variables E, V and N are allowed to exchange: the grand canonical ensemble, in which E and N are exchanged amongst the systems of the ensemble, one in which E and V and exchanged, and one in which V and N are exchanged. Finally, it might be thought that there would be one in which all three variables are allowed to exchange. The characteristic functions are listed in Table 4.2 together with the names of the ensembles and the natural variables, when appropriate.

The entries in Table 4.2 can be easily generated by assuming that in each sum, the most probable value of the summand contributes overwhelmingly more to the sum than any other term or sum of terms. For example, in the microcanonical ensemble $S(E, V, N) = k_B \ln \Omega(E, V, N)$. This implies

$$\Omega = e^{S/k_B} = e^{TS/k_B T} \tag{4.27}$$

In the grand canonical ensemble, for example, the sum in Table 4.2 may be rewritten as

$$pV = k_B T \ln \sum_E \sum_N e^{(TS-E+\mu N)/k_B T} \tag{4.28}$$

Since by definition, $G = H - TS$, and $H = E + pV$, and for a one component system Euler's theorem on homogeneous functions

Table 4.2.   Equivalent ensembles and communication dependencies. All these ensembles are equivalent, in the same sense that the microcanonical and canonical ensembles were shown above to be equivalent.

Characteristic Equation	Functional Dependence	Communication Variables	
$S = k_B \ln \Omega$	$E, V, N*$	None	Microcanonical
$A = -k_B T \ln \sum_E \Omega e^{-E/k_B T}$	$T, V, N*$	E	Canonical
$pV = k_B T \ln \sum_E \sum_N \Omega e^{-E/k_B T} e^{N\mu/k_B T}$	$T, V, \mu*$	E, N	Grand Canonical
$G = \mu N = -k_B T \ln \sum_E \sum_V \Omega e^{-E/k_B T} e^{-pV/k_B T}$	$P, T, N*$	E, V	Gibbs
$H = k_B T \ln \sum_N \Omega e^{\mu N/k_B T}$	E, V, μ	N	No name
$E - G = TS - pV = k_B T \ln \sum_V \Omega e^{-pV/l_B T}$	E, p, N	V	No name
$E = k_B T \ln \sum_N \sum_V \Omega e^{\mu N/k_B T} e^{-pV/k_B T}$	E, p, N	N, V	No name
$0 = k_B T \ln \sum_E \sum_N \sum_V \Omega e^{\mu N/k_B T} e^{-pV/k_B T} e^{-E/k_B T}$	T, μ, p	E, V, N	No name

*The natural variables are starred. The other ensembles have inconvenient functional dependencies, and are therefore not much used.

requires $G = \mu N$, immediately in thermodynamics,

$$pV = TS - E + \mu N \tag{4.29}$$

This is exactly the same as the expression in Table 4.2, if the most probable values of E and N overwhelm the contributions of all the other possibilities. Similar considerations lead to the evaluations shown in the first column of Table 4.2 for the other ensembles.

When the energy is Gaussian, what happens to functions of the energy? To answer this question, consider again the canonical ensemble. Knowing the mean square standard deviation of the energy, σ_E^2, allows calculation of the mean square standard deviation of any function of the energy, $G(E)$. In Gaussian distributions, appreciable values of the energy will occur only very near the average, and $E - \langle E \rangle$ will generally be small quantities. Then $G(E)$ may be expanded as

$$G(E) = G(\langle E \rangle) + \left(\frac{\partial G}{\partial E}\right)_{E=\langle E \rangle} (E - \langle E \rangle) + \cdots \tag{4.30}$$

Define $\Delta G \equiv G(E) - G(\langle E \rangle)$ and note that in general $\left(\frac{\partial G}{\partial E}\right)_{E=\langle E \rangle}$ does not vanish. Then

$$(\Delta G)^2 = \left(\frac{\partial G}{\partial E}\right)^2_{E=\langle E \rangle} (E - \langle E \rangle)^2 \tag{4.31}$$

Neglecting higher powers of $E - \langle E \rangle$ as trivially contributory. Averaging

$$\langle (\Delta G)^2 \rangle = \int \left(\frac{\partial G}{\partial E}\right)^2_{E=\langle E \rangle} (E - \langle E \rangle)^2 f(E) dE \tag{4.32}$$

The derivative is constant in the integration, and the integral is the mean squared standard deviation of the energy, Thus, in general

$$(\Delta G)^2 = \left(\frac{\partial G}{\partial E}\right)^2_{E=\langle E \rangle} \sigma_E^2 \tag{4.33}$$

Take, as a concrete example, $G(E) = \frac{1}{E}$. Then $\frac{\partial G}{\partial E} = -\frac{1}{E^2}$, and

$$\sigma_{1/E}^2 = \left\langle \left(\Delta \frac{1}{E}\right)^2 \right\rangle = \left(-\frac{1}{\langle E \rangle^2}\right)^2 \sigma_E^2 \tag{4.34}$$

This can be rewritten as

$$\frac{\sigma_E}{\langle E \rangle} = \frac{\sigma_{1/E}}{1/\langle E \rangle} \tag{4.35}$$

That is, the relative standard deviation is the same for the energy and the function of energy.

Now ask, what is the probability distribution for $1/E$ given that it is Gaussian for E? For E, $P(E)$ is

$$P(E) = \frac{e^{-(E-\langle E \rangle)^2/2\sigma_E^2}}{\sqrt{2\pi}\sigma_E} \tag{4.36}$$

It will suffice to treat the argument of the exponential. Before averaging,

$$(\Delta G)^2 = \left(\frac{\partial G}{\partial E}\right)^2_{E=\langle E \rangle} (E - \langle E \rangle)^2$$

and this becomes

$$\left(\frac{1}{E} - \frac{1}{\langle E \rangle}\right)^2 = \frac{1}{\langle E \rangle^4}(E - \langle E \rangle)^2$$

Then, since $\sigma_E^2 = \langle E \rangle^4 \sigma_{1/E}^2, \langle E \rangle^4$ cancels from the expression when both sides are divided by $2\,\sigma_E^2$, and there results

$$\frac{(E - \langle E \rangle)^2}{2\sigma_E^2} = \frac{\left(\frac{1}{E} - \frac{1}{\langle E \rangle}\right)^2}{2\sigma_{1/E}^2} \tag{4.37}$$

and a Gaussian probability for the energy implies a Gaussian probability for the function of the energy.

4.4. Fluctuations in N

The grand canonical ensemble is the most convenient ensemble to use in studying fluctuations in N, since the natural variables are T, V and μ. These are the ones that define variations in the thermodynamic function pV, since $dpV = SdT + pdV + Nd\mu$. The partition function in this ensemble may be written

$$\Xi = \sum_E \sum_N \Omega(E, V, N) e^{-E/k_B T} e^{\mu N/k_B T} \qquad (4.38)$$

The activity $\lambda = e^{\mu/k_B T}$, and in terms of this the partition function may be rewritten as

$$\Xi = \sum_N \lambda^N \sum_E \Omega(E, V, N) e^{-E/k_B T} \qquad (4.39)$$

Or, more simply,

$$\Xi = \sum_N \lambda^N Q(T, V, N) \qquad (4.40)$$

Differentiating the logarithm of the partition function with respect to the activity produces

$$\left(\frac{\partial \ln \Xi}{\partial \lambda} \right)_{T,V} = \frac{\sum_N N \lambda^{N-1} Q(T, V, N)}{\sum_N \lambda^N Q(T, V.N)} \qquad (4.41)$$

Multiplying both sides by the activity λ produces the average value of N

$$\left(\frac{\partial \ln \Xi}{\partial \ln \lambda} \right)_{T,V} = \frac{\sum_N N \lambda^N Q(T, V, N)}{\sum_N \lambda^N Q(T, V.N)} = \langle N \rangle \qquad (4.42)$$

Differentiating again,

$$\left(\frac{\partial N}{\partial \ln \lambda} \right)_{T,V} = \langle N^2 \rangle - \langle N \rangle^2 = \left(\frac{\partial^2 \ln \Xi}{\partial \ln \lambda^2} \right)_{T,V} = \langle \Delta N^2 \rangle = \sigma_N^2 \quad (4.43)$$

Finally, as $\ln \lambda = \frac{\mu}{k_B T}$, there results

$$\left(\frac{\partial \langle N \rangle}{\partial \ln \lambda} \right)_{T,V} = \left(\frac{\partial \langle N \rangle}{\partial \mu} \right)_{T,V} k_B T = \frac{k_B T}{\left(\frac{\partial \mu}{\partial \langle N \rangle} \right)_{T,V}} = \langle \Delta N^2 \rangle = \sigma_N^2$$

$$(4.44)$$

where $\left(\frac{\partial \mu}{\partial \langle N \rangle} \right)_{T,V}$ is the reversible work to transfer matter from one system of the ensemble to an adjacent system. These fluctuations will ordinarily be small since the work of transfer of one particle will generally be quite large. But in special circumstances the work may be small to vanishingly small, and then the fluctuations become correspondingly very large. Two examples come immediately to mind.

1. Fluctuations are large near a critical point, which produces the experimental phenomenon of critical opalescence. Critical opalescence may be observed optically, as at the interface between vapor and liquid CO_2 retained in heavy walled glass capillaries at or very near to the critical point. It may also be observed in neutron scattering experiments conducted on crystalline solids maintained near a critical temperature in substances with several stable convertible crystalline forms.

2. Colloid particle occupation numbers undergo large fluctuation since $k_B T$ is of the order of the relatively small work of transfer of colloid particles from one region of suspension volume to an adjacent one. If the colloid particles are regarded as residing in cells, the cell walls are only conceptual and provide no hindrance to the passage of matter. Mainly, in this case, N is relatively small and as a result, $\Delta N/N$ is therefore large.

In either event, fluctuations may become large but do not become infinite, for there is a coupling between the several regions that is provided by the need to conserve mass.

If there is only one phase present, fluctuations will indeed be extremely small. This may be illustrated in a simple way by calculating the relative fluctuation $\Delta N/N$ for an ideal gas. Gibbs' Eq. (97) from Ref. 1 may be written once for each phase

$$-V dp + S dT + N d\mu = 0 \qquad (4.45)$$

Holding temperature T constant, and defining the specific volume $v = V/N$, there remains

$$d\mu = v dp \qquad (4.46)$$

Holding volume constant as well as temperature, thermodynamically

$$\left(\frac{\partial \mu}{\partial N}\right)_{T,V} = v \left(\frac{\partial p}{\partial N}\right)_{T,V} \qquad (4.47)$$

The ideal gas is partially defined by the equation of state, $pV = Nk_BT$, and therefore $(\frac{\partial p}{\partial N})_{T,V} = \frac{k_BT}{V}$. Therefore

$$\left(\frac{\partial \mu}{\partial N}\right)_{T,V} = \frac{k_BT}{N} \tag{4.48}$$

Then the mean square fluctuation in N is

$$\langle \Delta N^2 \rangle = \sigma_N^2 = \langle N^2 \rangle - \langle N \rangle^2 = \frac{k_BT}{\left(\frac{\partial \mu}{\partial N}\right)_{T,V}} = N \tag{4.49}$$

and the relative fluctuations are

$$\frac{\sigma_N^2}{N^2} = \frac{1}{N} \tag{4.50}$$

and that is of the order of 10^{-23} for one mole of the ideal gas. Fluctuations per mole are expected to be really small when the particles are of atomic dimensions.

In summary, fluctuations in E, N and V are given by

$$\langle \Delta E^2 \rangle = k_BT \times C_vT \tag{4.51}$$

$$\langle \Delta N^2 \rangle = k_BT \times \frac{1}{\left(\frac{\partial \mu}{\partial \langle N \rangle}\right)_{T.V}} \tag{4.52}$$

$$\langle \Delta V^2 \rangle = k_BT \times \frac{1}{-\left(\frac{\partial p}{\partial V}\right)_{T,N}} = k_BT\kappa V \tag{4.53}$$

where κ is the isothermal compressibility. Derivation of the third relation is left as an exercise (Hint: Use the Gibbs ensemble). Note that in every case, k_BT appears as a driving force for fluctuation, and as such, it dominates the theory of Brownian motion.

4.5. Cross Terms

In ensembles with two or more fluctuating variables, typically in the grand canonical or the Gibbs ensembles, cross fluctuations arise. In a grand canonical ensemble with two components the partition function is

$$\Xi(T, V, \mu_1, \mu_2) = \sum_E \sum_{N_1} \sum_{N_2} \Omega(E, V, N_1, N_2) e^{\beta(E - \mu_1 N_1 - \mu_2 N_2)}$$

(4.54)

where $\beta = -1/k_B T$. The averaging process produces

$$\langle \Delta N_1 \Delta N_2 \rangle = \langle (N_1 - \langle N_1 \rangle)(N_2 - \langle N_2 \rangle) \rangle = \langle N_1 N_2 \rangle - \langle N_1 \rangle \langle N_2 \rangle$$

(4.55)

These averages can be calculated using various derivatives of the logarithm of the partition function. Specifically

$$\left(\frac{\partial \ln \Xi}{\partial \mu_1} \right)_{T,V,\mu_2} = \beta \langle N_1 \rangle$$

(4.56)

$$\left(\frac{\partial \ln \Xi}{\partial \mu_2} \right)_{T,V,\mu_1} = \beta \langle N_2 \rangle$$

(4.57)

$$\left(\frac{\partial^2 \ln \Xi}{\partial \mu_1^2} \right)_{T,V,\mu_2} = \beta^2 \langle N_1^2 \rangle$$

(4.58)

$$\left(\frac{\partial^2 \ln \Xi}{\partial \mu_2^2} \right)_{T,V,\mu_1} = \beta^2 \langle N_2^2 \rangle$$

(4.59)

$$\frac{\partial^2 \ln \Xi}{\partial \mu_1 \partial \mu_2} = \beta \left(\frac{\partial N_1}{\partial \mu_2} \right)_{T,V,\mu_1} = \beta \left(\frac{\partial N_2}{\partial \mu_1} \right)_{T,V,\mu_2}$$

(4.60)

Finally, it is left as an exercise to show that

$$\frac{\partial^2 \ln \Xi}{\partial \mu_1 \partial \mu_2} = \beta^2 \langle N_1 \rangle \langle N_2 \rangle$$

(4.61)

starting from the partition function. The beginner is cautioned that $\langle \Delta N_1 \Delta N_2 \rangle$ is not equal to $\langle \Delta N_1 \rangle \langle \Delta N_2 \rangle$, and that in fact $\langle \Delta N_1 \rangle \langle \Delta N_2 \rangle$ vanishes identically, which the reader should show. On the other hand, $\langle \Delta N_1 \Delta N_2 \rangle$ does not vanish, but is finite.

Problems

Independently of one another, Stockmayer[2] at MIT in Cambridge and Kirkwood[3] at Yale in New Haven produced theoretical treatments based on fluctuations in solutions. Both papers solve the same problem and arrived in the editorial offices of the Journal of Chemical Physics in the same mail. The editor arranged for the two papers to be published back to back in the same issue. While Stockmayer[2] elected to use a classical partition function for a modified grand canonical ensemble in treating his light scattering problem, all of his results may be derived from the following quantum mechanical partition function Ξ:

$$\Xi = \sum_{E,V,\{m_i\}} \Omega(E, V, \{m_i\}) e^{-(E+pV-\sum_{i=2}^{c} \mu_i m_i)/k_B T} \tag{4.62}$$

in which $\{m_i\}$ denotes the set $m_2, m_3, \ldots, m_c$ from which m_1 has been removed. The m_i are the concentrations of the ith species in a solution in which m_1 is the concentration of the solvent. The other symbols have their usual meanings. A more precise definition of m_1 and the m_i is given in Stockmayer's paper, which should be consulted while doing this problem.

a. Show that for i, j and k not equal to 1, the following mean fluctuations arise

$$\langle \Delta m_j \Delta m_k \rangle = k_B T \left(\frac{\partial \langle m_j \rangle}{\partial \mu_k} \right)_{p,T,m_{i \neq j}, m_1} \tag{4.63}$$

$$\langle \Delta V^2 \rangle = -k_B T \left(\frac{\partial \langle V \rangle}{\partial p} \right)_{T,\mu,m} \tag{4.64}$$

$$\langle \Delta V \Delta m_i \rangle = -\left(\frac{\partial \langle m_i \rangle}{\partial p} \right)_{T,\mu,m_{j \neq i}, m_1} \tag{4.65}$$

b. Assuming Stockmayer's equation (2.1), show that the turbidity τ is

$$\tau = \frac{32\pi^2 B k_B T}{3\lambda^4} \left\{ \frac{\kappa}{V} \left(\frac{\partial n}{\partial \ln \rho} \right)_m^2 + \sum_{i=2}^{c} \sum_{j=2}^{c} \left(\frac{\partial n}{\partial m_i} \right)_{p,T,m} \right.$$

$$\left. \times \left(\frac{\partial n}{\partial m_j} \right)_{p,T,m} \left(\frac{\partial m_i}{\partial \mu_j} \right)_{p,T,\mu_{i \neq j}, m_1} \right\} \tag{4.66}$$

where c is the number of components.

c. Noting that in general, $\mu_\iota = \mu_i\,(p, T, m_1, m_2, \ldots, m_c)$, show that in a solution composed of one polymer solute and two liquid solvents

$$\left(\frac{\partial m_3}{\partial \mu_2}\right)_{p,T,m_2,m_1,\mu_3} = \left(\frac{\partial m_2}{\partial \mu_3}\right)_{p,T,m_3,m_1,\mu_2} \tag{4.67}$$

$$-\left(\frac{\partial m_3}{\partial m_2}\right)_{p,T,\mu_3} = \frac{\left(\frac{\partial \mu_3}{\partial m_2}\right)_{p,T,m_3}}{\left(\frac{\partial \mu_3}{\partial m_3}\right)_{p,T,m_2}} = \frac{a_{23}}{a_{33}} \tag{4.68}$$

What is the similar relation for the ratio a_{23}/a_{22}?

d. Derive Stockmayer's Equation (3.5), and state what the omission in this equation is.

e. By considering the case of a polydisperse polymer and one solvent, show that the turbidity formula is not sensitive to molecular weight distribution.

Hint: In completing the above work, it will be found necessary to show that

$$\left(\frac{\partial n}{\partial T}\right)_{p,m} = -\left(\frac{\partial n}{\partial \ln \rho}\right)_m \left(\frac{\partial \ln V}{\partial T}\right)_{p,m} \tag{4.69}$$

$$\left(\frac{\partial n}{\partial p}\right)_{T,m} = -\left(\frac{\partial n}{\partial \ln \rho}\right)_m \left(\frac{\partial \ln V}{\partial p}\right)_{T,m} \tag{4.70}$$

References

1. J. W. Gibbs, Collected Works, Vol. 1, "On the equilibrium of heterogeneous substances", Yale Univ. Press, New Haven, Ct, reprinted 1948.
2. W. H. Stockmayer, J. Chem. Phys. **18**, 58 (1950).
3. J. G. Kirkwood and R. J. Goldberg, J. Chem. Phys. **18**, 54 (1950).

Chapter 5

A Variational Principle, Phase Space, Classical Averaging and Ω

5.1. Variational Principle

Variational techniques may be used in statistical mechanics, just as these are in quantum mechanics, and the variational theorem given below provides a firm theoretical basis for these calculations. The variational theorem in statistical mechanics resembles that in quantum mechanics in that an approximate free energy is minimized by variation of certain naturally occurring adjustable parameters, just as is done in quantum mechanics. For the free energy, the physical basis of the minimization is essentially the not negative definite character of the entropy change produced by addition of perturbative corrections to the approximate Hamiltonian, while in quantum mechanics the mathematical basis of the minimization of the energy is the fact that the energy calculated with a trial function and perturbed Hamiltonian forms an upper bound for the correct energy. Entropic considerations do not occur in quantum mechanics as applied to single systems, as is evident from the fact that a single system is not an ensemble of systems.

A proof of the classical statistical mechanical variation theorem was first given by J. Willard Gibbs[1] as a part of his investigation of the way energy may be distributed over the systems of a canonical ensemble. The theorem was independently rediscovered by Girardeau,[2] who, in a useful paper, cast it in modern form and notation, together with several other theorems also first proved by Gibbs. In Russian literature the relation is referred to as the Gibbs–Bogoliubov relation, evidently because Bogoliubov independently discovered the theorem. The fact remains that Gibbs did it first before quantum mechanics was invented. Quantum statistical theorems of similar content were given by Peierls[3] and are discussed in the book by Huang.[4] Other theorems given by Gibbs have been revisited in modern notation by Falk and Adler.[5] The proof of the variational theorem given here is based on that given by Gibbs as modified by Fixman.[6]

The theorem asserts that the Helmholz free energy A for a perturbed system is less than or equal to that for the unperturbed system corrected by a perturbative energy difference according to the inequality

$$A \leq A_0 + E - E_0 \tag{5.1}$$

Here E is the energy of the perturbed system and E_0 is that of the unperturbed. The Hamiltonian that defines E is the same as that for E_0 with a perturbation H' added, and is

$$H = H_0 + H' \tag{5.2}$$

The appropriate Schrödinger equations are $H\psi = E\psi$ and $H_0\psi_0 = E_0\psi_0$. The averaging implied by the brackets $\langle\ \rangle_0$ is accomplished with the energies of the unperturbed system. Take $\beta = -1/k_BT$, and presume the perturbation to be so slight that the density of states is unaltered to within infinitesimals of second order. Then, averaging with respect to the density of states G,

$$e^{\beta A_0} = \int e^{\beta E_0} dG \tag{5.3}$$

$$e^{\beta A} = \int e^{\beta E} dG \tag{5.4}$$

where the volume element $dG = dG(E) = dG(E_0) = \Omega(E_0, V, N)dE_0$. That is, the density of states is sensibly unaltered and therefore no change in volume element is required. The definition of averaging with respect to the unperturbed energies E_0 is

$$\langle X(E_0)\rangle_0 = \frac{\int X(E_0)\, e^{\beta E_0}\, dG}{\int e^{\beta E_0}\, dG} \tag{5.5}$$

X is an arbitrary function of the energy. The proof begins with Equations (5.3) and (5.4). From these, the crux equations follow

$$1 = \int e^{\beta(E-A)}\, dG = \int e^{\beta(E_0-A_0)}\, dG \tag{5.6}$$

Since both A and A_0 are constant in the integration over the density of states. By definition of average

$$\langle X(E_0)\rangle_0 = \int X(E_0) e^{\beta(E_0-A_0)}\, dG \tag{5.7}$$

for any function of E_0 that can be averaged. The ensemble entropy change that accompanies the alteration in accessible states per unit energy interval that results from introduction of the perturbation H' is

$$\Delta S = (\langle \Delta E\rangle_0 - \Delta A)/T \tag{5.8}$$

That is

$$\Delta S = -k_B \beta (\langle E - E_0\rangle_0 - A + A_0) \tag{5.9}$$

This rearranges to

$$\Delta S/k_B = -\beta\langle E - E_0 + A_0 - A\rangle_0 \tag{5.10}$$

Since A and A_0 are both constants under the averaging defined in Equation (5.5). From Equation (5.4), one can write

$$1 = \int e^{\beta(E-A)}\, dG \tag{5.11}$$

Both E_0 and A_0 are constants under the averaging procedure. Adding and subtracting their exponentials leaves the equality unaltered

$$1 = \int e^{\beta(E-E_0+A_0-A)} e^{\beta(E_0-A_0)}\, dG \tag{5.12}$$

This is another way of writing

$$1 = \langle e^{\beta(E - E_0 + A_0 - A)} \rangle_0 \qquad (5.13)$$

Adding and subtracting 1 on the right hand side of Equation (5.10) for the entropy change leaves the value of the change unaltered, but results in

$$\Delta S/k_B = -\beta \langle E - E_0 + A_0 - A \rangle_0$$
$$= \langle -\beta(E - E_0 + A_0 - A) + e^{\beta(E - E_0 + A_0 - A)} - 1 \rangle_0 \qquad (5.14)$$

This expression is incapable of assuming a negative value, for β is defined as $-1/k_B T$ and $-\beta$ is therefore a positive quantity in ensembles lacking energy level structures bound both from below and from above. That is, negative temperatures are barred from this proof. The expression in brackets is of the form $x + e^{-x} - 1$, and is convex upward about $x = 0$. For example, if $x = 1$, the function becomes $1/e$ or about 0.36788, which is positive. If $x = -1$, the value is $e^2 - 2$, or about 5.389. Finally if $x = 0$ the function vanishes. But if the expression cannot become negative, neither can the entropy change, and the inescapable conclusion is that for negative β

$$\langle E - E_0 + A_0 - A \rangle_0 \geq 0 \qquad (5.15)$$

Since A and A_0 are constant with respect to the averaging, this may be rewritten as

$$A \leq A_0 + \langle E - E_0 \rangle_0 \qquad (5.16)$$

and that is Gibbs' theorem.

Rewriting $\langle E - E_0 \rangle_0$ as $\langle H' \rangle_0$ shows that the theorem compares the Helmholtz free energy of the corrected system with that of the approximate system partially corrected by the perturbation. All averages are computed using the known energies E_0 of the approximate or unperturbed system. This is a computational convenience of considerable utility. The main idea of the theorem is that the free energy corrected simply by a first order perturbation forms an upper bound to the true free energy. The inequality introduced in Equation (5.15) provides the basis for a variational approach to the true free energy

A, accomplished by allowing E_0 or H_0 to depend on parameters with respect to which the first order corrected free energy may be minimized.

5.2. Phase Space

Phase space is spanned by $\boldsymbol{q}$, the set of generalized coordinates $(q_0, q_1, \ldots, q_{N-1})$ and their canonically conjugate momenta $\boldsymbol{p}$, the set $(p_0, p_1, \ldots, p_{N-1})$ defined by $p_i = \frac{\partial L}{\partial \dot{q}_i}$, where the Lagrangian function $L(\dot{\boldsymbol{q}}, \boldsymbol{q}) = T(\dot{\boldsymbol{q}}, \boldsymbol{q}) - V(\boldsymbol{q})$. Hamilton's formulation of classical mechanics is convenient to use since $H(\boldsymbol{p}, \boldsymbol{q}) = T(\boldsymbol{p}, \boldsymbol{q}) + V(\boldsymbol{q})$ in problems that involve only forces that can be derived from a potential. The great advantage to the Hamiltonian for statistical mechanical purposes is that the Hamiltonian function is equal to the total energy. The relevance of these remarks will shortly become evident.

The harmonic oscillator provides a simple illustration of the nature of phase space. The potential energy for a one dimensional oscillator that consists of a mass m moving along the x axis under the action of a spring of force constant k is $V = \frac{1}{2}kx^2$, where x is the displacement of the mass from its equilibrium position. The Hamiltonian function is

$$H = \frac{p^2}{2m} + \frac{1}{2}kx^2 = E \tag{5.17}$$

and the time rates of change of the coordinate x and momentum p are

$$\frac{\partial H}{\partial x} = -\dot{p} = kx \tag{5.18}$$

and

$$\frac{\partial H}{\partial p} = \dot{x} = \frac{p}{m} \tag{5.19}$$

Combining these gives the classic equation of motion

$$\dot{p} + kx = 0 \tag{5.20}$$

Since $p = m\dot{x}$, this becomes

$$m\ddot{x} + kx = 0 \tag{5.21}$$

The solution is well known

$$x = x_0 \sin(2\pi t + \varphi) \tag{5.22}$$

where x_0 is the amplitude and φ the phase. These are the constants of integration of this second order differential equation. Substitution into the equation of motion yields the classical harmonic frequency

$$\nu = \frac{1}{2\pi}\sqrt{\frac{k}{m}} \tag{5.23}$$

in units of sec^{-1}. More importantly, substitution into the Hamiltonian shows the energy is

$$E = \frac{p^2}{2m} + \frac{1}{2}4\pi^2\nu^2 m x^2 \tag{5.24}$$

At maximum excursion, the momentum p vanishes and the energy is momentarily

$$E = 2\pi^2\nu^2 m x_0^2 \tag{5.25}$$

At the equilibrium position, the mass is travelling at its fastest and the momentum is a maximum while $x = 0$. The energy momentarily is

$$E = \frac{p_{max}^2}{2m} \tag{5.26}$$

The first law of thought is that a thing is always identically itself. The energy is the energy is the energy, and therefore

$$\frac{p_{max}^2}{2m} = 2\pi^2\nu^2 m x_0^2 \tag{5.27}$$

and immediately $p_{max} = 2\pi\nu m x_0$. Next, cast the energy equation into standard form, The energy is

$$E = 2\pi^2\nu^2 m x_0^2 = \frac{p^2}{2m} + 2\pi^2\nu^2 m x^2 \tag{5.28}$$

Division by $2\pi^2\nu^2 m x_0^2$ gives the relation

$$1 = \frac{p^2}{4\pi^2\nu^2 m^2 x_0^2} + \frac{x^2}{x_0^2} \tag{5.29}$$

From the above work, the maximum momentum is $p_{max} = 2\pi\nu m x_0$. Hence

$$1 = \frac{p^2}{p_{max}^2} + \frac{x^2}{x_0^2} \tag{5.30}$$

This is the standard form for an equation describing an ellipse in a space spanned by variables p and x. The state of the oscillator at any time t may be obtained by the position of a point representative of the values of p and x on the ellipse. This point is called the representative point, or sometimes the phase point. Now inquire, what is the area enclosed by the ellipse? To get this. solve for p and integrate pdx from zero to x_0 along the x axis. This gives the area of one quadrant of the ellipse, and the area A enclosed by the ellipse is then

$$A = 4 \int_0^{x_0} p dx \tag{5.31}$$

A little algebra shows that

$$p = p_{max}\left(1 - \frac{x^2}{x_0^2}\right)^{1/2} \tag{5.32}$$

and the area becomes

$$A = 4p_{max} \int_0^{x_0} \left(1 - \frac{x^2}{x_0^2}\right)^{1/2} dx \tag{5.33}$$

The integral is elementary (use the substitution $\sin\Theta = \frac{x}{x_0}$) and the result is

$$A = 4p_{max}x_0 \int_0^{\pi/2} \cos^2\Theta d\Theta = 4p_{max}x_0 \left.\frac{1}{2}(\Theta + \sin\Theta\cos\Theta)\right|_0^{\pi/2}$$

$$= \pi p_{max}x_0 \tag{5.34}$$

But $p_{max} = 2\pi\nu m x_0$, and making this substitution yields

$$A = \pi(2\pi\nu m x_0)x_0 = \frac{2\pi^2\nu^2 m x_0^2}{\nu} = \frac{E}{\nu} \tag{5.35}$$

Quantum mechanical simple harmonic oscillators obey the energy relation

$$E = h\nu \left(v + \frac{1}{2} \right) \quad v = 0, 1, 2, 3 \cdots \infty \qquad (5.36)$$

This has a profound effect on the structure of phase space, for the area enclosed by one elliptical trajectory is

$$A_v = h \left(v + \frac{1}{2} \right) \qquad (5.37)$$

and that enclosed by an elliptical trajectory for vibrational quantum number $v + 1$ is

$$A_{v+1} = h \left(v + 1 + \frac{1}{2} \right) \qquad (5.38)$$

The difference between these two areas is

$$A_{v+1} - A_v = h \qquad (5.39)$$

that is, the annular volume between successive vibrational trajectories has area h. Furthermore, the representative point can describe an ellipse only when the enclosed area of the ellipse is given by A_v. If the energy is set equal to a value that forces the representative point into the annular area between quantized trajectories, immediately the wavefunction for the oscillator interferes destructively with itself: The wavefunction cannot satisfy the boundary conditions unless the energy is one of the quantum levels. In other words, the representative point cannot access all of phase space, but is restricted to one or another of the quantized annuli. Phase space is subdivided into cells of dimensionality h, one for each product $dpdq$. The sole exception is the area enclosed by the ellipse for the ground state ($v = 0$), in which case the enclosed area is $h/2$.

To determine the partition function q classically one expects

$$q_{classical} = \int_{-\infty}^{\infty} \int_{-\infty}^{\infty} e^{-\frac{H(p,x)}{k_B T}} dpdx \qquad (5.40)$$

The Hamiltonian for the simple harmonic oscillator may be written

$$H(p, x) = \frac{p^2}{2m} + 2\pi^2 \nu^2 m x^2 \tag{5.41}$$

and the classical integral is

$$q_{classical} = \int_{-\infty}^{\infty} \int_{-\infty}^{\infty} e^{-\frac{p^2}{2mk_BT} - \frac{2\pi^2\nu^2 m x^2}{k_BT}} \, dp \, dx \tag{5.42}$$

The definite integral can be found in tables, and is

$$\int_{-\infty}^{\infty} e^{-a^2 y^2} \, dy = \frac{\sqrt{\pi}}{a} \tag{5.43}$$

There are two independent integrals of this form multiplied together to be evaluated. The p integral is

$$\int_{-\infty}^{\infty} e^{-\frac{p^2}{2mk_BT}} \, dp = \sqrt{2\pi m k_B T} \tag{5.44}$$

while the x integral is

$$\int_{-\infty}^{\infty} e^{-\frac{2\pi^2\nu^2 m x^2}{k_BT}} \, dx = \frac{\sqrt{\pi k_B T}}{\sqrt{2\pi^2\nu^2 m}} = \frac{\sqrt{k_B T}}{\sqrt{2\pi m \nu}} \tag{5.45}$$

The product of these two integrals is

$$q_{classical} = \frac{k_B T}{\nu} \tag{5.46}$$

But when the quantum mechanical partition function is evaluated at high temperature, the result is

$$q = \frac{k_B T}{h\nu} \tag{5.47}$$

and evidently in the classical formulation a factor of $1/h$ has been left out. The classical error is, of course, that classically the representative point is allowed access to every point in phase space, and is not constrained by Heisenberg uncertainty to swim about on quantized annuli. The problem with the classical formulation is that the volume element $dpdx$ is too large, and it should have been $dpdx/h$. Even that does not fix all the problems the classical formulation has. The result is only valid at temperatures large in comparison with the characteristic vibrational temperature, here $\Theta_{vib} = h\nu/k_B$ (the frequency is in $\sec^{-1}$ units, not the spectroscopic preference of cm^{-1} units).

Evidently, whenever the limiting high temperature form of the partition function is all that is required, the problem may be correctly formulated as

$$q_{\text{classical}} = \frac{1}{h^n} \int \cdots 2n \text{ iterated } \cdots \int e^{-\frac{H(p,q)}{k_B T}} dpdq \qquad (5.48)$$

where there are n momenta p and conjugate coordinates q, and the integrals run over the accessible space and all possible momenta.

5.3. Classical Averaging and Ω

In the canonical ensemble, the partition function is given by a sum over the energies of the systems of the ensemble, and as a result, no two systems need have exactly the same energy. In this ensemble, energy fluctuates. In the microcanonical ensemble, the energy is specified, and every system of the ensemble has exactly the same energy. The energy does not fluctuate in this ensemble. It is important to keep in mind what fluctuates and what does not in any ensemble. In general, if it is summed over, it fluctuates.

In the canonical ensemble, the Helmholtz free energy for non-localized molecules is given by

$$A = -k_B T \ln \frac{q^N}{N!} \qquad (5.49)$$

By definition, $A = E - TS$, and if it is allowable to mix ensemble equations, $S = k_B \ln \Omega$. Then

$$k_B T \ln \Omega = E + k_B T \ln \frac{q^N}{N!} \qquad (5.50)$$

Solving for Ω yields

$$\Omega = \frac{q^N}{N!} e^{\frac{E}{k_B T}} \qquad (5.51)$$

Classically, for a system consisting of N gaseous molecules each of the same n nuclei, the value should be calculable by writing

$$\Omega_{\text{non-loc}} = \left(\frac{1}{h^{3n}} \int \cdots 6n - \text{ fold iterated } \cdots \int e^{-\frac{H(p,q)}{k_B T}} dpdq \right)^N$$

$$\times e^{\frac{E}{k_B T}} / N! \tag{5.52}$$

where $H(p, q)$ is the Hamiltonian for a single molecule of the gas. Every molecule in the ensemble has the same Hamiltonian. Designating each molecule by a subscript i, and the coordinates and conjugate momenta similarly, allows writing the equation in the form

$$\Omega_{\text{non-loc}} = \left(\frac{1}{h^{3n}} \int \cdots 6n \cdots \int e^{-\frac{H_1(p_1,q_1)}{k_B T}} dp_1 dq_1 \right)$$

$$\times \left(\frac{1}{h^{3n}} \int \cdots 6n \cdots \int e^{-\frac{H_2(p_2,q_2)}{k_B T}} dp_2 dq_2 \right) \cdots e^{\frac{E}{k_B T}} / N! \tag{5.53}$$

or, more compactly,

$$\Omega_{\text{non-loc}} = \frac{1}{h^{3nN} N!} \int \cdots 6nN \cdots \int e^{-\frac{\sum_i H_i(p_i,q_i)}{k_B T}} \prod_i dp_i dq_i e^{\frac{E}{k_B T}} \tag{5.54}$$

The energy is in fact $E = \sum_i H_i(p_i, q_i)$, and as these have opposite signs in the integrand, the result is apparently

$$\Omega_{\text{non-loc}} = \frac{1}{h^{3nN} N!} \int \cdots 6nN \cdots \int \prod_i dp_i dq_i \tag{5.55}$$

However, this cannot be correct, for it assumes that the energy E is specified to a precise value, and in the canonical ensemble, the energy fluctuates. As written, the equation is analogous to trying to find the volume of a rectangle, or the area of a line. The product $dp_i dq_i$ represents a volume in phase space and the specification of a precisely defined energy E forces the representative point to swim about on the hypersurface defined by E. As E fluctuates, the hypersurfaces change, and a more nearly correct statement would be to write the equation with a restriction confining the energy to values lying in the vicinity of $H(p, q)$. That is

$$\Omega_{\text{non-loc}} \left(E - \Delta E \leq H\left(p, q\right) \leq E + \Delta E \right)$$

$$= \frac{1}{h^{3nN} N!} \int \cdots 6nN \cdots \int \prod_i dp_i dq_i \qquad (5.56)$$

This is what is understood as the number of *a priori* equally probable accessible states Ω in the quantum mechanical ensemble averaging postulate

$$\langle f_{\text{obs}} \rangle = \frac{\sum_i f_i}{\Omega} \qquad (5.57)$$

the sum running from 1 to Ω. Here, Ω is a sort of fuzzy volume in phase space, the size of which is determined by the size of the fluctuation in the energy, itself dependent on the size of the systems of the ensemble. Small systems have large fluctuations, and the ensemble average is correspondingly less well defined than is the case for large systems. The classical average is somewhat less well defined

$$f_{\text{obs, class}} = \frac{\int f\left(p, q\right) dp dq}{\int dp dq} \qquad (5.58)$$
$$E - \Delta E \leq H\left(p, q\right) \leq E + \Delta E$$

It is postulated that this classical average is equal to the observational long time average

$$f_{\text{obs, class}} = \lim_{\tau \to \infty} \frac{1}{\tau} \int_0^\tau f(t) dt \qquad (5.59)$$

The classical ergodic hypothesis is that these two averages are equal in the limit of vanishing fluctuations in energy. Since fluctuations are unavoidable in principle, it is not likely that this deterministic view may ever be proven.

On the other hand, the quasi-ergodic hypothesis is less restrictive, and asserts essentially that the amount of time a representative point spends in a volume of phase space is proportional to the volume. Proof of this hypothesis is beyond the scope of this book.

Problems

1. Repeat the derivation of the variational principle in the canonical ensemble $(A \leq A_0 + \langle E - E_0 \rangle_0)$ for a grand ensemble canonical with respect to E and N, and show that for this ensemble, the pressure obeys the inequality

$$p \geq p_0 - \frac{\langle E - E_0 \rangle_0}{V}$$

where p is the actual pressure and p_0 is the uncorrected. When $\beta = +1/k_B T$, averages may be written for this ensemble as

$$\langle X(E, N) \rangle_0 = \frac{\sum_N \lambda^N \int X(E, N) e^{-\beta E_0} dG(E_0, V, N)}{\sum_N \lambda^N \int e^{-\beta E_0} dG(E_0, V, N)}$$

Averages may also be written

$$\langle X(E, N) \rangle_0 = \sum_N \lambda^N \int X(E, N) e^{-\beta(E_0 + p_0 V)} dG(E_0, V, N)$$

where $\lambda = e^{\beta \mu}$ is the absolute activity.

2. Consider an ensemble of points of mass m moving in one dimension between limits of 0 and L. Let a typical mass have momentum p and position x, where $0 < x < L$. No mass exists outside these limits. The interval $0 < x < L$ is potential free.

   a. Make a sketch for the representative point for one mass moving in phase space with energy E.

   b. Show that the phase space for one mass point is subdivided into cells of dimensionality h.

   c. On a separate diagram, sketch some of the regions of phase space that are accessible to the representative point. E may assume various values.

   d. Use classical statistics to obtain the partition function q.

   e. In an ensemble of N such non-interacting mass points, the thermodynamic energy function is defined in part by the relation $dE = T dS - f dL$, where f is the magnitude of the force exerted by the particles on either end of the x interval. The definition of the energy function is completed by specification of a zero of energy. Obtain an expression for the force f in terms of N, T and L. Show that as L increases boundlessly, the force diminishes to vanishing.

3. A one dimensional oscillator of mass m and displacement x is governed by a potential V that includes anharmonicity, of the form

$$V = ax^2 - bx^3$$

where b is of the order of $0.02a$ or less. Obtain the classical partition function q and from this the vibrational contribution to the specific heat at constant volume for N such oscillators, correct to a term linear in T.

4. A gaseous diatomic molecule has a harmonic frequency of 500 cm^{-1} with no anharmonicity. Let N_i be the number of gaseous molecules in vibrational quantum state i. The fraction of molecules f_v in a specific quantum state v or higher is given by

$$f_v = \frac{\sum_{i=v}^{\infty} N_i}{N}$$

Calculate f_v at 300, 500 and 1000 K for $v = 2, 3, 4, 5,$ and 6.

References

1. J. Willard Gibbs, "Collected Works", Vol. II, Yale University Press, New Haven, 1948, Chapter XI.
2. M. D. Girardeau, J. Chem. Phys. **40**, 899–900 (1964).
3. R. E. Peierls, Phys. Rev. **54**, 918 (1938).
4. K. Huang, "Statistical Mechanics", J. Wiley & Sons, New York, 1963, p. 220.
5. H. Falk and E. Adler, Phys. Rev. **168**, 185 (1968).

Chapter 6

Gases

6.1. Introduction

Gases differ from crystals in one important aspect. In a crystal, each molecule or atom is localized, its position in space fixed and addressable by coordinates measured with respect to some origin. When the crystal evaporates, the molecules or atoms lose their addresses and become completely delocalized, a fact that has a profound effect on the partition function, and on all the thermodynamic functions derivable from the partition function.

In the canonical ensemble, the partition function Q is a function of the natural variables, and the variation in the Helmholz free energy

$$dA = -SdT - pdV + \mu dN \tag{6.1}$$

shows that the entropy S, pressure p and chemical potential μ are all derivable from a Helmholz free energy expressed as a function of the variables T, V and N by differentiation, as

$$S = -\left(\frac{\partial A}{\partial T}\right)_{V,N} \tag{6.2}$$

$$p = -\left(\frac{\partial A}{\partial V}\right)_{T,N} \tag{6.3}$$

$$\mu = \left(\frac{\partial A}{\partial N}\right)_{T,V} \tag{6.4}$$

Any other thermodynamic functions may be written as algebraic combinations of these three functions, the natural variables and the Helmholz free energy. For example, since by definition $A = E - TS$, the energy $E = A + TS$, and so on. The connection between the partition function and thermodynamics is provided by the fundamental equation in the canonical ensemble

$$A(T, V, N) = -k_B T \ln Q(T, V, N) \tag{6.5}$$

All that is necessary is to evaluate Q in a differentiable form that describes some physical situation with adequate precision. The crystal is somewhat more easily visualized than the gas, and so the analysis starts with a crystal composed of non-chemically bonded atoms, such as neon or argon at a temperature high enough that quantum effects may be ignored but of course low enough that a stable crystal will form. The quantum effects will be considered in a later chapter.

For the crystal, the number of ways of placing N molecules on N numbered sites is $N!$ if the molecules are all in different states. The first molecule may be put on any of the N sites, and the second on any of the $N-1$ remaining sites. The total number of ways of placing the first two molecules on two sites is $N(N-1)$, since for each possible placement of the first molecule there are $N-1$ possible placements of the second. The third molecule can be put in any of $N-2$ sites, and the number of ways of putting three molecules on three of the N sites is $N(N-1)(N-2)$. This continues until the number of molecules is exhausted, and the total number of ways of placing N molecules on N numbered sites is $N(N-1)(N-2)(N-3)\ldots 2 \times 1 = N!$, when the molecules are all in different states. When the molecules are identical and all in the same state, there is only one way to place N such molecules on N sites. Suppose the first three molecules selected are in the same state and are otherwise identical. Then the first may be put on in N ways, the second in $N-1$ ways. But it would not matter whether molecule a were put on first and then molecule b, or b first and then a, for the same configuration results. Then the number of ways of putting the first two molecules on N sites is $N(N-1)/2$. When the third is put on, there are $N-2$ ways to do it, but now the same configuration results whether a, b or c

is chosen first and the other two are chosen from what is left, and there are 3! ways of choosing the molecules in order. Then there are $N(N-1)(N-2)/3!$ ways of placing three identical molecules on N numbered sites. If n_i molecules are in the ith state, and $N = \sum_i n_i$, then there are $\frac{N!}{\prod_i n_i!}$ ways of placing the N molecules on N sites. If one of the n_i, say n_1, is the only non-zero n_i, then $n_1 = N$ and the number of ways becomes $\frac{N!}{N!0!0!\cdots} = 1$ since $0! = 1$.

For a gas, the partition function in the canonical ensemble might be expected to be

$$Q_{gas} = \sum_E \Omega(E)e^{-E/k_B T} \tag{6.6}$$

The energy E is composed of the energy levels ε_i of the individual molecules in the gas, each with occupation numbers n_i, and the number N of molecules in a system of the canonical ensemble is

$$N = \sum_i n_i \tag{6.7}$$

In this equation, n_i is the number of molecules in the ith state with energy ε_i and the sum is over states. Similarly, the energy E of one system of the canonical ensemble is

$$E = \sum_i n_i \varepsilon_i \tag{6.8}$$

For the gas, then, the expectation is that

$$Q_{gas} = \sum_{(n)} \frac{N!}{\prod_i n_i!} e^{-\sum_i \frac{n_i \varepsilon_i}{k_B T}} \quad \text{subject to} \quad \sum_i n_i = N \tag{6.9}$$

if all the molecules in the gas are stationary. According to the multinomial theorem, this evaluates as

$$Q_{gas} = \left(\sum_i e^{-\varepsilon_i/k_B T} \right)^N \tag{6.10}$$

and, defining $q = \sum_i e^{-\varepsilon_i/k_B T}$, Q_{gas} is apparently equal to q^N. But this cannot be entirely correct. The problem is that there were $N!$ ways of putting N distinct molecules onto N sites, and when the

N molecules evaporated, the $N!$ ways disappeared because the sites did. As it stands, the partition function grossly over counts states, and to correct for the over count, the function should be divided by the lost $N!$. In terms of q, then, the corrected partition function for the gas in the canonical ensemble is

$$Q_{gas} = \frac{q^N}{N!} \tag{6.11}$$

The argument for dividing by $N!$ to correct for the over count of states is not very rigorous, and was in fact first suggested by Gibbs, who couldn't prove it either. In fact, it is only true as a very good approximation. The approximation is so good that it is better than nearly any conceivable measurement, and it is generally accepted as entirely adequate for engineering or chemical purposes. A more rigorous argument must await a discussion of the quantum nature of real molecules.

The partition function for the crystal is, in terms of q, simplicity itself, provided the crystal may be regarded as a localized but independent collection of identical entities

$$Q_{crystal} = q^N \tag{6.12}$$

since the argument given above need not be used for the crystal. In fact, nothing in the crystal is truly independent, and equation (6.12) is actually a practical but non-rigorous approximation that often works well enough in practice. These preliminaries lead immediately to applications, and the first of these is the statistical consequences of translation, a property of molecules in gases but not in crystals.

6.2. Translation

The classical translational energy is all kinetic. For one molecule of mass m, the kinetic energy T is

$$T = \frac{1}{2}m\dot{x}^2 + \frac{1}{2}m\dot{y}^2 + \frac{1}{2}m\dot{z}^2 \tag{6.13}$$

where the dot denotes time differentiation. The linear momentum p_i is defined by

$$p_i = \frac{\partial T}{\partial \dot{q}_i} \tag{6.14}$$

where $q_i = x$, y or z. The momentum in the x direction, for example, is $p_x = m\dot{x}$. The classical Hamiltonian is a function of the coordinates and the corresponding momenta (often called the conjugate, or even canonically conjugate, momenta), and in this case the Hamiltonian is equal to the classical energy E

$$E = \frac{1}{2m}(p_x^2 + p_y^2 + p_z^2) \tag{6.15}$$

Solution of the quantum mechanical problem when the molecule is regarded as contained in a box with impenetrable walls of height, width and depth a, b and c, gives energy levels

$$\varepsilon = \frac{h^2}{8m}\left(\frac{n_x^2}{a^2} + \frac{n_y^2}{b^2} + \frac{n_z^2}{c^2}\right) \tag{6.16}$$

A particular translational state is specified by giving non-zero positive integer values for the n_i, that is, $n_i = 1, 2, 3, \ldots$. Each molecule in the system has the same set of translational energy levels, and the partition function for translation q is

$$q = \sum_{n_x=1}^{\infty} \sum_{n_y=1}^{\infty} \sum_{n_z=1}^{\infty} e^{-\frac{h^2}{8mk_BT}\left[\left(\frac{n_x}{a}\right)^2 + \left(\frac{n_y}{b}\right)^2 + \left(\frac{n_z}{c}\right)^2\right]} \tag{6.17}$$

The power of the method depends on the evaluation of these sums. The first thing to notice is that the function q is actually a product of three nearly identical sums.

$$q = \sum_{n_x=1}^{\infty} e^{-\frac{h^2 n_x^2}{8ma^2 k_BT}} \sum_{n_y=1}^{\infty} e^{-\frac{h^2 n_y^2}{8mb^2 k_BT}} \sum_{n_z=1}^{\infty} e^{-\frac{h^2 n_z^2}{8mc^2 k_BT}} \tag{6.18}$$

Fortunately, h is very small in comparison to the other parameters, and this fact allows the Euler-MacLauren sum formula (see later in this chapter) to be truncated at the leading (integral) term without incurring detectable error for all finite temperatures. The energy levels are therefore so close together as to form essentially a continuum (Heisenberg uncertainty broadening $\Delta E \Delta t$ may be regarded as blending adjacent levels into one another for Δt sufficiently long and a box of laboratory dimensions), and the partition function may be written, to an extremely good approximation at temperatures above the millikelvin at which quantum effects first appear

$$q = \int_0^\infty e^{-\frac{h^2 n_x^2}{8ma^2 k_B T}} dn_x \int_0^\infty e^{-\frac{h^2 n_y^2}{8mb^2 k_B T}} dn_y \int_0^\infty e^{-\frac{h^2 n_z^2}{8mc^2 k_B T}} dn_z$$

(6.19)

In these integrals, the lower limit 1 has been replaced by 0 since the energy difference between the energy level origin and the lowest possible level (all the n_i are 1) is very small in comparison with the rest of the energy levels. The resulting integral is the product of three nearly identical integrals, and can be found in tables.

$$q = \int_0^\infty e^{-\alpha_x^2 x^2} dx \int_0^\infty e^{-\alpha_y^2 y^2} dy \int_0^\infty e^{-\alpha_z^2 z^2} dz$$

(6.20)

For the x integral,

$$\alpha_x^2 = \frac{h^2}{8ma^2 k_B T}$$

(6.21)

and similarly for the other two, with appropriate width b and depth c changes. The integrals have the value $\sqrt{\pi}/2\alpha_i$ ($i = x, y$ or z) and q becomes

$$q = \frac{(2\pi m k_B T)^{3/2}}{h^3} abc$$

(6.22)

The volume of the box $V = abc$. The partition function in the canonical ensemble is $Q = \frac{q^N}{N!}$ or

$$Q = \frac{(2\pi m k_B T)^{3N/2}}{N! h^{3N}} V^N$$

(6.23)

It is characteristic of statistical mechanics that once the partition function is evaluated, it literally starts raining results. This treatment provides the ideal gas law, the Sackur–Tetrode expressions for the translational entropy and the translational component of the chemical potential, as worked out below.

6.3. The Ideal Gas Equation of State

The Helmholz free energy is $A = -k_B T \ln Q$. For translation alone

$$A = -N k_B T \ln \left(\frac{2\pi m k_B T}{h^2} \right)^{\frac{3}{2}} + k_B T \ln N! - N k_B T \ln V \quad (6.24)$$

In the V differentiation, only the last term contributes. Using $p = -(\frac{\partial A}{\partial V})_{T,N}$ there results

$$p = N k_B T \left(\frac{\partial \ln V}{\partial V} \right)_{T,N} \quad (6.25)$$

The derivative is elementary, and immediately the ideal gas law equation of state ensues.

$$p = N k_B T / V \quad (6.26)$$

6.4. Sackur–Tetrode Translational Entropy

Since the Helmholz free energy for translation may be written

$$A = -N k_B T \ln q\,(T) + k_B T \ln N! \quad (6.27)$$

the temperature derivative is $S = -(\frac{\partial A}{\partial T})_{V,N}$ and the entropy of the gas is therefore

$$S = N k_B \ln q - k_B \left(N \ln N - N \right) + N k_B T \left(\frac{\partial \ln q}{\partial T} \right)_{V,N} \quad (6.28)$$

The logarithm of q is

$$\ln q = \frac{3}{2} \ln \left(\frac{2\pi m k_B}{h^2} \right) + \frac{3}{2} \ln T + \ln V \quad (6.29)$$

and the temperature derivative is just $3/2T$. The expression for S becomes

$$S = N k_B \ln \left[\left(\frac{2\pi m k_B T}{h^2} \right)^{3/2} V \right] - N k_B \ln N + N k_B + \frac{3}{2} N k_B \quad (6.30)$$

In this form, the relation is not practical for use. The volume dependence is the sticking point. However, that can be eliminated by introducing the ideal gas equation of state. It is easy to show that

$$S = Nk_B \left\{ \frac{5}{2} \ln T - \ln P + \frac{5}{2} + \ln \left[\left(\frac{2\pi m}{h^2} \right)^{3/2} k_B^{5/2} \right] \right\} \qquad (6.31)$$

For the molar entropy, set the universal gas constant $R = Nk_B$. Doing so implies that N is Avogadro's number, N_0. Practical calculations require P to be expressed in suitable units. For example, if k_B is chosen in ergs $\deg^{-1}$molecule^{-1}, then P must be in units of dynes cm^{-2}. The mass m is given in grams per molecule. Usually, chemists like to express the molar mass in grams per mole and pressure in some unit such as atmosphere or torr or bar, and the corresponding conversions must be inserted in the entropy expression for practical use. To illustrate the process, the units selected are atmosphere for pressure and grams per mole for molar mass. One atmosphere is by definition 1.01325×10^6 dynes cm^{-2}, and if P is in dynes cm^{-2} and p is in atmosphere, $P = p \times 1.01325 \times 10^6$. The mass of one molecule m is the molar mass in grams per mole divided by Avogadro's number, in molecules per mole, that is, $m = M/N_0$, where M is the molar mass in grams per mole. Making these changes,

$$S = R \left[\frac{5}{2} \ln T - \ln p + \frac{3}{2} \ln M + \ln \left[\left(\frac{2\pi k_B}{N_0 h^2} \right)^{\frac{3}{2}} \frac{k_B}{1.01325 \times 10^6} \right] + \frac{5}{2} \right] \tag{6.32}$$

The numerical entries can be evaluated, within limits imposed by the margins of error in the universal constants N_0, h and k_B, and are approximately

$$\ln \left[\left(\frac{2\pi k_B}{N_0 h^2} \right)^{\frac{3}{2}} \frac{k_B}{1.01325 \times 10^6} \right] + \frac{5}{2} = -1.1648_{487} \qquad (6.33)$$

This allows a practical equation for the translational entropy to be written as

$$\frac{S}{R} = \frac{5}{2} \ln T - \ln p + \frac{3}{2} \ln M - 1.1648_{487} \qquad (6.34)$$

Table 6.1. Translational entropy for Argon gas at 1 bar pressure in the hypothetical state of an ideal gas.

T/K	S/R	S/eu	$S/\mathrm{JK^{-1}mol^{-1}}$
10	10.136	20.142	84.277
20	11.869	23.586	98.685
30	12.883	25.600	107.113
40	13.602	27.030	113.093
50	14.160	28.138	117.731
60	14.616	29.044	121.521
70	15.001	29.810	124.725
80	15.335	30.473	127.501
90	15.629	31.058	129.949
100	15.893	31.582	132.139
110	16.131	32.055	134.120
120	16.348	32.487	135.929
130	16.549	32.885	137.593
140	16.734	33.253	139.133
150	16.906	33.596	140.567
160	17.068	33.917	141.909
170	17.219	34.218	143.169
180	17.362	34.502	144.357
190	17.497	34.770	145.481
200	17.625	35.025	146.547

In this equation, p is in atmospheres. If p is to be in bars, 1 atmosphere is 1.01325 bar, which will slightly modify the numerical term. Table 6.1 shows the Sackur–Tetrode translational entropy calculated for Argon in the hypothetical state of an ideal gas at 1 bar pressure.

The translational component of the chemical potential is left as an exercise for the reader.

6.5. Rotation

If the molecules in the gas have a little more structure than simply neon or argon, such as a diatomic molecule, or a linear polyatomic molecule, the energy levels are usually given as a function of a spectroscopic constant B that is inversely proportional to the moment of inertia about an axis perpendicular to the internuclear axis and passing through the center of gravity. The quantum number for rotation is usually chosen to be J. The quantum number for rotation J assumes values $0, 1, 2, 3, \ldots$. Unlike the translational quantum numbers n_x, n_y and n_z, J is allowed to assume the value zero. In first approximation, the $2J + 1$ degenerate rotational energy level $F(J)$ is, in the cm^{-1} units preferred by spectroscopists,

$$F_e(J) = B_e J(J + 1) \tag{6.35}$$

where the rotational constant B_e in cm^{-1} units is given by

$$B_e = \frac{h}{8\pi^2 I_e c} \tag{6.36}$$

and the vacuum velocity of light c is $2.99792458 \times 10^{10}$ cm sec^{-1}. In this expression, I_e is the equilibrium moment of inertia. The diatomic molecule is regarded as held together by a potential energy function that has a minimum at an internuclear distance r_e, called the equilibrium distance. The potential may be approximated by a parabola with a minimum at r_e. For a diatomic molecule with two different masses m_1 and m_2 and equilibrium internuclear distance r_e, the moment of inertia I_e is

$$I_e = \mu r_e^2 \tag{6.37}$$

where the reduced mass μ is given by

$$\frac{1}{\mu} = \frac{1}{m_1} + \frac{1}{m_2} \tag{6.38}$$

Multiplication of B_e by $hc/k_B = 1.4388$ cm K converts B_e into degrees Kelvin, with the symbol Θ_{rot}, a characteristic temperature, that is

$$\Theta_{rot} = \frac{B_e hc}{k_B} \tag{6.39}$$

The rotational partition function q_{rot} is, for the equilibrium case,

$$q_{rot} = \sum_{J=0}^{\infty} (2J + 1) e^{-\frac{\Theta_{rot}}{T} J(J+1)} \tag{6.40}$$

There is, unfortunately, no way to evaluate the sum in closed algebraic form. However, the Euler–MacLaurin sum formula allows approximate evaluation at temperatures above Θ_{rot}. At temperatures below Θ_{rot}, the sum should be evaluated term by term until no further appreciable contribution is made. Except for temperatures very near Θ_{rot}, the series usually converges rapidly, and relatively few terms need to be kept. The Euler–MacLaurin sum formula can be quite tediously formidable. In general, it is as follows. If

$$\sum_{j=m}^{j=n} f(j) = f(m) + f(m+1) + f(m+2) + \cdots + f(n) \qquad (6.41)$$

then the sum may be approximately evaluated as

$$\sum_{j=m}^{j=n} f(j) = \int_m^n f(x)dx + \frac{1}{2}[f(n) + f(m)]$$

$$+ \sum_{k \geq 1} (-1)^k \frac{B_k}{(2k)!} \left[\left(\frac{d^{2k-1} f}{dx^{2k-1}} \right)_{x=m} - \left(\frac{d^{2k-1} f}{dx^{2k-1}} \right)_{x=n} \right]$$

$$(6.42)$$

The B_k are the Bernoulli numbers, $B_1 = \frac{1}{6}, B_2 = \frac{1}{30}, B_3 = \frac{1}{42}$, $B_4 = \frac{1}{30}, B_5 = \frac{5}{66}$ etc. When this theorem is appled to evaluating q_{rot} at temperatures just greater than the characteristic rotational temperature Θ_{rot}, the result is

$$q_{rot} = \frac{T}{\Theta_{rot}} \left(1 + \frac{\Theta_{rot}}{3T} + \frac{\Theta_{rot}^2}{15T^2} + \frac{4\Theta_{rot}^3}{315T^3} + \cdots \right) \qquad (6.43)$$

This works well for heteronuclear diatomic molecules for $T > \Theta_{rot}$. For higher temperatures, the higher terms in the series contribute increasingly less, until in the limit of very high temperature only the integral term is left. It is easy to show that

$$q_{rot} = \int_0^\infty (2J+1) e^{-\frac{\Theta_{rot}}{T} J(J+1)} dJ = \frac{T}{\Theta_{rot}} \qquad (6.44)$$

In the case of homonuclear diatomic molecules, the influence of nuclear spin makes its appearance in the experimental pure rotational Raman spectrum, as the nuclear spins alter the degeneracies of successive rotational energy levels differently. The altered degeneracy has a dramatic effect on the partition function by in effect making two distinct and non-interconverting molecular species in most cases. In the case of oxygen, the extreme case is encountered. The nuclear spin is zero, and, since the ground electronic state is a triplet sigma g minus $(^3\Sigma_g^-)$, the need to satisfy the Pauli principle allows only odd numbered J values to exist. Half of the energy levels considered for heteronuclear diatomic molecules do not exist, and the sum that forms the partition function has half its terms missing. For a diatomic molecule of oxygen-16 in the high temperature limit

$$q_{rot} = \sum_{J_{odd}} (2J+1)\, e^{-\frac{\Theta_{rot}}{T} J(J+1)} = \frac{T}{2\Theta_{rot}} \tag{6.45}$$

The influence of nuclear spin will be discussed in greater detail in the next chapter.

6.6. Vibration

The vibrational motions of the nuclei of molecules are quantized. For diatomic molecules, in first approximation the vibrational energy is given in terms of a vibrational quantum number v that assumes non-negative integer values $0, 1, 2, \ldots$ as non-degenerate levels

$$E_{\text{vib}} = \left(v + \frac{1}{2}\right) h\nu \tag{6.46}$$

Here, ν is the classical oscillation frequency. In units of $\sec^{-1}$ the frequency is

$$\nu = \frac{1}{2\pi}\sqrt{\frac{k}{\mu}} \tag{6.47}$$

where k is the Hooke's law force constant (essentially, k is the second derivative of the potential function evaluated at r_e) and μ is the reduced mass of the two nuclei. Spectroscopists measure ν in cm^{-1} units calculated by dividing ν by the velocity of light in vacuo expressed in cm $\sec^{-1}$ units. In statistical mechanics, it is convenient to introduce a characteristic vibrational temperature Θ_{vib} as

$$\Theta_{vib} = \frac{hc}{k_B}\nu \tag{6.48}$$

where ν is in cm^{-1} units. The vibrational partition function q_{vib} is

$$q_{vib} = \sum_{v=0}^{\infty} e^{-\left(v+\frac{1}{2}\right)\frac{\Theta_{vib}}{T}} \tag{6.49}$$

Factoring the zero point energy contribution $e^{-\frac{\Theta_{vib}}{2T}}$ gives

$$q_{vib} = e^{-\frac{\Theta_{vib}}{2T}} \sum_{v=0}^{\infty} \left(e^{-\frac{\Theta_{vib}}{T}}\right)^v \tag{6.50}$$

The sum is of the form $\sum_{n=0}^{\infty} x^n = \frac{1}{1-x}$, provided that $x^2 < 1$. This will practically always be the case, and when it is, the harmonic part of the vibrational partition function is

$$q_{vib} = \frac{e^{-\frac{\Theta_{vib}}{2T}}}{1 - e^{-\frac{\Theta_{vib}}{T}}} \tag{6.51}$$

when the vibrational energy levels are referred to a zero of energy at the bottom of the potential energy well, requiring retention of the zero point energy contribution. However, when the zero of energy is taken as the lowest energy level (that for $v = 0$), the harmonic part of the vibrational partition function becomes

$$q_{vib}^0 = \frac{1}{1 - e^{-\frac{\Theta_{vib}}{T}}} \tag{6.52}$$

In this approximation, these contributions to the energy are simply additive, and the total molecular energy is

$$E_{tot} = E_{trans} + E_{rot} + E_{vib} \tag{6.53}$$

Correspondingly, the partition function q_{tot} that includes contributions from all three energy sources is

$$q_{tot} = q_{trans} q_{rot} q_{vib} \tag{6.54}$$

However, the bond between the two atoms involved in a diatomic molecule is not rigid, but stretches or compresses a little under stress.

Additionally, the vibrational energy is not purely harmonic, but has anharmonic contributions as well, and these smaller parts have to be accounted for in addressing the real thermodynamic functions of real molecules.

The fact that the bond between the two nuclei in a diatomic molecule is not rigid also affects the rotational energy levels. As the molecule rotates, the nuclei are subject to centrifugal forces that slightly stretch the bond, the more so the faster the molecule rotates. This causes the moment of inertia to increase slightly, effectively reducing the rotational constant B. The consequence is that the rotational energy levels are altered. For diatomic and small linear molecules, analysis shows that the rotational energy levels are governed by the relation

$$F(J) = B_v J(J+1) - D_v J^2 (J+1)^2 + H_v J^3 (J+1)^3 \qquad (6.55)$$

The subscript v emphasizes the fact that the spectroscopic constants B_v, D_v, H_v are functions of the vibrational state. The constants D_v and H_v are called centrifugal distortion constants. The constant H_v is known for very few diatomic molecules, and is many orders of magnitude smaller than the centrifugal distortion constant D_v. The constant D_v is several orders of magnitude smaller than the rotational constant B_v.

Simultaneously, the vibrational energy levels are altered by the anharmonicity constants $\omega_e x_e$, $\omega_e y_e$, and $\omega_e z_e$. The vibrational energy levels are given in cm^{-1} units as

$$G(v) = \omega_e \left(v + \frac{1}{2} \right) - \omega_e x_e \left(v + \frac{1}{2} \right)^2$$

$$+ \omega_e y_e \left(v + \frac{1}{2} \right)^3 + \omega_e z_e \left(v + \frac{1}{2} \right)^4 \qquad (6.56)$$

The expansion cannot be carried further without violating the conditions of the Born–Oppenheimer approximation that permits separation of nuclear and electronic energies. In practice, there are no molecules for which the anharmonicity constant $\omega_e z_e$ is known to a higher precision than its value, and it is rarely encountered. The dominant majority of the anharmonicity is in the $w_e x_e$ term.

6.7. Anharmonicity

It is often convenient to refer the energies of the vibrational states to the lowest level, rather than to the bottom of the energy well. When this is done, the vibrational energy becomes

$$G_0\left(v\right) = w_0 v - w_0 x_0 v^2 + w_0 y_0 v^3 + w_0 z_0 v^4 \qquad (6.57)$$

which is an algebraically much simpler expression, representing exactly the same vibrational energy levels. It immediately follows that the constants are related as

$$w_0 = w_e - w_e x_e + \frac{3}{4} w_e y_e \qquad (6.58)$$

$$w_0 x_0 = w_e x_e - \frac{3}{2} w_e y_e \qquad (6.59)$$

$$w_0 y_0 = w_e y_e \qquad (6.60)$$

Similar relations may be written to include the z term. Since this is so poorly known, it is omitted here as of no practical importance. The anharmonicities generally rapidly decrease in magnitude with increasing power of the vibrational quantum number. For $H^{35}Cl$, $w_e x_e$ is about 52 cm^{-1} while $w_e y_e$ is 0.056 cm^{-1}, and the z component is too small to be measured. The method given below was first given by Mayer and Mayer in their textbook on statistical mechanics, and may be expanded to include higher and smaller terms. The results are given by Wooley.[1] Wooley also treated numerous other small perturbations on the energy of diatomic and small polyatomic molecules, and determined the contributions these make to the partition function and thus to the thermodynamic properties of these compounds. For simplicity, only the $w_e x_e$ component of the anharmonicity will be retained in the following analysis. The simplified relations needed are $w_0 = w_e - w_e x_e$ and $w_0 x_0 = w_e x_e$. It is spectroscopic convention to measure all energies in cm^{-1} units, and a conversion to Joules has to be done with care. The SI units for the speed of light in vacuo are inconvenient for this purpose, and it is preferable to use the cgs unit. The rigorous insistence on SI units throughout requires the inconvenient and unaesthetic transport of factors of 100 from equation to equation. If the vacuum speed of light is taken as $2,99792458 \times 10^{10}$ cm sec^{-1}, and Planck's constant is measured in Joule·sec, vibrational energies E_v and E_v^0 in Joules are given by

$$E_v = hcG(v) \tag{6.61}$$
$$E_v^0 = hcG_0(v) \tag{6.62}$$

These equations provide energies in Joules per molecule in terms of the spectroscopically derived parameters ω_e and $\omega_e x_e$.

Now, both of these energy expressions suffer from the same defect, namely, neither of the parameters ω_e or $\omega_e x_e$ is the observed frequency. Instead, these are derived quantities. In terms of these quantities or their zero counterparts, the observed fundamental frequency ν_{obs} is either of the relations

$$\nu_{\text{obs}} = \omega_e - 2\omega_e x_e \tag{6.63}$$
$$\nu_{\text{obs}} = \omega_0 - \omega_0 x_0 \tag{6.64}$$

Exactly the same observed frequency is described by both equations. The first refers to an energy zero selected at the bottom of the potential energy well. The second refers to a zero of energy selected to be the ground vibrational state. The following analysis works in terms of experimentally observed quantities, rather than in terms of derived quantities. For reasons of convenience, which will soon become apparent, the following definitions are advanced.

$$U = \frac{\Theta_v}{T} = \frac{hc\nu_{\text{obs}}}{k_B T} = \frac{hc\omega_0}{k_B T}(1 - x_0) = \frac{hc\omega_e}{k_B T}(1 - 2x_e) \tag{6.65}$$

where ν_{obs} is the experimentally observed frequency and $hc/k_B = 1.4388$ cm K is called the second radiation constant. Also define

$$X = \frac{\text{Anharmonicity}}{\nu_{obs}} = \frac{\omega_0 x_0}{\nu_{obs}} = \frac{\omega_e x_e}{\nu_{obs}} = \frac{x_e}{1 - 2x_e} = \frac{x_0}{1 - x_0} \tag{6.66}$$

These definitions allow a convenient expression for the energy referred to the ground state as the zero of energy

$$\frac{E_v^0}{k_B T} = U v - X U v (v - 1) \tag{6.67}$$

This may be shown as follows. Start from $E_v^0 = hcG_0(v)$ and write

$$\frac{E_v^0}{k_B T} = \frac{\omega_0 hc}{k_B T} v - \frac{\omega_0 x_0 hc}{k_B T} v^2 \tag{6.68}$$

Spectroscopists regard the symbol $\omega_0 x_0$ as a single symbol. It is convenient to imagine that in this analysis, however, $\omega_0 x_0 = \omega_0 \times x_0$, for then the expression becomes

$$\frac{E_v^0}{k_B T} = \frac{\omega_0 hc}{k_B T} \left(v - x_0 v^2 \right) \tag{6.69}$$

The value of this relation is unaltered if $x_0 v$ is added to and subtracted from the quantity in parenthesis.

$$\frac{E_v^0}{k_B T} = \frac{\omega_0 hc}{k_B T} (v - x_0 v^2 + x_0 v - x_0 v) \tag{6.70}$$

Doing so allows the selective factoring needed

$$\frac{E_v^0}{k_B T} = \frac{\omega_0 hc}{k_B T} (1 - x_0) v - \frac{\omega_0 x_0 hc}{k_B T} v (v - 1) \tag{6.71}$$

Using the definition of U given above, this can also be expressed

$$\frac{E_v^0}{k_B T} = U v - \frac{\omega_0 hc}{k_B T} v (v - 1) x_0 \tag{6.72}$$

Now, $X = \frac{x_0}{1 - x_0}$, and therefore $x_0 = X(1 - x_0)$ Making this substitution yields

$$\frac{E_v^0}{k_B T} = U v - \frac{\omega_0 hc}{k_B T} v (v - 1) X (1 - x_0) \tag{6.73}$$

But U as defined above is embedded in the term quadratic in v. Thus

$$\frac{E_v^0}{k_B T} = U v - X U v (v - 1) \tag{6.74}$$

which is the relation that was to be shown. To progress, note that $X = \frac{\omega_0 x_0}{\omega_0 - \omega_0 x_0}$ is quite small. For $H^{35}Cl$, for example, the anharmonicity is about 52 cm^{-1} while the observed vibrational frequency is about 2886 cm^{-1}. The numerical value of X in that case is $52/2886 = 1.8 \times 10^{-2}$. This is small enough to allow the approximation scheme developed below to function properly. Note that since X is small, X^2 is even smaller. The vibrational partition function, including the anharmonicity, may be written in the present approximation

$$q_{vib} = \sum_{v=0}^{\infty} e^{-Uv + XUv(v-1)} \tag{6.75}$$

This may be written

$$q_{vib} = \sum_{v=0}^{\infty} e^{-Uv} e^{XUv(v-1)} \tag{6.76}$$

The fact that X is small and successive powers of X are rapidly smaller suggests that it may be profitable to expand the second exponential and drop all but the linear term

$$q_{vib} = \sum_{v=0}^{\infty} e^{-Uv}[1 + XUv(v - 1) + \cdots] \tag{6.77}$$

This may be rewritten

$$q_{vib} = \sum_{v=0}^{\infty} e^{-Uv} + XU \sum_{v=0}^{\infty} v(v - 1) e^{-Uv} \tag{6.78}$$

All that need be done now is to evaluate the three sums. The starting point is a minor theorem from functions of a complex variable. If z is complex, the sum that represents the fraction $1/(1 - z)$ is $\sum_{n=0}^{\infty} z^n$. The sum is uniformly convergent and thus possesses derivatives of all orders, and represents the function $1/(1 - z)$ everywhere inside a circle of radius $|z| = 1$. These are precisely the conditions under which the vibrational partition function is to be evaluated. Thus, differentiating $1/(1 - z)$ twice and multiplying the result by z^2 produces the desired relation. The first derivative is

$$\frac{1}{(1-z)^2} = \sum_{n=0}^{\infty} n z^{n-1} \tag{6.79}$$

Differentiating again and multiplying the result by z^2 produces

$$\frac{2z^2}{(1-z)^3} = \sum_{n=0}^{\infty} n(n-1) z^n \tag{6.80}$$

Identifying z with e^{-U} and n with v allows the vibrational partition function to be evaluated

$$q_{\text{vib}} = \frac{1}{1-e^{-U}} + XU \frac{2e^{-2U}}{(1-e^{-U})^3} \tag{6.81}$$

This may be factored as

$$q_{\text{vib}} = \frac{1}{1-e^{-U}} \left[1 + \frac{2XU e^{-2U}}{(1-e^{-U})^2} \right] \tag{6.82}$$

This is of the form

$$q_{\text{vib}} = q_{\text{vib,harmonic}}[1 + \text{small correction term}]$$

This expression has the advantage that the variable U is defined in terms of the observed fundamental vibrational frequency, and thus has anharmonicities of all orders built in, rather than in terms of the hypothetical harmonic frequency ω_0 with some of the anharmonicity removed algebraically. This analysis is the justification for the recommendation that observed vibrational frequencies be used in the vibrational partition function. Often, the small correction term will be so small that there is no point to carrying it through thermodynamic calculations, except possibly at the highest temperatures. Usually the highest temperatures will be so high that the real molecule would pyrolyze.

In most applications, it is the logarithm of the partition function that is required, the single prominent exception being in the treatment of chemical equilibrium, where the partition function itself is needed, as will appear in a later chapter of this book. For example, the vibrational contribution to the Helmholz free energy in the canonical ensemble is

$$A_{\text{vib}} = -Nk_BT \ln q_{\text{vib}} \tag{6.83}$$

Thus

$$\ln q_{\text{vib}} = \ln q_{\text{vib, harmonic}} + \ln \left[1 + 2XU \frac{e^{(-2U)}}{(1 - e^{(-U)})^2} + \cdots \right] \tag{6.84}$$

Since $\ln(1 - z) = z - \frac{1}{2}z^2 + \frac{1}{3}z^3 + \cdots$, if z is very small ($z \ll 1$) then all but the leading term in the series may be ignored and $\ln(1 - z) \cong z$. Then $\ln q_{\text{vib}}$ may be approximated by

$$\ln q_{\text{vib}} = \ln \frac{1}{1 - e^{-U}} + 2XU \frac{e^{-2U}}{(1 - e^{-U})^2} + \text{smaller terms} \tag{6.85}$$

In this approximation, involving retention of only the linear term in the expansion of the anharmonicity contribution, the correction term appears simply additively. It vanishes at low temperatures, and makes no appreciable contribution to the thermodynamic functions until relatively high temperatures are attained. In the case of HCl, the contribution becomes appreciable near or above about 500 K. If the anharmonicity is small, as is common for molecules not involving hydrogen, the contribution may be ignored and X may be set equal to zero. Then the effect of anharmonicity is contained entirely in the value of the experimental vibrational frequency used to define the variable U, and the entire logarithmic vibrational contribution to the Helmholtz free energy, and thus to the other thermodynamic functions, is just

$$\ln q_{\text{vib}} \cong \ln \frac{1}{1 - e^{-U}} = -\ln(1 - e^{-U}) \tag{6.86}$$

An example of early computation of thermodynamic functions from spectroscopic data is given for several triatomic molecules, including SO_2, by McBride and Gordon.[2] The calculation is routinely included in output from the Gaussian suite of *ab initio* molecular orbital programs. The calculation was routinely done in most of the versions, such as Gaussian 03.[3] McBride and Gordon pointed out that one of the problems with such computation is that new spectroscopic measurement often brings new values for the spectroscopic parameters, requiring recalculation. This should present no special problem with modern computers, as all one needs to do is design a reservoir for the spectroscopic constants. The computational method will not change, just the constants, and a rerun costs little but time on an ordinary desktop computer. In fact, changing a spectroscopic constant by a few standard deviations and comparing the results of the two runs (one with no change, one with change) could give insight into the relative importance of the parameter changed. It will be found that at low temperatures the correction terms disappear, but at high temperatures begin to make their presence felt.

Problems

1. Calculate the entropy of Xenon in the hypothetical state of an ideal gas at 10 degree intervals from 10 K to 200 K and tabulate the results.
2. Use the Euler–MacLaurin sum formula to calculate the coefficient of the fifth term (the next term) in the expansion of the rotational partition function

$$q_{\text{rot}} = \frac{T}{\Theta_{\text{rot}}} \left(1 + \frac{\Theta_{\text{rot}}}{3T} + \frac{\Theta_{\text{rot}}^2}{15T^2} + \frac{4\Theta_{\text{rot}}^3}{315T^3} + \cdots \right)$$

3. Calculate the entropy of $H^{35}Cl$ in the hypothetical state of an ideal gas at 50 degree intervals from 100 to 1000 degrees Kelvin, and tabulate the results.
4. The rotational energy levels in cm^{-1} units are given by

$$F(J) = BJ(J+1) - DJ^2(J+1)^2$$

to an adequate precision. If $\Theta = \frac{Bhc}{k_B T}$ and $\varphi = \frac{Dhc}{k_B T}$, give an approximate rotational partition function valid at high temperatures. Note that D is much smaller than B.

References

1. H. W. Wooley, J. Research Natl. Bur. Standards **56**, 105 (1956).
2. B. J. McBride and S. Gordon, J. Chem. Phys. **35**, 2198 (1961).
3. M. J. Frisch, G. W. Trucks, H. B. Schlegel *et al.*, Gaussian 03 , Revision D.01; Gaussian, Inc.: Wallingford, CT, 2004.

Chapter 7

Homonuclear Diatomic Molecules

7.1. Spectroscopic Constants

Homonuclear diatomic molecules are just like heteronuclear diatomic molecules in the structure of their energy levels. The typical term in both cases is

$$T(v, J) = \omega_e \left(v + \frac{1}{2} \right) + B_v J(J+1) + \text{small terms} \qquad (7.1)$$

The small terms carry the corrections to the equilibrium values shown here, and are, for vibration

$$-\omega_e x_e \left(v + \frac{1}{2} \right)^2 + \omega_e y_e \left(v + \frac{1}{2} \right)^3 + \omega_e z_e \left(v + \frac{1}{2} \right)^4$$

and for rotation

$$-D_v J^2 (J+1)^2 + H_v J^3 (J+1)^3$$

In these relations the value of the rotational constants depends in principle on the value of the vibrational quantum number. The reason is that the vibrational, rotational and even the electronic motions do not really separate, but are connected by usually small perturbative couplings. Ignoring the electronic contributions, the reason the rotational motions are coupled with the vibrational is classically clear enough. As the molecule vibrates, the internuclear distance changes periodically, causing the moment of inertia to fluctuate anharmonically. This coupling increases with vibrational state, and is accommodated by a linear dependence of the rotational constants on the vibrational quantum number or state. The centrifugal distortion constant is similarly affected. The result of a power series analysis is

$$B_v = B_e - \alpha_e \left(v + \frac{1}{2} \right) \tag{7.2}$$

where α_e is a small constant that can be calculated from the rotational and vibrational constants using the formula quoted below, and the other symbols have the following meaning

$$B_e = \frac{h}{8\pi^2 I_e c}$$

$$I_e = \mu r_e^2$$

$$\mu = \frac{m_1 m_2}{m_1 + m_2}$$

$$\alpha_e = \frac{6\sqrt{\omega_e x_e B_e^3}}{\omega_e} - \frac{6 B_e^3}{\omega_e}$$

In addition, Kratzer's formula allows a reasonable estimate of the equilibrium centrifugal distortion constant D_e. Kratzer's formula states

$$D_e = \frac{4 B_e^3}{\omega_e^2} \tag{7.3}$$

This is good enough to estimate D_e certainly within an order of magnitude, and usually much better. However, given D_e and B_e, it may not be used to estimate the harmonic vibrational frequency, which it will miss by as much as twenty percent or more. Very few situations require the calculation of α_e, as it is usually just measured as a by-product of the measurement of B_e.

The constant H_v is not known for most diatomic molecules, but may be estimated from the relation

$$H_v = \frac{2 D_e}{3 \omega_e^2} (12 B_e^2 - \alpha_e \omega_e) \tag{7.4}$$

In this relation, D_e is the equilibrium centrifugal distortion constant. Unfortunately, spectroscopists also use D_e to represent the dissociation energy measured from the bottom of the potential energy well to the dissociation limit. It is easy to show that it is related to the harmonic frequency ω_e and the first anharmonicity constant $\omega_e x_e$ approximately by

$$D_e = \frac{\omega_e^2}{4\omega_e x_e} \qquad (7.5)$$

7.2. Nuclear Spin and the Pauli Exclusion Principle

The statistics of homonuclear diatomic molecules use the energy levels described by the relations above in ways that are very different from the ways the heteronuclear diatomic molecules use them. The difference is based on the need to satisfy the Pauli Exclusion Principle. In order to understand this, it is first necessary to understand some elementary facts about nuclear structure. For this purpose, it is only necessary to regard the nuclei as constructed of protons and neutrons, which give rise to a nuclear spin I, which can be measured (the magnetic resonance methods depend on this) but not predicted in an elementary way. Thus ^{12}C and ^{16}O have no spin ($I = 0$), ^1H and ^{19}F have spin $1/2(I = 1/2)$, ^2H has spin 1 as has ^{14}N ($I = 1$), ^{35}Cl has spin 3/2 ($I = 3/2$) and so on.

The statement of the Pauli Principle most useful in this context is: The complete wave function for the molecule, including nuclear spin but excluding translation, must be antisymmetric under exchange of pairs of identical particles, specifically protons and neutrons.

The wave functions describing the quantum behavior of molecules are defined as follows. The electronic part of the wavefunction is Ψ_e. The vibrational-rotational product wavefunction combined with the electronic is called the total wavefunction, $\Psi_e \Psi_v \Psi_r = \Psi_{total}$. Finally the complete wavefunction is the product of the nuclear spin wavefunction and the total wavefunction

$$\Psi_{complete} = \Psi_e \Psi_v \Psi_r \Psi_{ns} = \Psi_{total} \Psi_{ns} \qquad (7.6)$$

Here the subscripts v, r and ns stand for vibrational, rotational and nuclear spin. The Pauli Exclusion Principle uses the complete wavefunction. This requires an examination of some of the properties of the nuclear spin wave functions, without knowing exactly what these functions might be. Notice that although the electronic, vibrational and rotational motions are treated separately, these motions are actually inextricably intertwined. This does not matter, since only the symmetry properties are needed, and these are correctly summarized in any properly constructed approximation.

Consider ^1H, which has a nuclear spin $I = 1/2$. This means that each nucleus has two states, α and β. In a magnetic field, one would expect the two states to correspond to alignment with and against the field direction. The possibilities for the diatomic molecule H_2 are

$$\alpha\alpha$$

$$\beta\beta$$

$$\alpha\beta$$

$$\beta\alpha$$

Both $\alpha\alpha$ and $\beta\beta$ are symmetric under exchange of nuclei, and are thus acceptable nuclear spin wave functions as they stand. But exchange of nuclei converts $\alpha\beta$ into $\beta\alpha$ and conversely, so neither is an acceptable spin wave function. However, the sum and the difference of these functions do meet the criterion. The functions then divide into two types, symmetric and antisymmetric under exchange.

$$\text{Symmetric}$$

$$\alpha\alpha$$

$$\beta\beta$$

$$\frac{1}{\sqrt{2}}(\alpha\beta + \beta\alpha)$$

$$\text{Antisymmetric}$$

$$\frac{1}{\sqrt{2}}(\alpha\beta - \beta\alpha)$$

The factor $\frac{1}{\sqrt{2}}$ is used to normalize the spin wavefunction. The symmetry arguments are dependent only on the commutativity of the symmetry operator R with the full molecular Hamiltonian, H. The translational wave function is not included since it is totally symmetric. The operator R can be, for example, the operator for a reflection through a plane or through a center of symmetry, or for expressing a rotation about an axis of symmetry by some angle such as π for a two-fold symmetry axis, or $2\pi/3$ for a three fold axis.. The molecular Hamiltonian, no matter how approximate, must display the full symmetry of the molecule. Then any such Hamiltonian H behaves as $RH = HR$, that is, H and R commute. For operators of order two (planes, inversion, two-fold symmetry axis), only two behaviors of the wavefunction are possible. Either $R\Psi_{total} = +\Psi_{total}$, which defines a symmetric state, or $R\Psi_{total} = -\Psi_{total}$, which defines an antisymmetric state. In the case of three- or higher-fold symmetry axes, the function may be sent into a linear combination of functions. However, such axes do not occur in diatomic molecules, and the more complex analysis required is only needed in polyatomic molecules.

When $RH = HR$, Ψ_{total} is said to be a simultaneous wave function (sometimes called an eigenfunction) of the operators H and R, with simultaneous eigenvalues, namely, the energy for H and usually an integer for R such as $+1$ or -1. In nature, symmetric states do not combine with antisymmetric states, and transitions between them do not occur. Of course, if the molecule is dissociated and then resynthesized, a new set of levels and populations will be generated. Graphite has the property of relaxing the restriction on combination at low temperatures. In the absence of such strong external perturbations, if a molecule is in a symmetric state at time 0, it will still be in a symmetric state at any time t later. This is not strictly true, and if t is long enough, conversion can occur. However, such times are long in comparison with experimental times. Thus, in a given thermodynamic state, a collection of molecules of kind A in symmetric states will act completely independently of molecules of kind A in antisymmetric states. It will be as if two distinct kinds of non-interconvertible molecules were present, and this will be reflected in both spectroscopic and thermodynamic properties of an ensemble that is composed of such molecules.

Application of the Pauli Principle determines which states are symmetric and which are antisymmetric. But to do this, the nature of an exchange of nuclei has first to be considered, and this requires consideration of the symmetry properties of the molecular wave function. It is convenient to consider first the electronic wave function, then the parity of rotational energy levels, then the exchange behavior, and finally the nature of the nuclear spin wave functions.

The electronic states in diatomic molecules are described by symbols that convey information pertaining to the angular momentum Λ of the electrons. If $\Lambda = 0$, the symbol is Σ. If $\Lambda = 1$, the symbol is Π. If $\Lambda = 2$, the symbol is Δ. For $\Lambda = 3$, the symbol is Φ, and so on. This symbol is decorated in the upper right with a plus (+) or minus (−), depending on the behavior of the electronic wave function on reflection through a plane of symmetry containing the two nuclei. If the wave function is unchanged by the reflection, the plus sign is used, and if it is transformed into its negative, the minus sign is used. Additional decoration is provided at the lower right of the symbol, where now a g (standing for the German word *gerade*, meaning symmetric) is used if the electronic wave function is unchanged on reflection through the center of symmetry, and if the wave function is transformed into its negative on this reflection, the symbol u (for *ungerade*, meaning antisymmetric) is used. An additional decoration appears in the upper left hand position that describes the multiplicity of the electron structure in the state. Most ground states are singlet, so a 1 appears there. Oxygen is the exception, as the ground electronic state is triplet, so a 3 appears there for that molecule. Most homonuclear diatomic molecules have ground electronic states described by $^1\Sigma_g^+$. The following table shows the possibilities. In the table, i is the operator for reflection through the center of symmetry, σ is the operator for reflection through the plane and the cross indicates application. In future, the cross will be omitted but understood, and writing two operators such as $C_2\sigma$ will mean to apply the reflection through the plane σ first, then perform the operation C_2, rotation about a two-fold axis of symmetry by 180 degrees, on the result of the first operation.

Table 7.1. Possible symmetry behavior
of Σ electronic states, for which $\Lambda = 0$.

Ψ_e	$i \times \Psi_e$	$\sigma \times \Psi_e$
Σ_g^+	$+$	$+$
Σ_u^+	$-$	$+$
Σ_g^-	$+$	$-$
Σ_u^-	$-$	$-$

Of course, in one molecule there will be only one ground electronic state, which will have only one of the behaviors shown.

Now consider parity and the exchange behavior of the total wave function. The total wave function is the wave function without the nuclear spin contribution. Rotational states are designated $+$ or $-$, according as the behavior of the total wave function under reflection through the center of symmetry, i. The property is called parity.

Rotational states are also designated a or s according as the behavior of the total wave function, under exchange of identical nuclei.

1. In order to get exchange symmetry, the parity must first be obtained, and to do that it is necessary to consider the nature of an inversion of all particles that form the molecule. An inversion i essentially consists of two steps, though performed as one, rather like the knight's move in chess. Every homonuclear diatomic molecule has an infinite number of 2-fold axes of symmetry orthogonal to the internuclear direction and passing through the center of mass of the molecule. Any one of them applied to the molecule causes a rotation of all its parts by 180 degrees. That is, the whole molecule rotates. This means

   A. the nuclei are exchanged
   B. a point at (x, y, z) is carried into one at $(-x, -y, z)$
   C. the electronic wave function does not change under the two-fold rotation, since it depends only on the coordinates of electrons measured with respect to the nuclei.

D. the product of vibrational and rotational wave functions may change sign, depending on the rotational quantum number J.
   1. If J is even, $C_2\psi_v\psi_r = +\psi_v\psi_r$
   2. If J is odd, $C_2\psi_v\psi_r = -\psi_v\psi_r$

   for any vibrational quantum number v.

2. The second step is the reflection of the electrons alone through a plane of symmetry that contains the nuclei. This has the following characteristics.

A. the nuclei are unchanged, for the plane contains them.
B. only the electronic wave function is involved in the reflection through the plane of symmetry. Its behavior is given by the right superscript sign + or − on the term symbol. Thus, if the state is Σ^- then $\sigma\psi_e = -\psi_e$ and so on, regardless of any other decoration the term symbol displays.
C. a point at $(-x, -y, z)$ is carried into a point at $(-x, -y, -z)$ in the electron cloud, thus completing the inversion.

Now use an inversion to obtain parity. Consider Σ_g^- as an example. In the reflection, $\sigma\psi_e = -\psi_e$ while ψ_e is unchanged under two-fold rotation, that is $C_2\psi_e = \psi_e$. Note that $\psi_v\psi_r$ is unchanged under the reflection σ, since the reflection applies to the electrons alone. By 1. D. above, $C_2\psi_v\psi_r = +\psi_v\psi_r$ if J is even, and $C_2\psi_v\psi_r = -\psi_v\psi_r$ if J is odd. Consider the two cases separately.

For J even

$$i\Psi_e\Psi_v\Psi_r = \sigma C_2\Psi_e\Psi_v\Psi_r = (\sigma\Psi_e)(C_2\Psi_v\Psi_r)$$
$$= (-\Psi_e)(+\Psi_v\Psi_r) = -\Psi_e\Psi_v\Psi_r \tag{7.7}$$

For J odd

$$i\Psi_e\Psi_v\Psi_r = \sigma C_2\Psi_e\Psi_v\Psi_r = (\sigma\Psi_e)(C_2\Psi_v\Psi_r)$$
$$= (-\Psi_e)(-\Psi_v\Psi_r) = +\Psi_e\Psi_v\Psi_r \tag{7.8}$$

Parity is the property of the total wave function $\Psi_e\Psi_v\Psi_r$ under reflection through the center of symmetry. The possibilities for $\Lambda = 0$ states are summarized in Table 7.2.

Table 7.2.   Parity of rotational levels for $\Lambda = 0$.

Electronic State	J	$C_2\Psi_v\Psi_r$	$\sigma\Psi_e$	$i\Psi_{evr}$
Σ_g^+	odd	$-$	$+$	$-$
	even	$+$	$+$	$+$
Σ_g^-	odd	$-$	$-$	$+$
	even	$+$	$-$	$-$
Σ_u^+	odd	$-$	$+$	$-$
	even	$+$	$+$	$+$
Σ_u^-	odd	$-$	$-$	$+$
	even	$+$	$-$	$-$

7.3. Exchange of Nuclei

Now consider the nature of an exchange of nuclei, which the foregoing allows.

An exchange of nuclei alone can be accomplished by the following stepwise process.

1. Reflect the entire molecule through the center of symmetry i. The result is given by parity.
2. Reflect only the electrons through the center of symmetry i. The result is given by the labels g or u on the term symbol for the electronic state.

Example. Let the electronic state be Σ_g^-. In the case of even J, $i\Psi_{evr} = -\Psi_{evr}$ from the table. Also $i\psi_e = +\psi_e$ from the subscript g on the term symbol. The product of these two operations is negative. Such a state is named antisymmetric, and associated with the label a. When J is odd, the rotational states have the opposite parity but the g character is retained. The result is that an exchange of nuclei when J is odd gives rise to a symmetric state. These levels are labeled s.

For Σ states, there are four possible behaviors, summarized in the table below. Note particularly the difference between Σ_g^+ and Σ_g^-. The ground electronic state of $^{16}O_2$ is $^3\Sigma_g^-$, while that of $^1H_2$ and most other common diatomic molecules is $^1\Sigma_g^+$. In the table, R_{ex} is the operator for the exchange of nuclei only.

	Σ_g^+		Σ_g^-		Σ_u^+		Σ_u^-	
	$R_{ex}\Psi_{evr}$	$i\Psi_{evr}$	$R_{ex}\Psi_{evr}$	$i\Psi_{evr}$	$R_{ex}\Psi_{evr}$	$i\Psi_{evr}$	$R_{ex}\Psi_{evr}$	$i\Psi_{evr}$
J_{odd}	a	$-$	s	$+$	s	$-$	a	$+$
J_{even}	s	$+$	a	$-$	a	$+$	s	$-$

States with $\Lambda = 1, 2, 3, \ldots$ do not typically occur as ground states for diatomic molecules. However, should one be encountered, it would be subject to Λ-type doubling, and all the rotational states would occur as doublets split by the electronic orbital angular momentum Λ. The treatment is similar to that given above, but is somewhat more complex, and has been described by G. Herzberg on page 238 of his book on diatomic molecules.[1]

7.4. Bose Einstein and Fermi Dirac Statistics

The form of the Pauli exclusion principle needed here states that only those wave functions that are antisymmetric in exchange of identical pairs of elementary particles are allowed in nature. The wave function in question is one that carries all the molecular information Ψ_{evr} and the nuclear spin wave function Ψ_{ns}, and that function is what is called the complete wave function Ψ in this book, that is

$$\Psi = \Psi_{evr}\Psi_{ns} \tag{7.9}$$

The nucleus of a hydrogen atom is a single proton, and exchange of nuclei in H_2 exchanges one pair of identical elementary particles. Hence, for H_2,

$$R_{ex}\Psi = -\Psi \tag{7.10}$$

for all J values.

For deuterium, the ground electronic state is exactly the same as that for hydrogen. But now, the nucleus of an atom of deuterium consists of a proton and a neutron, and exchange of nuclei causes the exchange of not one but two pairs of different but still elementary particles. The result could not be more different than the result for hydrogen.

$$R_{ex}\Psi = -(-\Psi) = +\Psi \tag{7.11}$$

The first minus sign may be regarded as representing the sign change due to exchange of protons, as in hydrogen. The other minus sign originates in the exchange of the neutrons in the deuterium nucleus. Extending these ideas, if the mass number of the nuclei involved in a homonuclear diatomic molecule is odd, the complete wave function must be antisymmetric under exchange of identical nuclei, and if the mass number is even, the complete wave function must be symmetric under the exchange. By convention, nuclei that follow the antisymmetric case are said to follow Fermi Dirac statistics, and those that follow the symmetric case are said to follow Bose Einstein statistics.

The wave functions for hydrogen were worked out above. For deuterium, the case is slightly more complicated since the nuclear spin is 1, and there are three projections on a preferred direction such as that established by a uniform magnetic field, one for +1, one for 0 and one for −1. The corresponding functions may be designated α, β, γ. The diatomic possibilities are

$$\alpha\alpha$$
$$\beta\beta$$
$$\gamma\gamma$$
$$\alpha\beta + \beta\alpha$$
$$\alpha\gamma + \gamma\alpha$$
$$\beta\gamma + \gamma\beta$$

$$\alpha\beta - \beta\alpha$$
$$\alpha\gamma - \gamma\alpha$$
$$\beta\gamma - \gamma\beta$$

Obviously there are six symmetric spin functions and three antisymmetric, whereas for hydrogen there were three symmetric and one antisymmetric spin function. The exchange behavior of the total wave function determines the rotational levels with which these functions must be associated. In the case of hydrogen, $R_{ex}\Psi = -\Psi$ for the complete wave function. The opposite result holds for deuterium. These facts produce the following alterations in the populations of the energy levels

	$R_{ex}\Psi_{evr}$	degeneracy	state	common name
$H_2\ ^1\Sigma_g^+$				
J odd	a	3	symmetric	ortho
J even	s	1	antisymmetric	para
$D_2\ ^1\Sigma_g^+$				
J odd	a	3	antisymmetric	para
J even	s	6	symmetric	ortho

Note the different behaviors of hydrogen and deuterium at low temperatures. When cooled to a sufficiently low temperature, a sample of hydrogen can be prepared as essentially pure para hydrogen, while deuterium would provide a sample of practically pure ortho deuterium. The preparation of nearly pure para hydrogen has been achieved and the liquid used as a solvent in infrared investigations.[2]

7.5. Spin Wave Functions

The number of spin wave functions can be found for arbitrary nuclear spin quantum number I. The method of writing down functions is to introduce $2I+1$ spin functions for each nucleus. There are then always $2I+1$ exchange symmetric spin functions of the form $\alpha\alpha$, $\beta\beta$ .... The rest of the spin functions are formed by taking linear combinations that are either symmetric or antisymmetric under exchange, such as $\alpha\beta + \beta\alpha$, $\delta\gamma - \gamma\delta$, and so on until the possible combinations are exhausted. In general, the first function (α, β, γ, δ, ...) may be

selected in $2I + 1$ different ways. The second selection must be different, and so may be made in any of $2I + 1 - 1$ different ways, a total of
$2I(2I + 1)$ ways. However, the selection $\alpha\beta$ is equivalent for counting
purposes to $\beta\alpha$, and the count is twice as large as it should be. The
number of antisymmetric functions and the number of symmetric
spin functions that are not simple products is then $I(2I + 1)$, Some
authors feel that $\rho = 2I + 1$ is a more compact expression. Since
$I = \frac{1}{2}(\rho - 1)$, the following expressions for the number of exchange
symmetric and antisymmetric spin states is given as follows

Number of antisymmetric spin states	$I(2I + 1)$	$\frac{1}{2}(\rho - 1)\rho$
Number of symmetric spin states	$2I + 1 + I(2I + 1)$ $= (I + 1)(2I + 1)$	$\frac{1}{2}(\rho + 1)\rho$
Total number of spin states	$(2I + 1)^2$	ρ^2

Consider $^{16}O_2$. The term symbol is $^3\Sigma_g^-$. The triplet splitting is
fairly small, and would not be resolved in a Raman experiment with
spectral slit widths near 1 cm^{-1}, a setting that is more than sufficient
to resolve the rotational Raman spectrum between a few cm^{-1} and
about 120 cm^{-1}. Since the nuclear spin of ^{16}O is zero and the number
of nucleons is even, the complete wave function follows Bose Einstein
statistics

$$R_{ex}\Psi = (-1)^{16}\Psi = +\Psi \qquad (7.12)$$

The only nuclear spin wave function is the totally symmetric function $\alpha\alpha$, and states with which it is associated are called ortho. There
are no para states for $^{16}O_2$. For Σ_g^- states, J even is associated with
antisymmetric behavior under exchange of nuclei, while J odd is associated with symmetric behavior. There is no antisymmetric nuclear
spin function, which would be required to make the exchange behavior of the complete wave function positive for J even. Thus, only
rotational states with J odd survive, and the pure rotational Raman
spectrum of $^{16}O_2$ is missing half its bands, appearing to rotate twice
as fast as it actually does. The main point is that states of even

J cannot satisfy the Pauli Principle and therefore do not exist in oxygen-16.

In Raman spectroscopy, the selection rule for rotation is $\Delta J = 0, \pm 2$ Experimentally, the first pure rotational transition for $^{16}O_2$ is found at $10B$, where B is the rotational constant, the next $8B$ further along at $18B$, the next several at intervals of $8B$, whereas in a heteronuclear diatomic molecule, the first pure rotational transition is found at $6B$, the next at $10B$, which is $4B$ further along, and the rest at intervals of $4B$. The ratio of $6B/8B = 3/4$, whereas that for $10B/8B = 5/4$, easily distinguishable. For oxygen, the experimentally observed ratio of Raman displacements is $5/4$. Other possible exchange behaviors are displayed in Table 7.3.

In passing from this topic, it is worth noting that similar, though far from identical, results are obtained for symmetric polyatomic molecules. The molecular wave function including nuclear spin must again satisfy the Pauli Principle. For example, methane, CH_4, has protons, which are fermions, and all of its wave functions must be antisymmetric under exchange of pairs of protons, while the deuterium substituted molecule CD_4 contains bosons (deuterium nuclei) in place of fermions and must therefore be symmetric under the same operations. The effects are generally of importance to spectroscopists, but less so to statistical mechanics, for the effects dominate only for small molecules at quite low temperatures. The matter was discussed by E. B. Wilson, Jr. (3,4) in 1935.

7.6. Statistical Weights of Rotational Energy Levels

It is customary to weight rotational levels with the usual $2J+1$ degeneracy and with the nuclear spin weights (the number of symmetric or antisymmetric spin functions). Then in the microcanonical ensemble, the partition function for a heteronuclear diatomic molecule including the spin weights is in first good approximation

$$q_{ns}q_{rot} = \rho_1\rho_2 \sum_{J=0}^{\infty}(2J+1)e^{-\frac{J(J+1)\Theta_{rot}}{T}} \qquad (7.13)$$

Table 7.3. Summary of possible behaviors under nuclear exchange of homonuclear diatomic molecules, with examples. EB stands for Einstein Bose and FD for Fermi Dirac.

Statistics	Example	I	ρ	ρ^2	Ψ_e	Nuclear spin states		Exchange Symmetry of		Allowed rotational states	$R_{ex}\Psi$
						number	name	Ψ_{ns}	Ψ_{evr}		
EB	$^{16}O_2$	0	1	1	$^3\Sigma_g^-$	1	Ortho	+	+	J_{odd}	+
FD	$^1H_2$	1/2	2	4	$^1\Sigma_g^+$	3	Ortho	+	−	J_{odd}	−
						1	Para	−	+	J_{even}	−
EB	$^2D_2$	1	3	9	$^1\Sigma_g^+$	3	Para	−	−	J_{odd}	+
						6	Ortho	+	+	J_{even}	+
FD	$^{35}Cl_2$	3/2	4	16	$^1\Sigma_g^+$	10	Ortho	+	−	J_{odd}	−
						6	Para	−	+	J_{even}	−
EB	$^{16}N_2$	2	5	25	$^1\Sigma_g^+$	10	Para	−	−	J_{odd}	+
						15	Ortho	+	+	J_{even}	+
FD	$^{127}I_2$	5/2	6	36	$^1\Sigma_g^+$	21	Ortho	+	−	J_{odd}	−
						15	Para	−	+	J_{even}	−
EB	$^{10}B_2$	3	7	49	$^3\Sigma_g^-$	28	Ortho	+	+	J_{odd}	+
						21	Para	−	−	J_{even}	+

in which the tiny contribution of centrifugal distortion is ignored, as is the similarly small contribution of rotational coupling with vibration.

Here the characteristic temperature $\Theta_{rot} = Bhc/k_B$ and B is usually taken as B_0, the rotational constant for the ground vibrational state. If I_1 is the nuclear spin of one nucleus and I_2 that of the other, $\rho_1 = 2I_1 + 1$ and $\rho_2 = 2I_2 + 1$. For temperatures close to but above Θ_{rot} the Euler–MacLaurin sum formula may be used to provide an approximate evaluation of the combined rotational and nuclear spin partition function, with the result

$$\rho_1\rho_2 q_{rot} = \rho_1\rho_2 \frac{T}{\Theta_{rot}} \left[\left(1 + \frac{\Theta_{rot}}{3T} \right) + \frac{1}{15} \left(\frac{\Theta_{rot}}{T} \right)^2 \right.$$
$$\left. + \frac{4}{315} \left(\frac{\Theta_{rot}}{3T} \right)^3 + \cdots \right] \qquad (7.14)$$

Clearly, at sufficiently high temperature, this becomes simply $\rho_1\rho_2 \frac{T}{\Theta_{rot}}$. For most molecules room temperature is high enough, though hydrogen compounds may require somewhat larger temperatures, and at room temperature may be somewhat more fairly represented by $\rho_1\rho_2 \frac{T}{\Theta_{rot}}(1 + \frac{\Theta_{rot}}{3T} + \frac{1}{15}(\frac{\Theta_{rot}}{T})^2)$, as for example hydrogen chloride. Even at temperatures near $700°C$, the linear term may still contribute a few percent, depending on the size of the characteristic temperature.

7.7. Rotational Partition Function for Spin 0

Homonuclear diatomic molecules with nuclear spin zero and Σ_g^- ground electronic states have only ortho rotational states of odd J. For such molecules, the product of the nuclear spin and rotational partition functions is

$$q_{ns}q_{rot} = \sum_{oddJ}^{\infty} (2J + 1)e^{-J(J+1)\frac{\Theta_{rot}}{T}} \qquad (7.15)$$

since there is only one exchange symmetric spin function, $\rho = 2I + 1 = 1$ and $1/2\rho(\rho + 1) = 1$, that is, the statistical weight of each existing rotational level due to nuclear spin is 1. If $T \gg \Theta_{rot}$, then as half the levels are missing, the sum is half as large and the product of partition functions is

$$q_{ns}q_{rot} = \frac{T}{2\Theta_{rot}} \tag{7.16}$$

Of course, at very low temperatures, only the first few terms in the expansion of the defining sum would be needed, as

$$q_{ns}q_{rot} = 3e^{-\frac{2\Theta_{rot}}{T}} + 7e^{-\frac{12\Theta_{rot}}{T}} + 11e^{-\frac{30\Theta_{rot}}{T}} + \cdots \tag{7.17}$$

The series should converge rapidly, since the absolute value of the exponent of the exponential increases at a much greater rate than the coefficient.

7.8. Rotational Partition Functions for Spins Other than 0

More generally, the problem of writing $q_{ns}q_{rot}$ is one of associating $\frac{1}{2}\rho(\rho+1)$ with either a sum over even J or a sum over odd J. There are 8 possible associations for Σ states, 4 for Fermions and 4 for Bosons. Consider first $^1\Sigma_g^+$ molecules. For even mass number, the nuclei are bosons and the exchange behavior of the complete wave function (that is, including the nuclear spin functions) is

$$R_{ex}\Psi_{complete} = +\Psi_{complete} \tag{7.18}$$

This means that ortho states (s) are associated with even J, while para states are associated with states of odd J. Thus for $^1\Sigma_g^+$ bosons in ortho states

$$q_{ns}q_{rot} = \frac{1}{2}\rho(\rho+1) \sum_{Jeven}^{\infty} (2J+1)e^{-J(J+1)\frac{\Theta_{rot}}{T}} \tag{7.19}$$

Evidently, for $^1\Sigma_g^+$ bosons in para states

$$q_{ns}q_{rot} = \frac{1}{2}\rho(\rho-1) \sum_{Jodd}^{\infty} (2J+1)e^{-J(J+1)\frac{\Theta_{rot}}{T}} \tag{7.20}$$

Table 7.4.   Symmetric (s) and antisymmetric (a) spin states in $\Lambda = 0$ homonuclear diatomic molecules.

Term symbol	$^1\Sigma_g^+$	$^1\Sigma_g^-$	$^1\Sigma_u^+$	$^1\Sigma_u^-$
J odd	a	s	s	a
J even	s	a	a	s

Fermions in $^1\Sigma_g^+$ ground electronic states behave in the opposite way, since for fermions

$$R_{ex}\Psi_{complete} = -\Psi_{complete} \tag{7.21}$$

And now ortho states are associated with J odd while para states are associated with J even. The governing equations are

$$q_{ns}q_{rot} = \frac{1}{2}\rho(\rho+1)\sum_{J odd}^{\infty}(2J+1)e^{-J(J+1)\frac{\Theta_{rot}}{T}} \tag{7.22}$$

$$q_{ns}q_{rot} = \frac{1}{2}\rho(\rho-1)\sum_{J even}^{\infty}(2J+1)e^{-J(J+1)\frac{\Theta_{rot}}{T}} \tag{7.23}$$

To determine the correct association for Σ_g^-, Σ_u^+, or Σ_u^-, refer to the behavior of the total wave function Ψ_{evr} under exchange of nuclei. The property is listed in the foregoing as a or s. Results are summarized in Table 7.4.

Upon analysis, it will be found that the same results hold for $^1\Sigma_g^+$ and $^1\Sigma_u^-$ on the one hand, and for $^1\Sigma_g^-$ and $^1\Sigma_u^+$ on the other. These are summarized in Table 7.5.

These spin considerations make a definitive appearance in the heat capacity of hydrogen gas. At the time that the ultraviolet spectrum, the heat capacity and the thermal conductivity of hydrogen gas was being investigated, none of the foregoing quantum mechanics was known. The first hint that there was something unusual about hydrogen came from the ultraviolet spectrum, which showed, when the vibrational bands were resolved to show the rotational structure, that alternate lines alternated in intensity, and the intensity patterns were as if there were two different kinds of hydrogen, one three times

Table 7.5. Electronic state and mass number influence on rotational state symmetry.

Term	$^1\Sigma_g^+, ^1\Sigma_u^-$				$^1\Sigma_u^+, ^1\Sigma_g^-$			
Nuclear type	Bosons		Fermions		Bosons		Fermions	
Mass Number	Even		Odd		Even		Odd	
Spin State	Ortho	Para	Ortho	Para	Ortho	Para	Ortho	Para
Spin Function	Sym	Asym	Sym	Asym	Sym	Asym	Sym	Asym
Weight	$1/2\ \rho(\rho+1)$	$1/2\ \rho(\rho-1)$	$1/2\ \rho(\rho+1)$	$1/2\ \rho(\rho-1)$	$1/2\ \rho(\rho+1)$	$1/2\ \rho(\rho-1)$	$1/2\ \rho(\rho-1)$	$1/2\ \rho(\rho+1)$
J	Even	Odd	Odd	Even	Odd	Even	Odd	Even

more abundant than the other. The first observations were made allegedly by Mecke in 1924 (Schrodinger did not write his equation until 1925, and Raman did not discover his effect until 1928). In 1927, Hund and Hori more definitively analyzed the rotationally resolved ultraviolet bands of hydrogen gas and found the 3:1 intensity variation of adjacent rotational bands iterated throughout the spectrum.

7.9. Heat Capacity of Hydrogen Gas

When normal hydrogen is cooled from room temperature to the boiling point of about 20 K, it would be expected that the molecules would all seek the lowest rotational state. In fact, in the absence of an ortho-para conversion catalyst, this is not what happens. Three times as many condense into a higher state ($J = 1$) as to the lower ($J = 0$). In 1927, Dennison[5] realized that in the absence of a catalyst, the rate of conversion of ortho to para was very slow, requiring months to accomplish, a rate that would provide negligible contributions to the heat capacity or thermal conductivity. By 1929, Bonhofer and Hartek[6] had discovered that activated charcoal was an efficient ortho-para conversion catalyst, and performed heat capacity measurements on equilibrium hydrogen for the first time, showing it was different from heat capacity measurements on hydrogen gas obtained at room temperature and cooled as a mixture of 1/4 para and 3/4 ortho hydrogen.

By 1930, Giauque[7] was able to calculate the percentage of hydrogen in the para state in equilibrium hydrogen, that is, hydrogen admitted to the presence of an ortho-para converter. His results are plotted in Figure 7.1. Notice that there is a considerable range of low temperature in which the sample is almost entirely para, and that at room temperatures the sample is essentially one quarter para, the rest ortho. In the figure, the squares are Giauque's calculated values, the lines are best guesses concerning where the function must go between calculated points.

Giauque was also able to calculate the low temperature heat capacities of pure para hydrogen, pure ortho hydrogen, the equilibrium mixture and the 1/4 para, 3/4 ortho mixture, as shown in Figure 7.2.

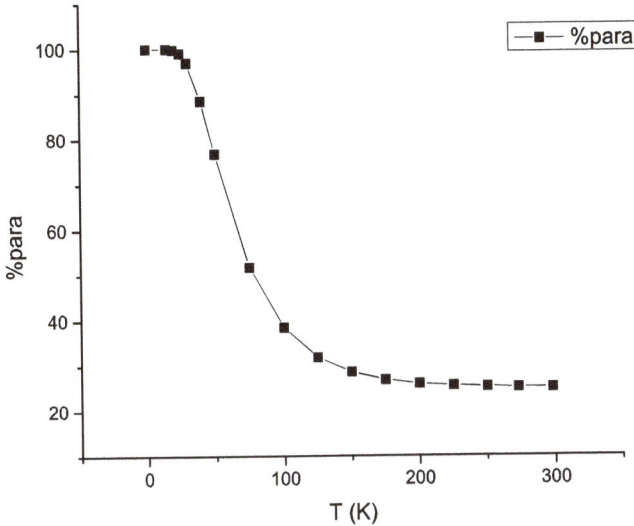

Fig. 7.1.   Percent para hydrogen in equilibrium hydrogen as a function of temperature, due to calculations by Giauque.[7]

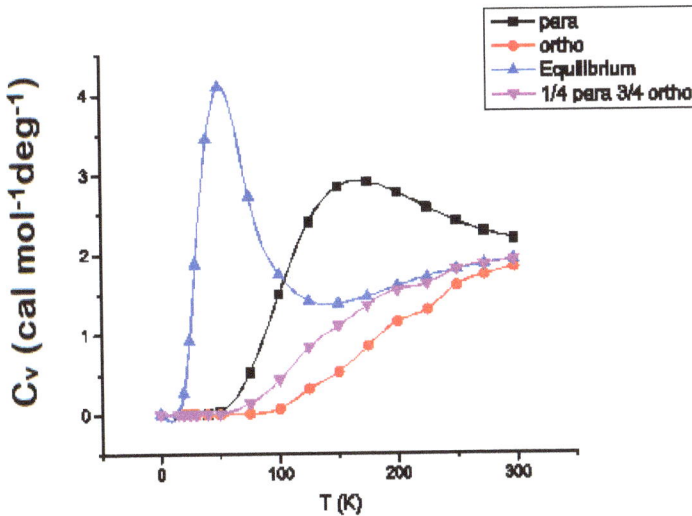

Fig. 7.2.   Calculated heat capacities of pure para, pure ortho, equilibrium and 1/4 para plus 3/4 ortho hydrogen mixtures. The calculations are due to Giauque,[7] who gave the results as a table of numbers. The symbols are his numbers. The heat capacity data obtained in the absence of an ortho-para conversion catalyst follow the 1/4 para plus 3/4 ortho mixture curve, while in the presence of the catalyst, follow the equilibrium curve.

The heat capacity of hydrogen was important in the development of quantum and statistical mechanics, for the agreement with experiment showed that the quantum hypotheses and statistical analyses were essentially correct.

7.10. Review

For Fermion nuclei in a $^1\Sigma_u^-$ electronic state of a homonuclear diatomic molecule, the ortho partition function for spin and rotation is

$$q_{ns}q_{rot} = \frac{1}{2}\rho(\rho+1) \sum_{J\,odd} (2J+1)e^{-J(J+1)\frac{\Theta_{rot}}{T}} \qquad (7.24)$$

The limiting high temperature form is

$$q_{ns}q_{rot} = \frac{1}{2}\rho(\rho+1)\frac{T}{2\Theta_{rot}} \qquad (7.25)$$

The para partition function for this molecule is

$$q_{ns}q_{rot} = \frac{1}{2}\rho(\rho-1) \sum_{J\,even} (2J+1)e^{-J(J+1)\frac{\Theta_{rot}}{T}} \qquad (7.26)$$

and the limiting high temperature form for para is

$$q_{ns}q_{rot} = \frac{1}{2}\rho(\rho-1)\frac{T}{2\Theta_{rot}} \qquad (7.27)$$

If high temperature equilibration occurred at some previous time in the sample history, the mole ratio o/p and the mole fraction of ortho are

$$\frac{o}{p} = \frac{\frac{1}{2}\rho(\rho+1)}{\frac{1}{2}\rho(\rho-1)} = \frac{\rho+1}{\rho-1} \qquad (7.28)$$

$$X_o = \frac{\frac{1}{2}\rho(\rho+1)}{\frac{1}{2}\rho(\rho+1) + \frac{1}{2}\rho(\rho-1)} = \frac{\rho+1}{2\rho} \qquad (7.29)$$

If the very slow conversion of ortho into para is hastened at all temperatures by an ortho-para conversion catalyst, equilibrium molecules result and the appropriate spin and rotational partition function is

$$q_{ns}q_{rot,eq} = \frac{1}{2}\rho(\rho+1)\sum_{Jodd}(2J+1)e^{-J(J+1)\frac{\Theta_{rot}}{T}}$$

$$+ \frac{1}{2}\rho(\rho-1)\sum_{Jeven}(2J+1)e^{-J(J+1)\frac{\Theta_{rot}}{T}} \quad (7.30)$$

The limiting high temperature form is

$$q_{ns}q_{rot,eq,hiT} = \rho^2\frac{T}{2\Theta_{rot}} \quad (7.31)$$

Problems

1. Apply the Euler–MacLaurin sum formula to the rotational partition function

$$q_{rot} = \sum_{J=0}^{\infty}(2J+1)e^{-J(J+1)\frac{\Theta_{rot}}{T}}$$

at temperatures T greater than Θ_{rot} and obtain the coefficient of the 4th power of the expansion parameter $\frac{\Theta_{rot}}{T}$. Answer: $\frac{1}{315}$

2. One mole of normal hydrogen ($\Theta_{rot} = 85$ K) is cooled from room temperature to 1 degree K in the absence of an ortho-para conversion catalyst. The cooled hydrogen is then admitted to the presence of an ortho-para conversion catalyst, such as graphite or activated charcoal, and rapidly equilibrated. The gas constant R is 1.98719 cal deg^{-1}mole^{-1}. The masses of hydrogen and deuterium are 1.00792522 and 2.01410222 amu (by definition $^{12}C = 12.00000000$ amu).

   a. How many calories of heat are evolved in this process?
   b. The experiment is repeated with one mole of hydrogen deuteride, HD. How many calories of heat are evolved in this experiment?
   c. The experiment is repeated with one mole of deuterium gas, D_2. Now how many calories of heat are evolved?

3. In Figure 7.2, the line for the heat capacity of ortho hydrogen has some inflection points that are probably not there. Repeat Giauque's calculation[7] at the temperature interval he used and plot the results on the same graph as Giauque's result. Keep in mind that you have a computer, while Giauque only had the multiplication table.

4. Hydrogen (H_2) has a rotational temperature of 85 degrees Kelvin. Calculate the rotational partition function from 100 K to 1000 K, ignoring terms that contribute less than 0.1% to the total, and ignoring centrifugal distortion.

References

1. G. Herzberg, "Molecular Spectra and Molecular Structure. I. Spectra of Diatomic Molecules", Ed. 2, D. van Nostrand, New York, 1950.
2. L. Andrews and X. F. Wang, "Simple Ortho-Para Hydrogen and Para-Ortho Deuterium conversion converter for matrix isolation spectroscopy", Rev. Sci. Instr. **75**, 3039–3045 (2004).
3. E. B. Wilson, Jr., J. Chem. Phys. **3**, 276 (1935).
4. E. B. Wilson, Jr., J. Chem. Phys. **3**, 818 (1935).
5. D. M. Dennison, Proc. Roy. Soc. London A115, 483 (1927).
6. K. F. Bonhofer and P. Hartek, Z. Physik. Chem. B, 4, 113 (1929).
7. W. F. Giauque, J. Am. Chem. Soc. 52, 4808, 4816 (1930).

Chapter 8

Polyatomic Molecules

8.1. Separation of Energy States

In principle, polyatomic molecules are not greatly different from diatomic molecules, as the same equations apply. However, polyatomic molecules usually have much larger moments of inertia than diatomic molecules, and as a result have characteristic rotational temperatures that are so low the full classical value of the rotational partition function is achieved at quite low temperatures. There are other differences, but in broad outline, the problems are just the same. The Born–Oppenheimer approximation permits first good approximate treatment of electronic problems separately from vibrational and rotational, and the customary assumption of non-relativistic quantum mechanics allows translation to be treated separately. In fact, the treatment of translation is identical with that for the diatomic molecule. The energy of the polyatomic molecule is, in this approximation

$$\varepsilon = E_{translational} + E_{electronic} + E_{vibrational} + E_{rotational} + E_{nuclear\ spin} \tag{8.1}$$

The molecular partition function q is

$$q = \sum_{accessible\ states} e^{-\varepsilon/k_B T} \tag{8.2}$$

In this first fairly good approximation, this appears as a product of partition functions

$$q = q_{trans}\, q_{elec}\, q_{vib}\, q_{rot}\, q_{ns} \tag{8.3}$$

137

The separation of translational energy is always good in sufficiently dilute gas, and at non-relativistic molecular speeds. The partition function evaluates as

$$q_{trans} = \left(\frac{2\pi M k_B T}{h^2}\right)^{3/2} V \tag{8.4}$$

In this equation, M is the molecular weight of the polyatomic molecule involved. When the gas density becomes large enough, this approximation no longer suffices, and the needed modifications will be discussed in Chapter 13.

8.2. Moment of Inertia

Owing to the larger number of atoms, the problems of rotation and vibration are more complicated than for the diatomic molecule. The statistical weights of rotational levels of molecules possessing symmetry and nuclei with nuclear spin (all nuclei have spin, even if it is zero) are modified by the need to satisfy the Pauli Principle, though this may be a rather complex way rather than the simple way shown by homonuclear diatomic molecules. As is the case for diatomic molecules, the effects of the statistical weight modifications can be taken into account by a symmetry number at temperatures high in comparison with the characteristic rotational temperature.

The rotations of polyatomic molecules fall into several different categories. For purposes of classification, the rotating molecule is considered rigid, and is called a "top". The different categories are best defined through the moment of inertia tensor. This tensor arises in the following way. If p is the vector angular momentum of the rotation at a vector angular velocity ω, and the inertial tensor is I, then

$$p = I\omega \tag{8.5}$$

in an arbitrary Cartesian axis system with origin at the center of mass of the molecule. This equation means

$$p_x = I_{xx}\omega_x + I_{xy}\omega_y + I_{xz}\omega_z \tag{8.6}$$

$$p_y = I_{yx}\omega_x + I_{yy}\omega_y + I_{yz}\omega_z \tag{8.7}$$

$$p_z = I_{zx}\omega_x + I_{zy}\omega_y + I_{zz}\omega_z \tag{8.8}$$

That is, p is a vector with components p_x, p_y and p_z, and ω is a vector with components ω_x, ω_y and ω_z. More compactly, the components of angular momentum may be written as a matrix product, either $p = I\omega$ or more explicitly as

$$p_i = \sum_j I_{ij}\omega_j \tag{8.9}$$

The inertial tensor is often written as a matrix

$$I = \begin{pmatrix} I_{xx} & I_{xy} & I_{xz} \\ I_{yx} & I_{yy} & I_{yz} \\ I_{zx} & I_{zy} & I_{zz} \end{pmatrix} \tag{8.10}$$

In an arbitrary Cartesian coordinate system, the products of inertia are defined as

$$I_{xy} = I_{yx} = \sum_i m_i x_i y_i \tag{8.11}$$

and similarly for the other products of inertia. The sum runs over the masses involved. The diagonal elements are somewhat different, and are

$$I_{xx} = \sum_i m_i(y_i^2 + z_i^2) \tag{8.12}$$

The other two are cyclically defined. Note that, since ordinary multiplication is commutative, the inertial tensor is a symmetrical tensor, since equivalent off diagonal elements such as I_{xz} and I_{zx} are equal.

Some authors regard the products of inertia as negative, and some as positive. Since a general rotation of the axes diagonalizes the matrix, leaving the principal moments on diagonal and zeros elsewhere, the sign does not matter. The appearance of the diagonalized matrix is

$$I = \begin{pmatrix} I_a & 0 & 0 \\ 0 & I_b & 0 \\ 0 & 0 & I_c \end{pmatrix} \tag{8.13}$$

I_a, I_b and I_c are called the principal moments of inertia, and are ordered, with I_a the smallest and I_c the largest. The axis system that defines the principal moments is called the principal axis system, with coordinates a, b and c. The principal moments may always be calculated using

$$I_a = \sum_i m_i(b_i^2 + c_i^2) \tag{8.14}$$

and similarly for I_b and I_c. To use this, select an axis. Any axis of symmetry will define a principal moment of inertia. In the case above, the a axis is defined, and the distances b_i and c_i are the perpendicular distances from the nuclei to the a axis.

Molecules are classified according to their principal moments of inertia in the ground state. There are three possibilities, the second of which has two possible realizations. These are

1. $I_a = I_b = I_c$. When all three principal moments are equal, the molecule is called a spherical top. The moment of inertia about any arbitrary axis passing through the center of mass will do to define the principal moment of inertia. Spherical tops lack a permanent dipole moment, and as a result have no dipole microwave or infrared rotational spectrum. A few examples are methane, CH_4, carbon tetrachloride, CCl_4, and sulfur hexafluoride, SF_6.

2. $I_a = I_b$, but I_c is different. There are two ways I_c could be different. It might be larger than $I_a = I_b$ in which case the molecule is called an oblate symmetric top, or it could be less than $I_a = I_b$, in which case the molecule would be called a prolate symmetric top. Oblate tops tend to be plate-like, and benzene (C_6H_6) is an extreme molecular example. The earth is an oblate spheroid. Prolate symmetric tops tend to be needle-like, and all diatomic and linear molecules are prolate symmetric tops. So is methyl acetylene, CH_3CCH, and many other long thin molecules. Generally, spectroscopists like to keep the c axis for the largest moment of inertia, and the b axis for the intermediate moment. This leads to definitions of spectroscopic rotational constants A, B and C that are different for different types of molecules. A very great deal is known about the details of the energy levels and allowed transitions. Most symmetric tops are so large and heavy that they

have very small characteristic rotational temperatures. The spectroscopic constants for oblate symmetric tops are B and C. These constants for prolate symmetric tops are A and B. These are more precisely defined below.

3. I_a, I_b and I_c are all different. Such molecules are called asymmetric tops.

8.3. Rotational Partition Functions

For rigid spherical tops, the rotational energy levels are the same as those for the heteronuclear diatomic molecule, but the degeneracies of the levels is not $2J + 1$ but $(2J + 1)^2$ instead. The characteristic rotational temperature $\Theta_{rot} = Bhc/k_B$, and for most spherical tops is so small that the full classical value of the partition function is attained at quite low temperatures. Formally, when centrifugal distortion is ignored, the rotational partition function is

$$q_{rot} = \sum_J (2J + 1)^2 e^{-J(J+1)\frac{\Theta_{rot}}{T}} \tag{8.15}$$

When the effect of nuclear spin is included by way of the introduction of a symmetry number, σ, at temperatures large compared with Θ_{rot}, the sum converges to

$$q_{ns}q_{rot} = \frac{\sqrt{\pi}}{\sigma} \left(\frac{T}{\Theta_{rot}}\right)^{3/2} \tag{8.16}$$

Symmetry numbers are discussed after symmetric tops, which also use them. If the spherical top is involved in very low temperature environments, such as interstellar space or in the gas in an adiabatic expansion, or is required at very high temperatures, a paper by McDowell[1] has given an evaluation of the rotational partition function that holds at very low as well as very high temperatures.

The rotational energies of symmetric tops are variously expressed, depending on the type. In general, two moments of inertia contribute, I_B and I_A for prolate tops, or I_B and I_C for oblate tops, which give rise to spectroscopic constants A, B or C as

$$A = h^2/(8\pi^2 I_a k_B c) \tag{8.17}$$

where c is the velocity of light in vacuo, measured in cm/s units. B and C are similarly defined. In these relations, A, B and C are measured in cm^{-1} units. The A axis is the highest fold symmetry axis in any prolate top while the B axis is an axis perpendicular to this and passing through the center of gravity. The energy expression, with centrifugal distortion omitted, then reads

$$F(J, K) = BJ(J + 1) + (A - B)K^2 \tag{8.18}$$

Both J and K are integers. J varies from 0 to ∞, while K is limited to the $2J + 1$ values lying between and including $+J$ and $-J$. The total angular momentum is always expressed by J. The partition function is correspondingly complicated. Keeping in mind that the characteristic rotational temperature is given by Bhc/k_B, the rotational partition function is

$$q_{rot} = \sum_{J} \sum_{K=-J}^{K=+J} (2J + 1)e^{-J(J+1)\frac{\Theta_B}{T}} e^{-K^2 \frac{(\Theta_A - \Theta_B)}{T}} \tag{8.19}$$

When the effect of nuclear spin is included, the high temperature form is

$$q_{ns}q_{rot} = \frac{\sqrt{\pi}}{\sigma} \left(\frac{T^3}{\Theta_A \Theta_B^2} \right)^{1/2} \tag{8.20}$$

It is worth a few words to define the roles of J and K in the rotating molecule. Consider first a prolate top such as methyl acetylene. The heavy atom axis is a three-fold axis of symmetry, and the B axis is perpendicular to it. Rotation about the B axis generates angular momentum characterized by J when there is no rotation about the three fold axis. However, when there is also rotation about the three fold axis, characterized by K, J becomes the total angular momentum, obtained by adding vectorially the two angular momenta, that about the B axis and that about the A axis. In such cases, J is not perpendicular to the A axis, but lies at an angle other than 90 degrees to it. For a given J, K can be 0, ± 1, ± 2, ... up to $\pm J$. But when $K = \pm J$, it can grow no more for a given J.

If the partition function at very high or very low temperatures is required, McDowell[2] has given evaluations that include the centrifugal distortion constants. These do not contribute at very low temperatures, but become appreciable at sufficiently high temperature.

Linear molecules are a special case of the prolate symmetric top. The A axis is the infinite fold symmetry axis on which all nuclei lie. There is no nuclear mass off the A axis, and the only contributors to the moment of inertia about the infinite fold axis are the electrons. These are very light. As a result, the A constant becomes very large, and the quantum number K represents the electronic orbital angular momentum. In order to change K, electronic energies are required, and as a result the sum over K disappears from the expression for the energy. This state of affairs is usually recognized by reassigning K to Λ, recognizing the quantum number Λ as representing electronic angular momentum. The term $A - B$ becomes essentially A, since B is very much smaller, and Λ does not change, making the whole a constant. The partition function is the same as that for heteronuclear diatomic molecules, unless there is also a center of symmetry. In that case, the symmetry number of 2 is appropriate. McDowell[3] has treated the special case of the linear molecule and included the effect of the quartic, hexic and octic centrifugal distortion constants over a very wide range of temperatures, from as low as 2 K to as high as 5000 K for acetylene, HCCH.

8.4. Symmetry Numbers

The classic symmetry number is the number of different ways a rigid symmetric top can be brought to the same spatial orientation by rotations alone. Some examples are given in Table 8.1.

Table 8.1. Symmetry numbers for some symmetric molecules.

CCl_4	12	CH_4	12	SF_6	24	SO_3	6
C_6H_6	12	H_2S	2	$H_2C{=}CH_2$	4	H_3CCH_3	6

Oblate symmetric tops follow the same statistics that prolate tops do. However, the A axis is called the C axis, and the term in the energy $(C - B)K^2$ is no longer positive, since $C < B$. High temperature results are the same, and usually are all that is needed, unless extremely low temperature work under near vacuum conditions is involved.

8.5. Asymmetric Top Partition Function

Asymmetric tops have three different moments of inertia, three different rotational constants and no simple treatment. However, in comparison with k_BT, the level separation is small for most molecules owing to the large moments of inertia of most asymmetric tops. The classical approximation is good enough for almost every purpose. It is, including the effects of nuclear spin,

$$q_{ns}q_{rot} = \frac{\sqrt{\pi}}{\sigma} \left(\frac{T^3}{\Theta_A\Theta_B\Theta_C} \right)^{1/2} \tag{8.21}$$

Derivation of this relation is given in Mayer and Mayer, pages 193 and following. The starting point is the rotational Hamiltonian written in terms of Euler's angles and the principal moments of inertia, and proceeds through complicated but straightforward integration to the result quoted above, not including the symmetry number σ.

Euler's angles are variously described in books on classical mechanics. Usually these are taken to be θ, φ, and ψ. These are defined by angles made between a set of Cartesian axes x, y and z fixed in space and the principal axes X, Y, and Z fixed in an asymmetric top with origins at the center of mass of the top. The angle made by z and Z is θ, which varies between 0 and π. The other two angles vary between 0 and 2π, and are defined as follows. The line in the xy plane that is perpendicular to the plane containing both the z and Z axes is called the nodal line. If $\mathbf{z}$ is a vector along the z axis and $\mathbf{Z}$ is a vector along the Z axis in the top then $\mathbf{z} \times \mathbf{Z}$ is a vector along the nodal line. The angle between the nodal line and the x axis is the angle φ. The two angles θ and φ determine the direction of the Z axis in space. The angle between the Z axis and the nodal line is the remaining angle ψ. The Hamiltonian for rotation of a rigid asymmetric top written in terms of these angles and the conjugate momenta is quite complicated, and is

$$
H = \frac{\sin^2 \psi}{2I_A} \left\{ p_\Theta - \frac{\cos \psi}{\sin \psi \sin \Theta} (p_\varphi - \cos \Theta p_\psi) \right\}^2
$$

$$
+ \frac{\cos^2 \psi}{2I_B} \left\{ p_\Theta + \frac{\sin \psi}{\cos \psi \sin \Theta} (p_\varphi - \cos \Theta p_\psi) \right\}^2 + \frac{p_\psi^2}{2I_C}
$$

$$\tag{8.22}$$

Since $\Theta_A \Theta_B \Theta_C$ is all that is needed, the product of the principal moments of inertia are sufficient. It is not necessary to know the principal moments I_a, I_b and I_c. It is a theorem in mechanics that the product of principal moments of inertia is equal to the determinant of the inertial tensor with elements expressed in an arbitrary Cartesian axis system such as

$$
I_a I_b I_c = \begin{vmatrix} I_{xx} & I_{xy} & I_{xz} \\ I_{yx} & I_{yy} & I_{yz} \\ I_{zx} & I_{zy} & I_{zz} \end{vmatrix}
$$

$$\tag{8.23}$$

Thus, any convenient selection of axes may be used. An example from the early literature shows how this theorem brought agreement between experiment and calculations, when at first such agreement was elusive.

In the case of H_2O, and HDO, the principal axes move around considerably, especially in HDO. In H_2O and D_2O, there is a two fold axis of symmetry on which the oxygen atom lies. This is not the case for HDO, where the asymmetric substitution of deuterium for hydrogen causes the principal axes to make a finite angle with the no longer extant two fold symmetry axis. The initial calculation assumed the same axis system for HDO as for H_2O, and on that basis calculated the equilibrium constant for the reaction

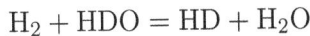

$$
H_2 + HDO = HD + H_2O
$$

But when the equilibrium constant was calculated, it was 40% in disagreement with the experimental result. Once it was realized that the principal axes had seriously shifted and the calculation of the equilibrium constant was repeated including the products of inertia, much closer agreement of the calculation with experiment resulted.

Asymmetric tops represent a class of molecules with no necessary symmetry and range from almost prolate to almost oblate tops. Nonetheless, Watson (4) was able to construct a high temperature asymptotic expansion of the rotational partition function for asymmetric tops.

8.6. Internal Rotation

Whenever methyl groups are encountered, the possibility of internal rotation must be considered. A classic case of nearly free internal rotation is exemplified by dimethyl acetylene, in which the heavy atoms (the carbons) are strictly linear in the ground electronic state. When the two methyl groups rotate in the same direction, the result is a rotation of the molecule as a whole about the A axis. However, there are two methyl groups in this compound, and one may counter-rotate against the other. The moment of inertia of the rotating methyl group is entirely determined by the off axis hydrogen atoms. If that is I_{CH3}, the counter-rotation is characterized by a reduced moment of inertia. If there are two counter-rotating tops attached to the same frame with moments of inertia about the same axis of rotation of I_1 and I_2, the reduced moment of inertia I_{red} is given by

$$I_{red} = \frac{I_1 I_2}{I_1 + I_2} \tag{8.24}$$

as will be shown later. Since in methyl acetylene I_1 and I_2 are both equal to I_{CH3}, for dimethyl acetylene the reduced moment of inertia about the axis of least inertia is

$$I_{red} = \frac{1}{2} I_{CH3} \tag{8.25}$$

Let the angle of rotation be measured by a variable φ and the canonically conjugate angular momentum by p_φ. In first good approximation, the corresponding classical energy contribution is expected to be

$$E_{rot} = \frac{p_\varphi^2}{2I_{red}} \tag{8.26}$$

provided the rotation is completely free. The classical rotational energy is then

$$E_{rot} = \frac{p_\varphi^2}{I_{CH3}} \tag{8.27}$$

The contribution the methyl groups make to the thermodynamic functions is collected in the classical partition function, defined in this case as

$$q_{\text{classical, unhindered}} = \frac{1}{\sigma h} \int_0^{2\pi} \int_{-\infty}^{\infty} e^{-\frac{p_\varphi^2}{I_{CH3} k_B T}} dp_\varphi d\varphi \tag{8.28}$$

where σ is a symmetry number, in this case, 3. The integration over φ is elementary and that over p_φ is tabulated. The result is

$$q_{\text{classical, unhindered}} = \frac{2\pi}{3h} (\pi I_{CH3} k_B T)^{1/2} \tag{8.29}$$

where k_B is Boltzmann's constant and h is Planck's constant. Since $T^{1/2}$ appears, the contribution to the specific heat is $R/2$. As long as the temperature is not taken too low, this matches the heat capacity data.

Now if the two acetylenic carbons are removed, the residue is two methyl groups attached to one another in the molecule ethane. When the counter-rotating motion is analyzed, however, it is found that each methyl group hinders the rotation of the other, The motion may be described by a hindering potential $V1(\varphi)$, and the methyl groups move under the action of this potential. The leading terms in a Fourier analysis of the potential energy are

$$V1(\varphi) \cong \frac{V_3}{2}(1 - cos(3\varphi) + \cdots) \tag{8.30}$$

This potential function is plotted in Figure 8.1, together with the potential function $V2(\varphi)$, which includes the next term in the Fourier

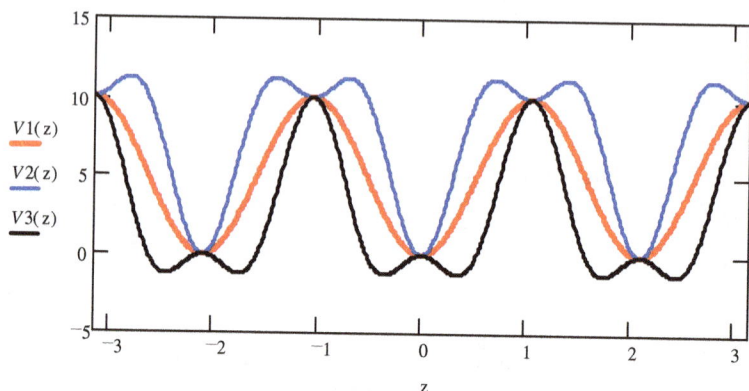

Fig. 8.1.   Plots of three-fold potential energies. Red curve is $V(\varphi)$. Blue is $V1(\varphi)$, Black is $V1(\varphi)$ but with the V_6 term negative. V_6 is half of V_3. If V_6 is small, say less than 20% of V_3, the only effect is to narrow the minimum slightly when positive, or to broaden it if negative. The minimum and maximum of $V(\varphi)$ is unaltered by any value of V_6.

series that represents the full three-fold symmetry of the methyl group. The function $V2(\varphi)$ is

$$V2(\varphi) \cong \frac{V_3}{2}(1 - \cos(3\varphi)) + \frac{V_6}{2}(1 - \cos(6\varphi)) + \cdots \qquad (8.31)$$

where V_6 appears as the amplitude of a small correction term. In ethane, the hydrogens on one methyl group are at minimum energy when located equidistant from the two nearest hydrogens on the other methyl group, essentially exactly between the two nuclei. A rotation of the methyl group by 180 degrees brings each hydrogen atom directly superposed on a hydrogen of the other methyl group, a position of maximum energy. Methyl groups are attached to many other frames, and if the frame is a benzene ring, the rotation by 180 degrees brings the methyl group to a position of minimum energy, the mirror image of the initial lowest energy position, and thus another lowest energy position. As a result, these methyl groups move in a six-fold potential rather than a three-fold one. In Figure 8.2, a six-fold potential is plotted. In this case, the leading term is V_6 rather than V_3, and the first correcting term is V_{12} rather than V_6. Note that

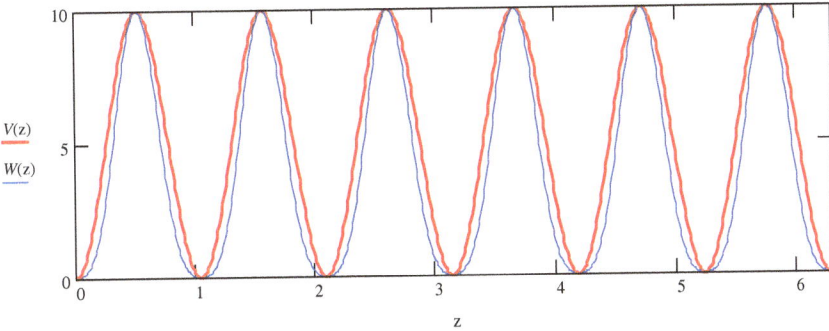

Fig. 8.2. Comparison of the potential functions $V(\varphi) = \frac{V_6}{2}(1 - cos(6\varphi))$ (wide red line) and $W(\varphi) = \frac{V_6}{2}(1 - cos(6\varphi)) - \frac{V_{12}}{2}(1 - cos(12\varphi))$ (thin blue line). V_{12} is 20% of V_6. φ ranges from 0 to 2π. If V_{12} is increased, the potential displays a double minimum that becomes more pronounced and deeper as V_{12} increases. The maximum is unaltered by any value of V_{12}.

in Figure 8.2, a negative correction term is applied rather than the positive term used in the potential function $V1(\varphi)$.

If V_3 in ethane (or some other compound in which it is appropriate) is large enough, the rotation is completely hindered and becomes essentially a vibration, with the methyl groups counter-vibrating. The motion appears to an external observer as a twisting motion, and spectroscopists call it a twist for that reason. The wavenumber ν is measured in cm^{-1} units, and in these units the partition function is

$$q_{\text{twist}} = \frac{e^{-hc\nu/2k_BT}}{1 - e^{-hc\nu/k_BT}} \tag{8.32}$$

Alas, neither $q_{classical, unhindered}$ nor q_{twist} fit the data for ethane, the truth lying somewhere between these two extremes. The potential has to be incorporated into the statistical mechanics to describe the heat capacity and other thermodynamic functions as a function of temperature. To do this properly, a little excursion into classical mechanics is required. Consider first the general rotational motion of a rigid body with principal moments of inertia I_x, I_y, I_z and angular velocities ω_x, ω_y, ω_z. The total rotational kinetic energy T is

$$T = \frac{1}{2}I_x\omega_x^2 + \frac{1}{2}I_y\omega_y^2 + \frac{1}{2}I_z\omega_z^2 \tag{8.33}$$

Now let the rotation axis of the internal rotor coincide with the principal z axis. Let I_φ be the moment of inertia of the internal rotor about the z axis. The rotation of the part of the molecule that is not internal rotor is called the frame and is described by a term $\frac{1}{2}(I_z - I_\varphi)\omega_z^2$ while the rotation of the internal rotor is described by a term $\frac{1}{2}I_\varphi(\omega_z + \omega_\varphi)^2$, where ω_φ is the angular velocity of the internal rotor. Then the kinetic energy for this special case (the rotor axis and the z axis are coincident, a circumstance that will not generally be obtained) is

$$T = \frac{1}{2}I_x\omega_x^2 + \frac{1}{2}I_y\omega_y^2 + \frac{1}{2}(I_z - I_\varphi)\omega_z^2 + \frac{1}{2}I_\varphi(\omega_z + \omega_\varphi)^2 \qquad (8.34)$$

The angular momentum is given by the derivative of the kinetic energy with respect to the velocity of the canonically conjugate coordinate. Thus, for example

$$p_z = \frac{\partial T}{\partial \omega_z} \qquad (8.35)$$

and

$$p_\varphi = \frac{\partial T}{\partial \omega_\varphi} \qquad (8.36)$$

Carrying out these differentiations produces

$$p_z = I_z\omega_z + I_\varphi\omega_\varphi \qquad (8.37)$$

and

$$p_\varphi = I_\varphi(\omega_z + \omega_\varphi) \qquad (8.38)$$

In order to write the Hamiltonian $H = T + V$, T must be expressed in terms of the momenta, not the angular velocities. Immediately

$$\frac{1}{2}I_\varphi(\omega_z + \omega_\varphi)^2 = \frac{p_\varphi^2}{2I_\varphi} \qquad (8.39)$$

However, the other term in T, $\frac{1}{2}(I_z - I_\varphi)\omega_z^2$, requires solving the two simultaneous equations for ω_z, The result is

$$\omega_z = \frac{p_z - p_\varphi}{I_z - I_\varphi} \qquad (8.40)$$

Then

$$\frac{1}{2}(I_z - I_\varphi)\omega_z^2 = \frac{p_z^2 - 2p_z p_\varphi + p_\varphi^2}{2(I_z - I_\varphi)} \tag{8.41}$$

and when these two results are combined, this part of the rotational Hamiltonian becomes

$$\frac{1}{2}(I_z - I_\varphi)\omega_z^2 + \frac{1}{2}I_\varphi(\omega_z + \omega_\varphi)^2$$

$$= \frac{p_z^2 - 2p_z p_\varphi + p_\varphi^2}{2(I_z - I_\varphi)} + \frac{p_\varphi^2}{2I_\varphi} \tag{8.42}$$

Combining these two terms produces

$$\frac{1}{2}(I_z - I_\varphi)\omega_z^2 + \frac{1}{2}I_\varphi(\omega_z + \omega_\varphi)^2$$

$$= \frac{p_z^2}{2(I_z - I_\varphi)} + \frac{p_\varphi^2 I_z}{2(I_z - I_\varphi)I_\varphi} - \frac{p_z p_\varphi}{I_z - I_\varphi} \tag{8.43}$$

The last term is a small coupling term that serves to couple the rotation of the frame to that of the top, and in first good approximation, it is ignored. Since the moment of inertia of the frame, I_F, about the common z axis is the difference between the moment of inertia of the entire molecule and that of the top, $I_z - I_\varphi$, $I_z = I_F + I_\varphi$, and the term in p_φ is of the form

$$\frac{p_\varphi^2 I_z}{2(I_z - I_\varphi)I_\varphi} = \frac{p_\varphi^2(I_F + I_\varphi)}{2I_F I_\varphi} \tag{8.44}$$

The reduced moment of inertia ($I_{red} = \frac{I_F I_\varphi}{I_F + I_\varphi}$) appears in the denominator. The moments of inertia of the frame and top are reciprocally additive to produce the reciprocal of the reduced moment. That is,

$$\frac{1}{I_{red}} = \frac{1}{I_F} + \frac{1}{I_\varphi} \tag{8.45}$$

This allows the classical kinetic energy for the top to be approximated by

$$T_\varphi = \frac{p_\varphi^2}{2I_{red}} \tag{8.46}$$

and the classical Hamiltonian for the top rotation to be written approximately as

$$H(\varphi, p_\varphi) = \frac{p_\varphi^2}{2I_{red}} + V(\varphi) \tag{8.47}$$

where $V(\varphi)$ is the potential function appropriate to the problem. For the methyl group moving in a three-fold symmetric potential, the leading term (often, the only term) is $\frac{V_3}{2}(1 - \cos(3\varphi))$, as in ethane. In toluene, the methyl group rotates in a six-fold symmetric potential, and the leading term is $\frac{V_6}{2}(1 - \cos(6\varphi))$, and so on. The only requirement for use of the above is that the axis of rotation of the frame must be that of the top. The more general case in which the top axis makes angles with the principal axes involves the direction cosines of these angles, and has been treated by Lin and Swalen.[5]

Both the top and the frame are regarded as rigid structures. In that case the appropriate quantum mechanics for the free top are those of the particle constrained to a circular track. The postulates of quantum mechanics state that the operator corresponding to the momentum is

$$p_x = \frac{h}{2\pi i} \frac{\partial}{\partial x} \tag{8.48}$$

and similarly for y and z. In the operator expression, i is the imaginary square root of unity. The postulates do not include an operator for p_φ. Fortunately, it is easy to show that for motion of a rigid body in a circle an effective operator for p_φ is

$$p_\varphi = \frac{h}{2\pi i} \frac{\partial}{\partial \varphi} \tag{8.49}$$

When the hindering potential is introduced, the appropriate time independent Schrodinger equation is formed by writing $H\Psi = E\Psi$ forming the operator H from the operator p_φ

$$-\frac{h^2}{8\pi^2 I_{red}}\frac{\partial^2\Psi}{\partial\varphi^2} + \frac{V_3}{2}(1 - \cos(3\varphi))\Psi = E\Psi \qquad (8.50)$$

This equation has energies that are drastically modified from those of the free rotor near the bottom of the potential energy wells, and is less so above the barrier between wells. At much higher energies, the energy levels of the free rotor are approached. The functions ψ were first studied by the mathematician Mathieu, and the equation is known as the Mathieu equation. The energies and functions are tabulated, and the functions can be approximated by truncated Fourier series. If the potential energy contains higher terms such as $\frac{V_6}{2}(1 - \cos(6\varphi))$, the greater complexity requires solutions to Hill's equation. However, these functions can again be well approximated by truncated Fourier series, and the energy levels calculated to spectroscopically useful precision. Tables of these energies have been published[6] and a rather useful computer program has been described that exploits the rapid convergence of Fourier series representations of Mathieu functions.[7] The first statistical use of the energies for hindered rotors was made much earlier by Pitzer and Gwinn.[8] The thermodynamic functions are tabulated in Pitzer's book in Appendix 18. Practical calculations are often carried out in terms of an F number, where F is measured in cm^{-1} units, and is defined in the energy units used for h as

$$F = -\frac{h^2(I_z - I_\varphi)}{8\pi^2 I_z I_\varphi} \qquad (8.51)$$

or, more compactly, $F = -\frac{h^2}{8\pi^2 I_{red}}$.

So far, only methyl groups have been considered. Hindered rotation is much more general than that. The simplest example of a lower symmetry is provided by 1,2-dichloroethane. In this case, although classic organic methods such as distillation, fractional crystallization and the like can only isolate one 1,2-dichloroethane, two forms, called conformers, in fact coexist. In one of these, the two chlorines are trans

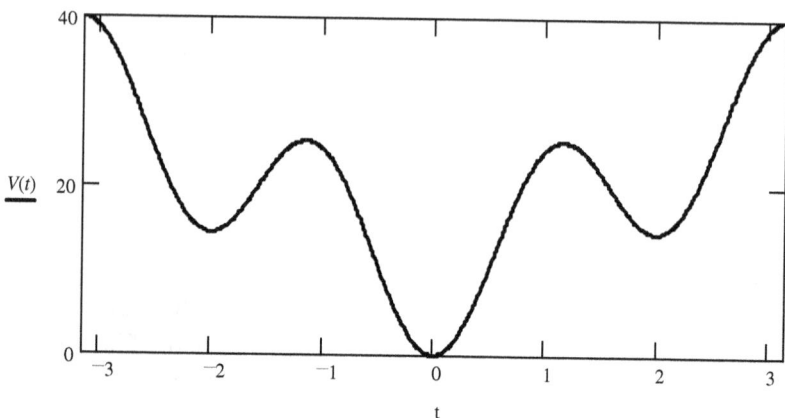

Fig. 8.3.   Potential function for gaseous 1,2-dichloroethane. The ante well is in the center of the figure. The potential function plotted is $\frac{V_1}{2}(1 - cos(\varphi)) + \frac{V_3}{2}(1 - cos(3\varphi))$, and $V_1 = V_3 = 20$ were taken for illustrative purposes.

to one another, and the form is called ante 1,2-dichloroethane. The other two forms are mirror images, and are both called gauche 1,2-dichloroethane. The chloromethyl groups move in a potential energy in the gas phase similar to that shown in Figure 8.3. The bottom of the gauche wells lies a few hundred cm^{-1} above the bottom of the ante well. However, when an early attempt was made by S. Mizushima to measure the energy difference by observing the temperature dependence of the Raman spectrum of the liquid phase of 1,2-dichloroethane, he found that there was no measurable difference. The reason for this peculiar behavior is that in the liquid, molecules are tightly packed, and each swims about in a sea of the others, not all of which are of the same kind. The trans form has a center of symmetry and thus cannot display an electric dipole moment. However, the gauche form lacks the center, and does in fact have a fairly large dipole moment. The result is that, since the forms are constantly interconverting, the dipole-dipole attractions of gauche conformers overwhelms the gas phase preference for the ante form. The net result of the dipole attraction is to increase the gauche population until both forms are about as likely in the liquid. In the gas, molecules are far apart and the dipole forces rarely act, so the ante form is the more stable there.

8.7. Vibration

The vibrations of polyatomic molecules are usually modeled as a set of mass points interconnected by a system of springs. In recent years, the anharmonicity is accommodated by using a force field calculated *ab initio* using some approximate basis and a computation method that at least partially includes electron correlation. The density functional B3LYP is a popular choice, and in an overnight run for small molecules with relatively small numbers of electrons, usually generates a reasonable force field at a fairly high level of basis, such as 6-311+g(2d,p). Some calculations have been done that include anharmonicity at a rather high level, matching the experimental spectrum to within 1 or 2 cm^{-1}, though this is unusual. Differences from experiment tend more often to be 10 to 50 cm^{-1}. The original algebra allowing treatment of the normal coordinate problem for relatively large molecules, such as benzene, was worked out long before modern computers and associated software in the years up to about 1935. The result is that a molecule with N nuclei vibrates with $3N - 6$ different characteristic frequencies unless linear. Linear molecules have $3N - 5$ characteristic frequencies. Each of these is anharmonic, and the bending vibrations of linear molecules are doubly degenerate, which requires an angular momentum due to vibration. The general vibrational energy level encountered experimentally is

$$G(v) = \sum_i \nu_i \left(v_i + \frac{d_i}{2} \right) + \sum_i \sum_j x_{ij} \left(v_i + \frac{d_i}{2} \right) \left(v_j + \frac{d_j}{2} \right)$$

$$+ \sum_i g_{ii} l_i^2 \tag{8.52}$$

Each vibration is quantized, and ν_i is the classical vibrational frequency of the ith mode. The vibrational quantum numbers are the $v_i = 0, 1, 2, 3 \ldots \infty$. The anharmonic constants are the x_{ij}, and the angular momentum constants are g_{ii}, while the quantum number associated with vibrational angular momentum due to doubly degenerate bending modes of linear molecules is l_i. This term is not present in non-linear molecules. The d_i are the degeneracies, and are often either 1 or 2.

In first good approximation the anharmonicity is neglected and for non-linear molecules the vibrational term is simply

$$G(\mathrm{v}) = \sum_i \nu_i \left(\mathrm{v}_i + \frac{d_i}{2}\right) \qquad (8.53)$$

The frequencies ν_i are measured in cm^{-1} units, and the effect of anharmonicity can be partially accommodated by using the experimental frequencies, rather than the anharmonicity corrected. Since the energy is a sum of algebraically identical functions, the partition function is a product of algebraically identical functions, which are the same as that for the diatomic molecule. If ν_i is the experimentally observed frequency of the ith mode measured in cm^{-1} units, then each vibration has a characteristic temperature $\Theta_{vib,i}$ that is 1.4388 cm K times greater numerically than the frequency. That is

$$\Theta_{vib,i} = \frac{hc}{k_B}\nu_i \qquad (8.54)$$

In terms of this parameter, the vibrational partition function is

$$q_{vib} = q_{vib,\,1}\, q_{vib,\,2}\, q_{vib,\,3} \cdots q_{vib,\,3N-6} \qquad (8.55)$$

where

$$q_{vib,i} = \frac{e^{-\frac{\Theta_{vib,i}}{2T}}}{1 - e^{-\frac{\Theta_{vib,i}}{T}}} \qquad (8.56)$$

For example, the energy E in the microcanonical ensemble is

$$E = Nk_BT^2 \left(\frac{\partial l \ln q}{\partial T}\right)_V \qquad (8.57)$$

Since each vibrational contribution is identical, it suffices to consider only one. In terms of a characteristic temperature Θ, the result is

$$E = \frac{Nk_B\Theta}{2} + Nk_B\Theta\frac{e^{-\frac{\Theta}{T}}}{1 - e^{-\frac{\Theta}{T}}} \qquad (8.58)$$

The first term is the zero point energy, perhaps more readily seen if N is identified as Avogadro's number, in which case Nk_B becomes R, the universal gas constant. An alternative expression of the vibrational contribution to the energy is

$$E = \frac{R\Theta}{2} + R\Theta \frac{1}{e^{\frac{\Theta}{T}} - 1} \tag{8.59}$$

The vibrational contribution to the molar heat capacity at constant volume (a condition of the microcanonical ensemble) is then

$$C_v = \left(\frac{\partial E}{\partial T}\right)_V \tag{8.60}$$

The result of the differentiation is called an Einstein function, since Einstein first obtained the function in one of his 1905 papers. It may be expressed in a variety of ways. The two most common are, in terms of a variable $u = \Theta/T$,

$$C_v = \frac{Ru^2 e^u}{(e^u - 1)^2} = \frac{Ru^2 e^{-u}}{(1 - e^{-u})^2} \tag{8.61}$$

Each vibration follows the same function, and the molar constant volume heat capacity of polyatomic molecules, including rotational contributions, is

$$C_v = 3R + R \sum_i \frac{u_i^2 e^{u_i}}{(e^{u_i} - 1)^2} \tag{8.62}$$

for non-linear molecules at temperatures large in comparison with the rotational characteristic temperature. Translation is regarded as fully excited at all temperatures. For linear molecules, $3R$ is replaced by $5R/2$ and the sum runs over $3N-5$ rather than $3N-6$ vibrations.

8.8. Normal Modes of Vibration

Usually, a normal coordinate problem is set up in Cartesian coordinates. The kinetic energy T and the potential energy V are of the form

$$2T = \sum_i m_i \dot{x}_t^2 \qquad (8.63)$$

$$2V = \sum_{i,j} k_{ij} x_i x_j \qquad (8.64)$$

The sums run over all the coordinates, and the dot denotes time differentiation. When the problem is finished, the results are the angular frequencies ω_k and the associated normal coordinates Q_k, in terms of which the kinetic and potential energy are brought to simultaneous diagonal form as

$$2T = \sum_k \dot{Q}_k^2 \qquad (8.65)$$

$$2V = \sum_k \omega_k^2 Q_k^2 \qquad (8.66)$$

Though this is a non-trivial algebraic problem, the path is now well worn, and most laboratories of vibrational spectroscopy have their own versions of the GF matrix method, described in the 1956 book by Wilson, Decius and Cross. To illustrate the process, consider the problem of two identical masses connected to one another by a Hooke's law spring of spring constant k_1, and let each mass be connected to a wall of infinite mass by a weaker spring of spring constant f, all aligned along the x axis. The appearance would be

Wall-----f----m₁-----k₁-----m₂-----f-----Wall

Let the displacement of mass m_1 be x_1 and that of mass m_2 be x_2. Then the kinetic energy T and the potential energy V for motion restricted to the x axis is

$$2T = m_1 \dot{x}_1^2 + m_2 \dot{x}_2^2 \qquad (8.67)$$

$$2V = f x_1^2 + k_1 (x_2 - x_1)^2 + f x_2^2 \qquad (8.68)$$

The LaGrangian $L = T - V$, and the equations of motion derive directly from this function. Since only velocities are involved in T

and only coordinates in V, the equivalent relation is

$$\frac{d}{dt}\frac{\partial T}{\partial \dot{x}_i} + \frac{\partial V}{\partial x_i} = 0 \tag{8.69}$$

In this case,

$$m_1\ddot{x}_1 + fx_1 - k_1(x_2 - x_1) = 0 \tag{8.70}$$

$$m_2\ddot{x}_2 + k_1(x_2 - x_1) + fx_2 = 0 \tag{8.71}$$

Rearranging these equations, and dropping the distinction between identical masses m,

$$\ddot{x}_1 + x_1\frac{(k_1 + f)}{m} + x_2\left(-\frac{k_1}{m}\right) = 0 \tag{8.72}$$

$$\ddot{x}_2 + x_1\left(-\frac{k_1}{m}\right) + x_2\frac{(k_1 + f)}{m} = 0 \tag{8.73}$$

Seeking oscillatory solutions, set $x_i \sim e^{j\omega_k t}$, for then $\ddot{x}_i = -\omega_k^2 x_i$ and the equations become

$$x_1\left[\frac{k_1 + f}{m} - \omega_k^2\right] + x_2\left[-\frac{f}{m}\right] = 0 \tag{8.74}$$

$$x_1\left[-\frac{f}{m}\right] + x_2\left[\frac{k_1 + f}{m} - \omega_k^2\right] = 0 \tag{8.75}$$

This is a system of two simultaneous linear equations in two unknowns. However, when either x_1 or x_2 is solved for, the result is zero, meaning that no motion takes place — a ridiculous result, since motion obviously can occur. The way out is to form the determinant of the coefficients of the variables and set that equal to zero in accordance with Cramer's rule. On expansion of the 2×2 determinant, an equation quadratic in ω_k^2 results. This equation is

$$\omega_k^4 - 2\left(\frac{k_1 + f}{m}\right)\omega_k^2 + \left(\frac{k_1 + f}{m}\right)^2 - \left(\frac{k_1}{m}\right)^2 = 0 \tag{8.76}$$

The quadratic formula and some algebra reduces the result to

$$\omega_k^2 = \left[\frac{k_1 + f}{m} \pm \frac{k_1}{m}\right] \tag{8.77}$$

If the plus sign is taken, the squared angular frequency is $(2k_1 + f)/m$, showing that the stretching frequency governed largely by k_1 is modified somewhat by the much smaller force constant f. On the other hand, if the negative sign is taken, the squared angular frequency is simply f/m. When the anchor to the wall is cut, f becomes zero, showing that that is the translation, which may be regarded as a non-genuine vibration. Indeed, both rotation and translation are regarded as non-genuine vibrations, even though each such degree of freedom has a perfectly good normal coordinate that describes its motion.

The frequency of the other vibration is the diatomic

$$\nu = \frac{1}{2\pi}\sqrt{\frac{2k_1}{m}} \tag{8.78}$$

The reduced mass of the two mass combination is $m/2$. The factor 2π arises since angular measure $\omega = 2\pi\nu$, where ν is in units of reciprocal seconds.

This problem is simple enough that it can be solved immediately by selection of suitable coordinates. The symmetry coordinates that solve the problem are

$$s_1 = x_1 + x_2 \tag{8.79}$$

$$s_2 = x_1 - x_2 \tag{8.80}$$

It develops that the normal coordinates are very similar to such trial symmetry functions, and are

$$Q_1 = \frac{1}{\sqrt{2m}}(x_1 + x_2) \tag{8.81}$$

$$Q_2 = \frac{1}{\sqrt{2m}}(x_1 - x_2) \tag{8.82}$$

In terms of these normal coordinates, the kinetic energy T and potential energy V are simultaneously diagonalized.

$$2T = \dot{Q}_1^2 + \dot{Q}_2^2 \tag{8.83}$$

$$2V = \frac{f}{m}Q_1^2 + \frac{2k_1 + f}{m}Q_2^2 \tag{8.84}$$

In more complicated cases, the normal coordinates will contain not only some of the masses, but also some of the geometry of the equilibrium molecule.

Problems

1. Methane, CH_4, has 9 fundamentals (a fundamental is a characteristic vibrational frequency). However, the symmetry is high, and the molecule is a spherical top that belongs to point group T_d, the tetrahedral group. Owing to the occurrence of three-fold symmetry axes, two of the fundamentals are triply degenerate, and one is doubly degenerate, leaving only four observable fundamentals. The observed spectroscopic activity and experimental frequency values are

$\nu_i(\gamma)$	cm^{-1}	Observed Activity	Degeneracy
$\nu_1(a_1)$	2914	Raman	1
$\nu_2(e)$	1526	silent*	2
$\nu_3(f_2)$	3020	Raman, infrared	3
$\nu_4(f_2)$	1306	infrared*	3

*Theoretically allowed in Raman, but too weak to detect experimentally.

The vibrations are all allowed in the Raman effect, but only ν_3 and ν_4 are allowed in the infrared. Then, neglecting contributions due to anharmonicity, centrifugal distortion, Coriolis coupling and other small energy terms, write the vibrational partition function for methane. Refer the energy zero to the bottom of the potential energy well.

2. Phosphorous vapor contains a tetratomic molecule P_4. The structure is that of an ideal tetrahedron. The $P - P$ distance is 2.21 Angstroms. P_4 belongs to point group T_d. It has 6 modes of vibration. These are the non-degenerate A_g totally symmetric stretch at 604 cm^{-1}, the doubly degenerate E bend at 381 cm^{-1} and the triply degenerate antisymmetric stretch at 506 cm^{-1}. All these vibrations are Raman active, and the F_2 vibration is strong in the

infrared. The A_g mode is polarized in the Raman effect but forbidden in the infrared, while the E vibration is depolarized in the Raman effect and silent in the infrared. The mass of a ^{31}P atom is 30.9737633 amu on a scale on which the mass of a ^{12}C atom is exactly 12 amu. Avogadro's number is 6.023045×10^{23} molecules per mole and the gas constant R is 1.98719 cal $deg^{-1}mol^{-1}$. The second quantization constant hc/k_B is 1.43877 cm deg. Planck's constant is 6.626176×10^{-27} erg sec and the velocity of light in vacuo is $2.99792458 \times 10^{10}$ cm sec^{-1}. The superscript zero indicates that the gas is ideal at one atmosphere and a temperature T. The zero of energy is E_0^0 and is the same for the thermodynamic functions H, A and G. Note that $C_p^0 = C_v^0 + R$ for one mole of ideal gas.

a. Calculate C_p^0 in units of cal deg^{-1} mol^{-1} from 100 to 1000 K at 100 degree intervals while in the hypothetical state of an ideal gas at one atmosphere pressure. Include the two temperatures 273.15 and 298.15 K as special cases.

b. Calculate the moment of inertia of P_4 in g cm^2 units and from this calculate both the spectroscopic constant B and the characteristic rotational temperature Θ_{rot}. Answer: $I = 251.17 \times 10^{-40}$ g cm^2, $B = 0.111453$ cm^{-1}, $\Theta_{rot} = 0.1604$ K.

c. Take the numerical constant in the Sackur Tetrode translational entropy to be 1.16512012. Calculate separately the translational, rotational and vibrational contributions to the entropy, then add these to get S^0 in cal mol^{-1} at the temperatures specified above.

d. Calculate the thermodynamic functions of P_4 as a function of temperature at 100 degree intervals from 100 to 1000 K. Include temperatures of 273.15 and 298.15 K in the table. The finished table will have columns headed T, $(H^0 - E_0^0)/T$, $-(G^0 - E_0^0)$, S^0, and C_p^0, in units of calories per mole except for C_p^0, which is in units of cal mol^{-1} K^{-1}. The P_4 molecules are regarded as in the hypothetical state of an ideal gas at one atmosphere pressure. Compare your results with those reported by G Thayarajan and Forest F. Cleveland, J. Mol. Spectrosc. **5**, 210–211 (1960). See also G. M. Rosenblatt, J. Mol. Spectrosc. **10**, 494 (1963). Note that these authors use F for the Gibbs free energy, whereas G is used above. The modern convention is to use Joules rather than calories for the energy unit. This problem is cast in calories to facilitate comparison with the literature values.

e. Thyagarajan and Cleveland actually published two papers on the thermodynamic functions of P_4, because they made a mistake in the first of the two, corrected in the second. What was the mistake? See G Thayarajan and Forest F. Cleveland, J. Mol. Spectrosc. **6**, 199 (1961).

3. OCS is a linear triatomic molecule. According to T. Wentink (J. Chem. Phys. **30**, 105 (1959)) the carbon–oxygen bond distance is 1.164 Angstrom and the carbon–sulfur bond distance is 1.559 Angstrom. The masses of the atoms are $^{16}O = 15.99491503$, $^{12}C = 12$, $^{32}S = 31.9720728$ amu. The doubly degenerate bending vibration occurs at 520.82 cm^{-1}, and the two non-degenerate stretching vibrations occur at 858.92 and 2062.18 cm^{-1}.

a. Calculate the heat capacity at constant pressure of one mole of OCS in the hypothetical state of an ideal gas at one atmosphere pressure at temperatures of 100, 200, 273.15, 298.15, 300, 400, and 500 K. Use R = 8.31447 J deg^{-1} mol^{-1}.

b. Calculate the moment of inertia about an axis perpendicular to the internuclear axis and passing through the center of gravity in amu A^2 and in g cm^2. Calculate the spectroscopic rotational constant B in cm^{-1} units and from this calculate the characteristic rotational constant Θ_{rot} in K. Answer: I = 83.119 amu A^2, I = 138.024 g cm^2, $B = 0.202811$ cm^{-1}, $\Theta_{rot} = 0.291805$ K.

c. Calculate the standard entropy of OCS in the hypothetical state of an ideal gas at one atmosphere and temperatures given above, in units of Joules per mole.

d. Visit the NIST Webbook and find the parameters for the Shomate equation that expresses the standard heat capacity and entropy in the temperature range $298 < T < 1200$ K, and compare your results obtained in parts a, b and c above. In the range of validity of the Shomate equation, the statistical mechanical results should be within 0.06 J mol^{-1} for the entropy, with similar results for the heat capacity.

4. Nitric oxide, NO_2, is a bent symmetric triatomic molecule that belongs to point group C_{2v}. The two-fold symmetry axis passes through the oxygen nucleus and bisects the ONO bond angle. Masses are 15.99491503 and 14.0030744 for oxygen and nitrogen respectively. The three rotational spectroscopic constants A, B and C are 8.003, 0.413, 0.404 cm^{-1}. The dimensionless constant in the translational entropy is -1.16511 closely enough. The symmetric NO stretch is at 1320.5 cm^{-1}, the bend is at 749.6 cm^{-1} and the antisymmetric NO stretch lies at 1517.0 cm^{-1}. The ground electronic state is doubly degenerate, and higher electronic states are inappreciably accessed at 500 K and below. Calculate the heat capacity C_p^0 and entropy S^0 in Joules per mol at 100 degree intervals from 100 to 500 K, given that the entropy S/R is 27.148 at 200 K and 28.886 at 300 K, while the heat capacity C_p/R is 4.002 at 100 K and 5.236 at 500 K.

References

1. R. S. McDowell, J. Quant. Spectros. Radiat. Transfer **38**, 337 (1987).
2. R. S. McDowell, J. Chem. Phys. **93**, 2801 (1990).
3. R. S. McDowell, J. Chem. Phys. **88**, 356 (1988).
4. J. K. G. Watson, Mol. Phys. **65**, 1377 (1988).
5. C. C. Lin and J. D. Swalen, Rev. Mod. Phys. **31**, 841 (1959).
6. J. R. Durig, S. M. Craven and W. C. Harris, "Determination of Torsional Barriers, from Far Infrared Spectra", in Vibrational Spectra and Structure, A Series of Advances, vol. 1, J. R. Durig, Ed., M. Dekker, New York, NY, 1975.
7. J. D. Lewis, T. B. Malloy, Jr., T. H. Chao and J. Laane, J. Mol. Struct.**12**, 427 (1972).
8. K. S. Pitzer and W. D. Gwinn, J. Chem. Phys. **10**, 428 (1942).

Chapter 9

Residual Entropy and Chemical Equilibrium

9.1. Residual Entropy

There are two kinds of entropy. One is due to a thermodynamic calculation from specific heat measurements, as illustrated in Figure 9.1. The other is the entropy as calculated from statistical mechanical principles discussed in earlier chapters. These should be equal at any specified temperature. Inspection of the entries in Table 9.1 shows that often the two are in fact equal within reasonable error, but there are some anomalies that lie outside any error estimate, and so must be real differences.

Two comments are in order. First, the zeros entered in the table merely mean that the agreement is within the estimated experimental error, not exactly zero. Second, in every case of disagreement between the calorimetric entropy, S_{calor}, and the spectroscopic entropy, S_{spec}, the spectroscopic entropy exceeds the calorimetric, and thus must contain a contribution that the calorimetric entropy somehow missed.

In thermodynamics, the defining relations for the energy and entropy are written as differential equations involving very small amounts of heat, dQ, and work, dW. For the entropy, the heat effect must be reversible, dQ_{rev}. These relations are

$$dE = dQ + dW \tag{9.1}$$

$$dS = \frac{dQ_{rev}}{T} \tag{9.2}$$

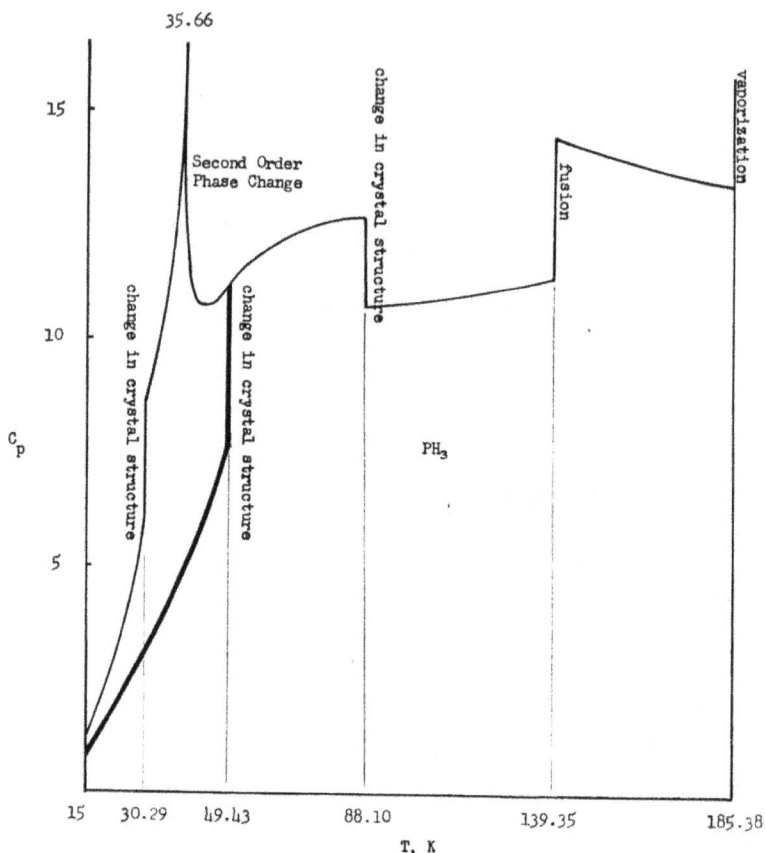

35.66

15

change in crystal structure

change in crystal structure

Second Order
Phase Change

change in crystal structure

change in crystal structure

fusion

vaporization

10

C_p

PH$_3$

5

| 15 | 30.29 | 49.43 | 88.10 | 139.35 | 185.38 |

T, K

C. C. Stephenson and W. F. Giauque, J. Chem. Phys. 5, 149 (1937). Third law
verification:

	Upper curve Form stable above 49.43 K		Lower curve Form stable below 49.43 K	
		eu		eu
0-15 K Debye extrapolation	hc ν/k=99	0.495	hc ν/k=114	0.338
Graphical	15-30.29 K	2.185	15-49.43 K	4.041
Transition	19.6/30.29	0.647	185.7/49.43	3.757
Graphical	30.29-49.43	4.800	S - S$_0$ =	8.14
	S-S$_0$ =	8.13		

Fig. 9.1. Specific heat of phosphine from 15 to 185 K.

where E and S are the energy and entropy respectively, and T is the
Kelvin temperature. These relations mean in particular that only
differences in energy and entropy may be measured. There is no
absolute entropy, nor absolute energy, since these functions are only
accessible by integration of these definitions, and must therefore each

Table 9.1. Entropies of selected small molecules at various temperatures. The column headed S_{spec} is the entropy calculated from spectroscopic data by statistical mechanical means. The column headed S_{calor} is the entropy calculated from thermodynamic measurements of the specific heat from some experimentally imposed low temperature limit. Entropy units (eu) are calories per degree per mole.

Molecule	T, K	S_{calor}	S_{spec}	Discrepancy
He	298.2	30.13	30.11	0
H_2	298.2	29.7	31.23	1.53
D_2	298.2	33.9	34.62	0.72
N_2	298.2	45.9	45.70	0
O_2	298.2	49.1	49.03	0
HCl	298.2	44.5	44.64	0
HBr	298.2	47.6	47.48	0
HI	298.2	49.5	49.4	0
CO	298.2	46.2	47.32	1.12
NO	121.36	43.0	43.75	0.75
H_2O	298.2	44.28	45.10	0.82
D_2O	298.2	45.89	46.66	0.77
H_2S	212.8	46.33	46.42	0
NH_3	239.68	44.13	44.10	0
PH_3	185.7	46.39	46.4	0
CH_4	111.5	36.53	36.61	0
CH_3D	99.7	36.72	39.49	2.77
CO_2	194.67	47.59	47.55	0
NNO	184.59	47.36	48.50	1.14
CS_2	318.39	57.48	57.60	0
OCS	222.87	52.56	52.66	0
SO_2	263.08	58.07	58.23	0
C_2H_4	169.4	47.36	47.35	0
CH_3Br	276.66	57.86	57.99	0

contain an arbitrary additive assignable constant of integration that may be chosen at pleasure once in the treatment of any problem. An apparent exception is the entropy in the microcanonical ensemble, $S = k_B \ln \Omega$, where Ω represents the number of accessible states, or really different equally probable ways of realizing the given thermodynamic condition. The defining relation for entropy implies that the statistical mechanical entropy $k_B \ln \Omega$ is really referred to a single definite state, $\Omega = 1$ as standard state, and the spectroscopic entropy is really a ΔS, given by

$$\Delta S = S_{spec} - 0 = k_B \ln \Omega - k_B \ln 1 \qquad (9.3)$$

The thermodynamic measurements are limited by the lowest temperature T^* to which the heat capacity apparatus can be driven. The heat capacity below this temperature can only be estimated, assuming no spikes in the heat capacity curve or other contributors other than the lattice modes of vibration. When this is the case, the heat capacity from $T = 0$ to T^* is estimated by the Debye approximation. Then the calorimetric entropy S_{calor} may be written

$$S_{calor} - S_0 = \int_0^{T^*} \frac{C_p^D}{T} dT + \int_{T^*}^{T_1} \frac{C_p^{(1)}}{T} dT + \frac{H_1}{T_1} + \int_{T_1}^{T_2} \frac{C_p^{(2)}}{T} dT + \frac{H_2}{T_2} + \ldots$$
$$(9.4)$$

Here, C_p^D is the Debye T^3 approximate heat capacity and the first term represents the conventional extrapolation estimate. If the spectroscopic and calorimetric entropies are equal, it implies that $S_0 = 0$. The spectroscopic entropy can be written

$$S_{spec} = S_{\text{translation}} + S_{\text{rotation}} + S_{\text{vibration}} + S_{\text{electronic}} \qquad (9.5)$$

The constant contribution of nuclear spin is deliberately omitted. Another important omission is the entropy of isotopic mixing. Both terms are constants that cancel when differences are taken, and for that reason are omitted from the definition. Additionally, the calorimetric entropy contains neither the nuclear spin nor the isotopic mixing contributions, since the use of the Debye approximation implicitly assumes that neither exists.

Thus, nuclear spin and isotopic mixing contributions are omitted from both S_{calor} and S_{spec}, and real differences between the two must be sought in other effects. In the early days of statistical mechanics, comparison between the two entropies was instrumental in establishing the validity of the statistical formalism. Today, the general agreement obviously establishes the theory, and it is reasonable to seek the explanation of the anomalies in other effects.

As noted above, in every case of a disagreement between S_{calor} and S_{spec}, S_{spec} is the greater. The reason is that S_{spec} is referred to a state at $T = 0$ with $\Omega = 1$, which defines a state of perfect molecular order. Hence, if there is any frozen in disorder in the physical crystal, S_0 will be greater than zero in the necessary amount. The result is that S_{calor} does not contain contributions from frozen in disorder, and S_{spec} does. Hence, S_{spec} is equal to or greater than S_{calor}.

Two simple cases in point are CO and NNO. Both molecules have a small electric dipole moment, but the size is small enough that crystal field forces far outweigh them. As a result, the molecules can fit into the crystal lattice in either of two essentially equivalent but distinguishable ways, CO as either CO or OC, and NNO as either NNO or ONN. In a crystal with N molecules, there would be 2^N ways of putting the N molecules on the N lattice sites. The associated entropy is $k_B \ln 2^N = N k_B \ln 2 = R \ln 2$, if N is Avogadro's number. The table is formed in entropy units, and $R = 1.98719$, $R \ln 2 = 1.377$. When the spectroscopic entropy is corrected by subtraction of $R \ln 2$, the corrected entropy for NNO is 47.12, and 45.93 eu for CO, bringing both spectroscopic entropies into agreement with the calorimetric.

Methane-d (CH_3D) is similar, but has four different orientations each molecule can assume in the crystal, leading to $R \ln 4 = 2.75$. The spectroscopic entropy corrected for the residual orientational entropy is 36.74 *eu*, which compares well with the calorimetric.

Nitric oxide, NO, might at first glance be thought to follow the same calculation as NNO and CO. But $R \ln 2$ is too large, essentially by a factor of 2. The reason is that NO dimerizes in the crystal, and there are two ways the dimer can fit into a lattice site. Then the residual entropy for NO is $k_B \ln 2^{N/2} = \frac{1}{2} R \ln 2 = 0.6887$, and the corrected spectroscopic entropy is 43.06 *eu*.

Light and heavy water, H_2O and D_2O, crystallize at atmospheric pressure and below in a form of ice in which each oxygen is tetrahedrally surrounded by four proton sites, two near the oxygen and two further away. The residual entropy of ice has been explained in a number of different ways, some more hand-waving than others. A simple explanation is given here, with some waving of the hands. If there were no hydrogens from the other water molecules in the ice structure, there would be six different ways of fitting a water molecule onto an oxygen site. But each hydrogen position has a 50:50 chance of being occupied, so the number of really different ways one molecule can be fit into the lattice is (since there are two hydrogens per water molecule) $6 \times 1/2 \times 1/2 = 3/2$. Then the residual entropy is $k_B \ln(3/2)^N = R \ln(3/2) = 0.8057$. The explanation for H_2O and D_2O is the same. The corrected spectroscopic entropy is 44.29 *eu* for H_2O and 45.85 *eu* for D_2O.

Another way to get $R \ln(3/2)$ for the residual entropy starts with N molecules in the crystal. Then there are $2N$ hydrogen atoms, and there are 2^{2N} ways of putting the hydrogen atoms into a lattice composed of oxygen atoms, one between each pair of oxygen atoms. This ignores the possibility that some of the water molecules would have formulas like H_3O or H_4O or HO or O, and the next step is to rule these out, allowing only H_2O. Consider one oxygen atom. It has 4 hydrogen positions, each of which may be near or far. There are then $2^4 = 16$ possible arrangements of 4 hydrogen atoms around one oxygen atom, but of these $4 \times 3/2 = 6$ arrangements allow 2 near and 2 far hydrogens. The reasoning is that there are 4 ways to select the first hydrogen atom position but only 3 to select the second, a total of 12 ways, not all different. Since placing a hydrogen in one position and another in a second position leads to the same configuration as placing the first hydrogen in the second position and the second hydrogen in the first position, division by 2 guarantees the number of ways is the number of really different ways. Then the residual entropy becomes $k_B \ln(2^{2N} (6/16)^N) = k_B \ln(3/2)^N = R \ln 3/2$. Whole papers have been published on the residual entropy of ice that use much more sophisticated methods, but these are beyond the introductory character of this book.

For hydrogen, H_2, the $J = 0$ state has a nuclear statistical weight of 1, as it is the lowest para state. If $J = 2$, the nuclear statistical weight is 3, as this is the lowest ortho state. Similarly, in deuterium, there are 6 ortho states in $J = 0$ and 3 para states in $J = 1$.

For hydrogen, high temperature equilibration produces a mixture of 1/4 para and 3/4 ortho molecules, which do not interconvert in the time of measurement of the heat capacity, and thus contribute their fractions to the calorimetric entropy. At very low temperatures, essentially all the molecules will have condensed into states of $J = 0$ or $J = 1$. In both H_2 and D_2, the rotational degeneracy is $2J + 1$, and thus is 1 for $J = 0$ and 3 for $J = 1$. Then in H_2,

$$S_{spec} - S_{calor} = \frac{1}{4}R\ln 1 + \frac{3}{4}R\ln 3 \qquad (9.6)$$

and the difference is $\frac{3}{4}R\ln 3 = 1.637$ *eu*.

The explanation for D_2 is very similar. Here, the 6 ortho and 3 para states make a total of 9 nuclear states, and high temperature equilibration produces 6/9 ortho and 3/9 para D_2. At extremely low temperatures, the molecules are essentially all in $J = 0$ or 1 states, ortho in $J = 0$ and para in $J = 1$. Then the difference between the spectroscopic and calorimetric entropies is

$$S_{spec} - S_{calor} = \frac{6}{9}Rl\ln 1 + \frac{3}{9}R\ln 3 \qquad (9.7)$$

and the difference is $\frac{1}{3}R\ln 3 = 0.7277$.

All of these results are entered in Table 9.2 for quick reference.

If heat capacity measurements are extended to sufficiently low temperatures, in very light molecules such as H_2 or D_2 the heat capacity curve may become lumpy or even display spikes. The contribution of ortho hydrogen[1] to $\int \frac{C_p}{T} dT$ at low temperatures is nearly $R\ln 3$, while the heat capacity for D_2 has been reported between 0.3 K and 13 K,[2] and appears to rise at the lowest temperatures. In Table 9.2, all of the molecules considered above are collected together with the explanation of the discrepancy between the spectroscopic and calorimetric entropies.

9.2. Chemical Equilibrium

Chemical equilibrium is of obvious interest to chemistry, since it controls the yield expected in any synthesis. Any chemical reaction can be written in the form

$$\sum \nu_i S_i = 0 \qquad (9.8)$$

where the S_i are the chemical substances involved and the ν_i are the stoichiometric coefficients. For example, the basic chemical change in

Table 9.2.   Summary of spectroscopic and calorimetric entropy disagreement, and of the reasons for them.

Molecule	T, K	S_{calor}	S_{spec}	Correction	S_{spec}-Corr.
He	298.2	30.13	30.11	0	
H_2	298.2	29.7	31.23	$(3/4)R\ln 3$	29.59
D_2	298.2	33.9	34.62	$(1/3)R\ln 3$	33.90
N_2	298.2	45.9	45.70	0	
O_2	298.2	49.1	49.03	0	
HCl	298.2	44.5	44.64	0	
HBr	298.2	47.6	47.48	0	
HI	298.2	49.5	49.4	0	
CO	298.2	46.2	47.32	$R\ln 2$	45.93
NO	121.36	43.0	43.75	$(1/2)R\ln 2$	43.06
H_2O	298.2	44.28	45.10	$R\ln(3/2)$	44.29
D_2O	298.2	45.89	46.66	$R\ln(3/2)$	45.85
H_2S	212.8	46.33	46.42	0	
NH_3	239.68	44.13	44.10	0	
PH_3	185.7	46.39	46.4	0	
CH_4	111.5	36.53	36.61	0	
CH_3D	99.7	36.72	39.49	$R\ln 4$	36.73
CO_2	194.67	47.59	47.55	0	
NNO	184.59	47.36	48.50	$R\ln 2$	47.12
CS_2	318.39	57.48	57.60	0	
OCS	222.87	52.56	52.66	0	
SO_2	263.08	58.07	58.23	0	
C_2H_4	169.4	47.36	47.35	0	
CH_3Br	276.66	57.86	57.99	0	

the Haber process for fixing nitrogen,

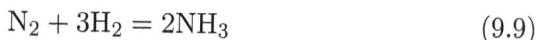

$$N_2 + 3H_2 = 2NH_3 \qquad (9.9)$$

can be written

$$2NH_3 - N_2 - 3H_2 = 0 \qquad (9.10)$$

and the coefficients are $-1, -3$, and $+2$. That is, reactants have negative coefficients and products have positive ones. This assignment of signs is clearly arbitrary. It is the convention in the United States, but not so in some other nations, which prefer the opposite selection. As the reaction progresses, the number of moles of each constituent

changes in accordance with a progress variable ξ, and the stoichiometry of the reaction. Let n_i represent the number of moles of substance
S_i. Then small changes under the stoichiometry are

$$dn_i = \nu_i d\xi \tag{9.11}$$

A variation in the Gibbs free energy for the reacting system can be
written

$$dG = -SdT + Vdp + \sum_i \mu_i dn_i \tag{9.12}$$

At constant temperature and pressure, the variation in the Gibbs
free energy is

$$dG = \sum_i \mu_i \nu_i d\xi \tag{9.13}$$

using equation (9.11) to replace dn_i. Integration would give the Gibbs
free energy as a function of the progress variable, and the slope of
this function is evidently

$$\frac{dG}{d\xi} = \sum_i \nu_i \mu_i \tag{9.14}$$

As equilibrium is approached, the Gibbs free energy approaches a
minimum. At the minimum, the slope vanishes, and the result is

$$\sum_i \nu_i \mu_i = 0 \tag{9.15}$$

That is, replacing the chemical substances with their chemical potential provides a correct equilibrium constraint on the values the chemical potentials may assume. This relation leads directly to the law of
mass action. The chemical potential of an ideal gas is

$$\mu_i = \mu_i^0(T) + RT \ln c_i \tag{9.16}$$

where c_i is the concentration of species i. Substituting this into equation (9.15) gives

$$\sum_i \nu_i \mu_i^0(T) + RT \sum_i \ln c_i^{\nu_i} = \sum_i \nu_i \mu_i^0(T) + RT \ln \prod_i c_i^{\nu_i} = 0. \tag{9.17}$$

The standard free energy change for the reaction is

$$\Delta G^0 = \sum_i \nu_i \mu_i^0. \tag{9.18}$$

The product of concentrations is represented by the symbol K and solving equation (9.17) for the product yields

$$\prod_i c_i^{\nu_i} = K = e^{-\frac{\Delta G^0(T)}{RT}}. \tag{9.19}$$

At constant temperature, a condition set at the start of this analysis, the quantity K is constant and characterizes the chemical equilibrium. For this reason it is called the equilibrium constant.

The statistical mechanic approach starts with a familiar relation from the canonical ensemble. In a pure gas, the partition function Q is given by

$$Q = \frac{q^N}{N!}, \tag{9.20}$$

and the Helmholz free energy by $A = -k_B T \ln Q$. Then the chemical potential is

$$\mu = \left(\frac{\partial A}{\partial N}\right)_{T,V} = -k_B T \ln q + k_B T \ln N. \tag{9.21}$$

To see how this works in practice, consider the chemical equilibrium among gases

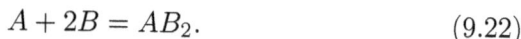

$$A + 2B = AB_2. \tag{9.22}$$

Using equation (9.15) with stoichiometric coefficients $\nu_A = -1, \nu_B = -2, \nu_{AB_2} = +1$, the equilibrium condition is

$$-\mu_A - 2\mu_B + \mu_{AB_2} = 0 = k_B T \ln\left(\frac{q_A}{N_A}\right) + 2k_B T \ln\left(\frac{q_B}{N_B}\right)$$
$$- k_B T \ln\left(\frac{q_{AB_2}}{N_{AB_2}}\right). \tag{9.23}$$

Transposing terms involving the partition functions q_i to the left hand side of the equality and cancelling $k_B T$, there results after taking antilogarithms

$$\left(\frac{q_{AB_2}}{q_A q_B^2}\right) = \left(\frac{N_{AB_2}}{N_A N_B^2}\right). \tag{9.24}$$

The arguments of the logarithms are obviously equal, and the ratio of the N's is an equilibrium constant K_N. Often it is preferable to express the equilibrium constant in concentrations rather than in molecules. Each q contains a volume dependence, since

$$q_i = q_{int,i} q_{translation,i} = q_{int,i} \frac{(2\pi m_i k_B T)^{3/2}}{h^3} V \tag{9.25}$$

where $q_{int,i}$ is the rotational and vibrational contributions to the molecular partition function q. Usually, a thermal wavelength Λ is used to collect compactly all the translational contributors except the volume, and the translational partition function is written as

$$q_{translation} = \frac{V}{\Lambda^3} \tag{9.26}$$

where $\Lambda = h/(2\pi m k_B T)^{1/2}$. Then each partition function can be written as

$$q = q_{int} \frac{V}{\Lambda^3} \tag{9.27}$$

and $q/V = q_{int}/\Lambda^3$, which is independent of V. Then each chemical potential can be written

$$\mu = -k_B T \ln\left(\frac{q}{V}\right) + k_B T \ln\left(\frac{N}{V}\right). \tag{9.28}$$

Define the concentration C_i as $C_i = N_i/V$. Then equation (9.24) becomes

$$\left(\frac{q_{int,AB_2}}{q_{int,A} q_{int,B}^2} \frac{\Lambda_A^3 \Lambda_B^6}{\Lambda_{AB_2}^3}\right) = \left(\frac{C_{AB_2}}{C_A C_B^2}\right) = K_C \tag{9.29}$$

where the antilogarithm has been used. While this seems simple, there is a complication. All the molecular partition functions must be evaluated relative to the same zero of energy. This is usually taken as the infinitely separated atoms. It leads to a temperature dependence of the equilibrium constant.

To illustrate the effect of a common energy zero, consider the equilibrium

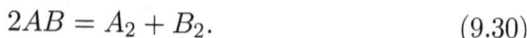

$$2AB = A_2 + B_2. \tag{9.30}$$

Each of these diatomic molecules has a dissociation energy, expressed either as D_e or D_0. The difference between the two is the zero point energy of the vibration. D_e measures the energy difference between the infinitely separated atoms and the bottom of the potential energy well. D_0 measures the energy difference between the infinitely separated atoms and the lowest vibrational energy level. Both are negative numbers. For example, D_0 for H_2 is -103.251 kcal mol^{-1}.

When the vibrational energy levels in one of the diatomic molecules is measured from the zero of energy taken as the infinitely separated atoms, the energy levels are

$$\varepsilon_v = D_0 + h\nu v. \tag{9.31}$$

The vibrational partition function is

$$q_{vib} = \sum_v e^{-\varepsilon_v/k_BT}. \tag{9.32}$$

Substituting equation (9.31) into equation (9.32) and evaluating the sum, the partition function becomes

$$q_{vib} = \frac{e^{-\frac{D_0}{k_BT}}}{1 - e^{-\frac{h\nu}{k_BT}}}. \tag{9.33}$$

The quantities $h\nu/k_BT = \Theta_{vib}/T$ and $-D_0$ are positive numbers. This applies to each of the diatomic molecules in the equilibrium shown in equation (9.30). The analog of equation (9.29) is

$$\left(\frac{q_{int,A_2} q_{int,B_2}}{q_{int,AB}^2} \frac{\Lambda_{AB}^6}{\Lambda_{A_2}^3 \Lambda_{B_2}^3}\right) = \left(\frac{C_{A_2} C_{B_2}}{C_{AB}^2}\right) = K_C \tag{9.34}$$

where $q_{int,A_2} = q_{rot,A_2} \frac{e^{-D_{0,A_2}/k_BT}}{1 - e^{-\Theta_{A_2}/k_BT}}$ and similarly for B_2 and AB. If the temperature is high enough, q_{rot,A_2} may be set equal to $T/2\Theta_{rot,A_2}$ and similarly for the rotational partition function for B_2.

It may be necessary to use the expansion $\frac{T}{2\Theta_{rot,A_2}}(1 + \frac{\Theta_{rot,A_2}}{3T} + \cdots)$ for hydrogen, H_2, or for hydrides such as HCl, HBr, HI, and so on.

The expression for K_c allows some simplification. First, note that

$$\frac{\Lambda_{AB}^6}{\Lambda_{A_2}^3 \Lambda_{B_2}^3} = \frac{m_{AB}^3}{m_{A_2}^{3/2} m_{B_2}^{3/2}} \tag{9.35}$$

the factors $\sqrt{2\pi k_B T}$ cancelling. Second, defining $\Delta E_0^0 = D_{0,A_2} + D_{0,B_2} - 2D_{0,AB}$ allows K_c to be written

$$K_C = \frac{m_{AB}^3}{m_{A_2}^{3/2} m_{B_2}^{3/2}} \frac{4\Theta_{rot,A_2}\Theta_{rot,B_2}}{\Theta_{rot,AB}^2} \left(\frac{\left(1 + \frac{\Theta_{rot,AB}}{3T}\right)^2}{\left(1 + \frac{\Theta_{rot,A_2}}{3T}\right)\left(1 + \frac{\Theta_{rot,B_2}}{3T}\right)} \right)$$

$$\times \frac{\left(1 - e^{-\Theta_{vibAB}/k_B T}\right)^2}{\left(1 - e^{-\Theta_{vibA_2}/k_B T}\right)\left(1 - e^{-\Theta_{vib,B_2}/k_B T}\right)} e^{-\Delta E_0^0/k_B T} \tag{9.36}$$

in which one or more of the Euler-MacLaurin rotational factors may contribute so little that it can be removed. Usually, the largest error arises in inaccurate dissociation energies. These appear in the argument of an exponential, and small errors are grossly magnified as a result.

The dissociation of a hydrogen atom provides an interesting illustration of these principles. The change in state is

$$H = H^+ + e^- \tag{9.37}$$

The equilibrium constant is

$$K(T) = \frac{C_{H^+} C_{e^-}}{C_H} = \frac{q_{H^+} q_{e^-}}{q_H} \tag{9.38}$$

The partition functions are for the proton

$$q_{H^+} = \frac{V}{\Lambda_{H^+}^3} \tag{9.39}$$

The proton has a nuclear magnetic moment and a spin of $1/2$, and thus a spin degeneracy of 2. By convention, nuclear spin degeneracies are omitted from statistical mechanical formulae. The electron has also a spin degeneracy of 2, and the partition function is

$$q_{e^-} = \frac{2V}{\Lambda^3} \tag{9.40}$$

Electron spin degeneracies are included, again by convention. The partition function for the hydrogen atom must take into account all the energy levels of the hydrogen atom, and is

$$q_H = \frac{V}{\Lambda_H^3} \sum_{n=1}^{\infty} 2n^2 e^{I/n^2 k_B T} \tag{9.41}$$

Hydrogen atoms have a degeneracy of $2n^2$, the factor of 2 deriving from the electron degeneracy and the n^2 from the mathematics (it is where the $p, d, f, g \dots$ levels are hiding). In general, the energies of the hydrogen atom are $\varepsilon = -I/n^2$, where I is the ionization energy and n is the principal quantum number. As the mass of the electron is about 1840 times lighter than the proton, the masses of the hydrogen atom and the proton are nearly the same. Then $\Lambda_H \cong \Lambda_{H^+}$ and the equilibrium constant can be written as

$$\begin{aligned}
K(T) &= \frac{1}{\Lambda_{H^+}^3} \frac{2}{\Lambda_{e^-}^3} \frac{\Lambda_H^3}{\sum_{n=1}^{\infty} 2n^2 e^{I/n^2 k_B T}} \\
&\cong \frac{(2\pi m_e k_B T)^{3/2}}{h^3} \frac{1}{\sum_{n=1}^{\infty} 2n^2 e^{I/n^2 k_B T}}
\end{aligned} \tag{9.42}$$

Note that the electron spin degeneracy 2 cancels exactly. If the proton or hydrogen atom spin degeneracy had been carried along, it too would have cancelled at this stage. When the quantum mechanical problem of the hydrogen atom is solved, it is assumed that the volume available to the atom is infinite, allowing the principal quantum number n to assume any value between 1 and infinity. Unfortunately, when the volume is infinite, the sum $\sum_{n=1}^{\infty} n^2 e^{I/n^2 k_B T}$ diverges for every finite temperature. The result is that the equilibrium constant is zero, and the hydrogen atom cannot dissociate. Of course hydrogen atoms do dissociate. In the laboratory, in any apparatus used to retain hydrogen atoms the volume available to the atoms is finite and usually relatively small, never infinite. Then there must be some principal quantum number n_{max} for which a hydrogen atom will just fit into the volume of the apparatus. Any larger principal quantum number would exceed the available volume. In this case, the sum does not extend to infinity, but terminates at the term in n_{max}. Since the sum is finite, it no longer diverges.

The magnitude of n_{max} can be estimated by noting that the radii of s states in spherical surroundings is $a_0 n^2$, where a_0 is the Bohr radius, the radius of the state with $n = 1$. This is about 0.5 Angstroms. An Angstrom is 10^{-8} cm. Place the hydrogen atom in a spherical container of radius 1 cm. Then $10^8 \cong 0.5 n_{max}^2$ and $n_{max} \cong 1.4 \times 10^4$. In any real apparatus, there will be a large but finite maximum principal quantum number.

Now inspect the magnitude of the terms in the series. At room temperature (300 K), the value of $k_B T$ in electron volts is about 0.026 eV. The dissociation energy of hydrogen is about 13 eV. The argument of the exponential is $I/n^2 k_B T \cong 13/.026 n^2 \cong 500/n^2$ This is so large that only the leading term in the sum contributes. For practical work, the equilibrium constant is

$$K(T) \cong \frac{(2\pi m_e k_B T)^{3/2}}{h^3} e^{-I/k_B T} \tag{9.43}$$

It may be objected that the argument above is not rigorous enough, and perhaps higher terms in the series really need to be carried. To show that the leading term entirely dominates the value of the series, first write it out in detail.

$$\sum_{n=1}^{n_{max}} n^2 e^{I/n^2 k_B T} = e^{I/k_B T} + 4e^{I/4k_B T} + 9e^{I/9k_B T} + \cdots$$

$$= e^{I/k_B T} \left(1 + 4e^{\frac{I}{k_B T}\left(\frac{1}{4}-1\right)} + 9e^{\frac{I}{k_B T}\left(\frac{1}{9}-1\right)} + \cdots \right) \tag{9.44}$$

or, more compactly,

$$\sum_{n=1}^{n_{max}} n^2 e^{I/n^2 k_B T} = e^{\frac{I}{k_B T}} \left(1 + \sum_{n=2}^{n_{max}} n^2 e^{\frac{I}{k_B T}\left(\frac{1}{n^2}-1\right)} \right) \tag{9.45}$$

Evidently

$$\sum_{n=2}^{n_{max}} n^2 e^{\frac{I}{k_B T}\left(\frac{1}{n^2}-1\right)} \ll \sum_{n=2}^{n_{max}} n^2 e^{-\frac{I}{k_B T}\left(\frac{3}{4}\right)} = e^{-\frac{I}{k_B T}\left(\frac{3}{4}\right)} \sum_{n=2}^{n_{max}} n^2$$

(9.46)

since $e^{-\frac{3I}{4k_B T}}$ exceeds every exponential $e^{\frac{I}{k_B T}\left(\frac{1}{n^2}-1\right)}$ in the sum on the left but the first, to which it is equal. The sum over n^2 on the right is difficult to evaluate in general (Jolly, 17). However, n_{max} is about 10^4 and the sum may be approximated as an integral. Integration suggests that the sum is of the order of $n_{max}^3/3$, neglecting $2^3/3$ in comparison with the upper limit of the integral. This is about 10^{12}. Since $I/k_B T$ is about 500, the exponential $e^{-\frac{3}{4}500} = e^{-375}$ and 10^{12} is much too small to drive the product $10^{12}e^{-375}$ to near unity. This is a very small number, and the quantity on the left of the inequality is even smaller. Hence only the leading term in the sum in the equilibrium constant in equation (9.42) need be retained.

Problems

1. Calculate the equilibrium constant for the gas phase reaction

$$2HI = H_2 + I_2$$

at 700 K, and compare your result with the experimental determination of Taylor and Crist.[3] Use the molecular properties given in the table below.

Gas	molecular weight	Θ_{rot}	Θ_{vib}	D_0, kcal mol^{-1}
HI	127.9	9.0	3200	70.5
H_2	2.016	85.4	6210	103.251
I_2	253.8	0.054	310	35.603

References

1. G. N. Lewis and M. Randall, Ed. 2, K. S. Pitzer, Ed., McGraw-Hill, New York, Ny, 1961, p. 598.
2. O. D. Gonzales, D. White and H. L. Johnston, J. Phys. Chem.**61**, 773 (1957).
3. A. H. Taylor and R. H. Crist, J. Am. Chem. Soc, **63**, 1377 (1941).

Chapter 10

Polymers

10.1. Introduction

Two general kinds of polymers may be encountered in practice, industrial or biological. Biological polymers are generally coded in an organism's DNA, and when synthesized are identical in length and monomer sequence. Industrial polymers have chains of different lengths and often have branches introduced during polymerization from the monomers by a variety of side reactions. Thus, industrial polymers are composed of a variety of different molecular weights. The average molecular weight of a polymer sample may be measured by the osmotic pressure, which yields the number average molecular weight, or by elastically scattered light, which yields the weight average molecular weight. In modern polymer laboratories, these two molecular weights are simultaneously measured by calibrated gel permeation chromatography. If the sample requires further characterization, the Z averaged molecular weights may be recovered by analysis of the results of sedimentation studies usually carried out by centrifugation.

10.2. The Polydispersity Index

The polydispersity index D is defined as the ratio of the weight average molecular weight M_w to the number average molecular weight M_n, and is used by polymer laboratories the world over to characterize the dispersion in molecular weights in polymer preparations. The number average molecular weight is defined according to the usual definition of averaging of a molecular property X

$$\langle X \rangle = \sum_i P_i X_i \tag{10.1}$$

where P_i is the probability of the property X_i. If there are n_i chains with molecular weight M_i, the probability a chain randomly selected from a sample of N chains has molecular weight M_i is $P_i = \frac{n_i}{\sum_i n_i}$. The number averaged molecular weight M_n is then

$$M_n = \langle M \rangle = \frac{\sum_i n_i M_i}{\sum_i n_i} \tag{10.2}$$

Since the weight of all the chains of molecular weight M_i is $w_i = n_i M_i$, the weight averaged molecular weight is a ratio of averages of the second power of the individual molecular weights to the average of the first power

$$M_w = \frac{\sum_i w_i M_i}{\sum_i w_i} = \frac{\sum_i n_i M_i^2}{\sum_i n_i M_i} = \frac{\langle M^2 \rangle}{\langle M \rangle} \tag{10.3}$$

Then, since $M_n = \langle M \rangle$, the polydispersity index defined as $D = \frac{M_w}{M_n}$ becomes

$$D = \frac{\langle M^2 \rangle}{\langle M \rangle^2} \tag{10.4}$$

The standard deviation is in general the root mean square deviation from the mean, and in this particular case may be used to characterize a molecular weight distribution. The mean squared deviation from the mean is

$$\langle \Delta M^2 \rangle = \langle (M_i - \langle M \rangle)^2 \rangle = \langle M^2 \rangle - \langle M \rangle^2 \tag{10.5}$$

From equation (10.4), $\langle M^2 \rangle = D \langle M \rangle^2$ and the relative mean squared deviation becomes

$$\frac{\langle \Delta M^2 \rangle}{\langle M \rangle^2} = D - 1 \tag{10.6}$$

The standard deviation relative to the mean molecular weight completely characterizes a Gaussian molecular weight distribution, and is

$$\frac{\sigma_M}{\langle M \rangle} = \sqrt{D - 1} \qquad (10.7)$$

For example, a Gaussian polymer with a polydispersity index of 1.03 would have a relative standard deviation of 0.17, and most of the chains in the sample would lie within one standard deviation from the mean. If the molecular weight distribution is not pure Gaussian, the polydispersive index and the other relations above can still be formed, since there is no assumption of Gaussian character in the sums that define them all. However further characterization depends on extracting the Z averages from sedimentation data, averages such as

$$\langle M^3 \rangle = \frac{\sum_i n_i M_i^3}{\sum_i n_i} \qquad (10.8)$$

and higher powers as needed.

Consider now binary polymer blends. In particular, consider a polymer blend formed by mixing n_1 molecules of a polymer with number average molecular weight M_1 with n_2 moles of another with mean molecular weight M_2, making a ratio $r = n_2/n_1$. The mean molecular weight of the blend is then

$$\langle M \rangle = \frac{\sum_i (n_{1i} + n_{2i}) M_i}{\sum_i (n_{1i} + n_{2i})} \qquad (10.9)$$

where n_{1i} is the number of molecules of kind 1 and molecular weight M_i, and similarly for molecules of kind 2. Let $c = 1/(1 + r)$. Then appropriate factoring in the denominator shows that

$$\langle M \rangle = c(\langle M_1 \rangle + r \langle M_2 \rangle) \qquad (10.10)$$

This is M_n for the blend. The weight average molecular weight M_w is

$$M_w = \frac{\langle M_1^2 \rangle + r \langle M_2^2 \rangle}{\langle M_1 \rangle + r \langle M_2 \rangle} \qquad (10.11)$$

The polydispersivity index D is

$$D = \frac{\langle M_1^2 \rangle + r\langle M_2^2 \rangle}{c(\langle M_1 \rangle + r\langle M_2 \rangle)^2} \qquad (10.12)$$

The mean squared deviation from the mean is

$$\langle \Delta M^2 \rangle = \langle M^2 \rangle - \langle M \rangle^2 = c(\langle M_1^2 \rangle + r\langle M_2^2 \rangle) - (c(\langle M_1 \rangle + r\langle M_2 \rangle))^2 \qquad (10.13)$$

The relative mean squared deviation is then again $D - 1$, and Equation (10.6) is recovered.

10.3. Dimensions of Polymer Chain Molecules

In solution or in the melt, polymer chains will only very rarely be found to be fully stretched out. A fair model of a polymer chain is a birds-nested rock climbing rope, but one writhing about under the onslaught of pelting by other chains or solvent molecules typically at Brownian frequencies. Evidently, in non-crystalline circumstanes the dimensions of polymer chains can be defined only in the mean. Several mean parameters have been used to characterize linear chains, particularly

1. the mean squared end to end distance, $\langle L^2 \rangle$.
2. the mean squared gyration radius, $\langle R^2 \rangle$.
3. mean squared Cartesian dimensions $\langle L_1^2 \rangle$, $\langle L_2^2 \rangle$ and $\langle L_3^2 \rangle$. In one investigation, Solc[1] found these stood in the ratio $\langle L_1^2 \rangle : \langle L_2^2 \rangle : \langle L_3^2 \rangle = 1 : 2.7 : 11.7$ for short chains of 50 and 100 bonds.

Consideration of these problems leads to the conclusion that Gaussian statistics, predicated on the central limit theorem, suffices except in such special circumstances as the condensation of chains into helical or other precisely determined array such as protein folding into biologically active native conformation. There are three main problems in the theory of polymer chain statistics. These are

1. Evaluation of the mean squared end to end distance and the gyration radius.

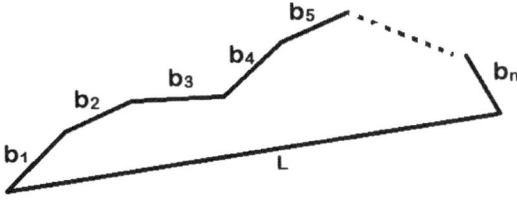

Fig. 10.1.   *b* vectors defining the *L* vector.

2. Phase transitions unique to polymer chains, such as the helix-coil condensation, protein folding or the polymer chain threading a pinhole.
3. Intersegment interaction effects resulting in excluded volume, altering configuration space available to a freely jointed Gaussian chain.

Many real linear polymer chains have nearly fixed bond angles along the chain, and have essentially free rotation about single bonds. Such chains have calculable purely numerical factors that modify the mean square end to end distance and gyration radius, which can be accommodated by increasing the statistical segment length. In this way the free flight results can be recovered for use unmodified by bond angle or rotation considerations. The model used to acquire free flight results is that of unform fixed bond lengths attached to universal joints lacking limit stops. Reference to Figure 10.1 should make this clear, and for a single polymer configuration, the end to end vector is defined as the vector sum

$$L = \sum_i b_i \tag{10.14}$$

Bold faced type is used to designate a vector. Thus, L is a vector, but L is its magnitude, and is a scalar. In Equation (10.14), b_i is a vector directed along a bond and is called a bond vector.

Bond vectors have the properties

$$|b_i|^2 = b^2 \tag{10.15}$$

$$b_i = b_{xi} u_x + b_{yi} u_y + b_{zi} u_z \tag{10.16}$$

In equation (10.16), the $\boldsymbol{u}_i$ are unit vectors along the Cartesian axes. These equations define the idealized free flight polymer chain.

Equation (10.1) emphasizes the central role played by the probability in producing mean values for molecular properties. To this end, write the probability that the end to end vector tip lie in the vector volume element $d\boldsymbol{L}$ as $P(\boldsymbol{L})d\boldsymbol{L}$. The volume element can be written

$$dL = dL_x dL_y dL_z = |L|^2 \sin \vartheta |dL| d\vartheta d\varphi, \qquad (10.17)$$

in Cartesian or spherical coordinates. $P(\boldsymbol{L})$ cannot be immediarely written down, but the individual bond vector probabilities can. In the freely jointed chain, each angle is equally probable. Further, the bond vector probability $p(\boldsymbol{b}_i)$ must vanish unless $|\boldsymbol{b}_i| = b$. These properties allow the individual bond vector probabilities to be written as proportional to a Dirac delta function

$$p(\boldsymbol{b}_i) = C\delta(|\boldsymbol{b}_i| - b) \qquad (10.18)$$

where C is the constant of proportionality. The delta function has the familiar properties $\delta(x) = 0$ if x is not equal to zero, and $\delta(x) = 1$ if x is equal to zero. It provides a method of evaluating a function at a specific value of its argument. For example, the value of $f(x)$ at $x = 0$ is

$$\int_{-\infty}^{\infty} f(x)\delta(x)dx = f(0) \qquad (10.19)$$

The latter property allows the constant C to be evaluated. Since the sum of all possible probabilities must be unity, the integral over all space of the bond vector probability must be one.

$$C \int \delta(b_i - b)db_i = 1 \qquad (10.20)$$

The vector volume element $db_i = b^2 \sin \vartheta d|\boldsymbol{b}_i| d\vartheta d\varphi$. The angles integrate to 4π leaving the b integration

$$4\pi C \int_0^{\infty} \delta(|\boldsymbol{b}_i| - \mathrm{b}) |\boldsymbol{b}_i|^2 \, d\,|\boldsymbol{b}_i| = 1 \qquad (10.21)$$

The value of the definite integral is b^2, which establishes the value of C as $1/4\pi b^2$.

The series of individual bond probabilities multiplies to produce the chain configuration probability just as the probability of throwing a specific number by throwing a pair of honest dice multiplies. For example, if one die reads 1, the second die could read any number between 1 and 6, and the same is true for each of the six numbers on the first die, for a total of $6 \times 6 = 36$ possible outcomes. In a polymer chain of n segments, designate the set of bond vectors $b_1, b_2, b_3 \ldots b_n$ as $\{b\}$. Then the probability that the configuration is in the volume element $d\{b\}$ is

$$P(\{b\})d\{b\} = \prod_{k=1}^{n} p(b_k)db_k \tag{10.22}$$

where the volume element $d\{b\} = \prod_{k=1}^{n} db_k$. Then the probability that the end to end vector tip lie in dL is obtained by integrating over all possible configurations

$$P(L) = \int \delta\left(L - \sum_{k=1}^{n} b_k\right) P(\{b\})d\{b\} \tag{10.23}$$

$P(L)dL$ must vanish independently of the orientation of L unless $L = \sum_{k=1}^{n} b_k$. In the vector relations, the meaning of the functional dependence is, for example, $\delta(A) = \delta(A_x)\delta(A_y)\delta(A_z)$, or other suitable coordinate system decompositions.

Substituting $P(\{b\})d\{b\}$ gives the probability as the integral of a product of delta functions In principle, the problem of writing the probability is solved, but the solution is of no practical use. The integral cannot be evaluated, and therefore averages of molecular properties cannot be calculated. Fortunately, it is possible to proceed through a Fourier transform to acquire something almost as good as the evaluated integral, namely, the characteristic function $K(k)$. The characteristic function can actually be calculated, and this function allows the calculation of averages. It is formed by writing

$$K(k) = \int \int \int_{-\infty}^{\infty} P(L)e^{ik \cdot L}dL \tag{10.24}$$

In equation (10.24), $\boldsymbol{k} \cdot \boldsymbol{L}$ is the vector dot, or inner, product. This equation shows $K(\boldsymbol{k})$ to be an average over the range of $\boldsymbol{L}$, of the imaginary exponential. Given $K(\boldsymbol{k})$, the Fourier inversion theorem allows evaluation of $P(\boldsymbol{L})$ as

$$P(\boldsymbol{L}) = \frac{1}{(2\pi)^3} \iiint e^{-i\boldsymbol{k}\cdot\boldsymbol{L}} K(\boldsymbol{k}) d\boldsymbol{k} \qquad (10.25)$$

This provides a useful formal expression for a delta function. Specifically

$$\delta(\boldsymbol{x}) = \frac{1}{(2\pi)^3} \iint e^{i\boldsymbol{k}\cdot\boldsymbol{x}} d\boldsymbol{k} \qquad (10.26)$$

Now the characteristic function can be evaluated. The trick is to first integrate over $\boldsymbol{L}$, making use of the properties of the delta function. First, substitute $P(\boldsymbol{L})$ into $K(\boldsymbol{k})$, with the result

$$K(\boldsymbol{k}) = \int d\boldsymbol{L} e^{i\boldsymbol{k}\cdot\boldsymbol{L}} \left\{ \prod_{j=1}^{n} p(\boldsymbol{b}_j) d\boldsymbol{b}_j \right\} \delta\left(\boldsymbol{L} - \sum_{j=1}^{n} \boldsymbol{b}_j \right) \qquad (10.27)$$

This integral vanishes unless $\boldsymbol{L} = \sum_{j=1}^{n} \boldsymbol{b}_j$. Then, upon making this substitution,

$$K(\boldsymbol{k}) = \int e^{i\boldsymbol{k}\cdot\sum_{j=1}^{n} \boldsymbol{b}_j} \prod_{j=1}^{n} p(\boldsymbol{b}_j) d\boldsymbol{b}_j \qquad (10.28)$$

Each bond probability $p(\boldsymbol{b}_j) = \frac{1}{4\pi b^2}\delta(|\boldsymbol{b}_j| - b)$ and on substituting this into $K(\boldsymbol{k})$, there results

$$K(\boldsymbol{k}) = \prod_{j=1}^{n} \int e^{i\boldsymbol{k}\cdot\boldsymbol{b}_j} p(\boldsymbol{b}_j) d\boldsymbol{b}_j \qquad (10.29)$$

The interchange of product and integral is rigorous since the product is finite. Here the exponential property $e^{a+b} = e^a e^b$ has also been used. $K(\boldsymbol{k})$ appears as a product of identical integrals, differing only in the orientation of the $\boldsymbol{b}_j$ vector. It is sufficient to consider only one of these integrals.

$$I = \frac{1}{4\pi b^2} \int e^{i\mathbf{k}\cdot\mathbf{b}_j} \delta(|\mathbf{b}_j| - b) d\mathbf{b}_j \tag{10.30}$$

The inner product $\mathbf{k} \cdot \mathbf{b}_j = kb\cos\Theta$, and the volume element $d\mathbf{b}_j = |\mathbf{b}_j|^2 \sin\Theta d\Theta d\varphi d|\mathbf{b}_j|$. Noting that the derivative $de^{ikb\cos\Theta} = -ikb\sin\Theta e^{ikb\cos\Theta} d\Theta$, the integral becomes

$$I = \frac{1}{4\pi b^2} b^2 \int_0^\pi \int_0^{2\pi} e^{ikb\cos\Theta} \sin\Theta d\Theta d\varphi \tag{10.31}$$

as the delta function selects b in the b integration. The φ integration gives just 2π and the remaining Θ integration is evaluated as

$$\int_0^\pi e^{ikb\cos\Theta} \sin\Theta d\Theta = \left.\frac{e^{ikb\cos\Theta}}{-ikb}\right|_0^\pi = -\frac{e^{-ikb}}{ikb} + \frac{e^{ikb}}{ikb} = 2\frac{\sin kb}{kb} \tag{10.32}$$

The integral I then becomes

$$I = \frac{\sin kb}{kb} \tag{10.33}$$

According to equation (10.29), the characteristic function is a product of n of these integrals, that is,

$$K(\mathbf{k}) = \left(\frac{\sin kb}{kb}\right)^n \tag{10.34}$$

The probability $P(\mathbf{L})$ is then

$$P(\mathbf{L}) = \frac{1}{(2\pi)^3} \int e^{i\mathbf{k}\cdot\mathbf{L}} \left(\frac{\sin kb}{kb}\right)^n d\mathbf{k} \tag{10.35}$$

and this formulation is exact. Unfortunately, there is no general technique for evaluating the integral other than in the limit of very large n. When n is very large, however, the mathematical property of the function $\frac{\sin x}{x}$ can be used to allow an approximate evaluation. The absolute value of this function must lie between 0 and 1 no matter what value x assumes, and any value less than 1 raised to a very large power tends toward zero. Only the value 1 survives unchanged. The algebraic trick that allows approximate evaluation is to use only values of kb that are very small, allowing series expansions of common functions to be used. The first few terms in the series expansions of the functions $\sin(x)$ and $ln(1-x)$ are

$$\sin x = x - \frac{x^3}{3!} + \cdots \tag{10.36}$$

$$ln(1-x) = -x + \cdots \tag{10.37}$$

From the series expansion for $\sin(x)$, a series expansion for $\frac{\sin x}{x}$ is evidently

$$\frac{\sin x}{x} = 1 - \frac{x^2}{6} + \cdots \tag{10.38}$$

Then, for kb sufficiently small, the function $\frac{\sin kb}{kb}$ can be approximated in the following way

$$\frac{\sin kb}{kb} = e^{ln\left(\frac{\sin kb}{kb}\right)} \tag{10.39}$$

Then, using equation (10.38), this becomes

$$\frac{\sin kb}{kb} = e^{ln\left(1 - \frac{k^2 b^2}{6} + \cdots\right)} \tag{10.40}$$

From equation (10.37)

$$\frac{\sin kb}{kb} \cong e^{-\frac{k^2 b^2}{6}} \tag{10.41}$$

This approximation allows writing the probability function for $\boldsymbol{L}$ as

$$P(\boldsymbol{L}) \cong \frac{1}{(2\pi)^3} \int e^{-i\boldsymbol{k}\cdot\boldsymbol{L} - n\frac{k^2 b^2}{6}} d\boldsymbol{k} \tag{10.42}$$

To get the whole thing, consider just one third of it. Decompose $P(\boldsymbol{L})$ into Cartesian components. Then one of these is

$$P(L_x) = \frac{1}{2\pi} \int e^{-ik_x L_x - \frac{nb^2 k_x^2}{6}} dk_x \tag{10.43}$$

and similarly for L_y and L_z. Working just with the argument of the exponential, factoring nb^2 out and completing the square yields a squared function of L_x and a term independent of k_x. Thus the argument of the exponential integrand is

$$-ik_x L_x - \frac{nb^2 k_x^2}{6}$$

$$= -\frac{nb^2}{6}\left(k_x^2 + \frac{6ik_x L_x}{nb^2} + \frac{1}{4}\left(\frac{6iL_x}{nb^2}\right)^2 - \frac{1}{4}\left(\frac{6iL_x}{nb^2}\right)^2\right)$$

$$(10.44)$$

This allows collecting all the functional dependence of the argument on k_x as a square of a function temporarily defined as $y(k_x)$.

$$y(k_x) = k_x + \frac{1}{2}\left(\frac{6iL_x}{nb^2}\right) \qquad (10.45)$$

Noting that $dy = dk_x$, the integral may now be written

$$P(L_x) = e^{-\frac{3L_x^2}{2nb^2}}\int_{-\infty}^{\infty} e^{-\frac{nb^2 y^2}{6}}\,dy \qquad (10.46)$$

The integral is tabulated and evaluation yields $P(L_x)$ explicitly, at least in the limit of long chains.

$$P(\boldsymbol{L}) = \left(\frac{3}{2\pi nb^2}\right)^{3/2} e^{-\frac{3L^2}{2nb^2}} \qquad (10.47)$$

This probability allows averages over molecular conformations to be calculated. For example,

$$\langle L^2 \rangle = \int L^2 P(\boldsymbol{L})d\boldsymbol{L} \qquad (10.48)$$

This average evaluates as nb^2. Note that the probability that $|\boldsymbol{L}|$ lies in $d|\boldsymbol{L}|$ is $4\pi L^2 P(\boldsymbol{L})d\,|\boldsymbol{L}|$.

The result nb^2 might have been anticipated from the definition of $\boldsymbol{L}$ as the vector sum $\boldsymbol{L} = \sum_{j=1}^{n} \boldsymbol{b}_j$.

Then the average of L^2 could be estimated as follows

$$\langle L^2 \rangle = \langle \boldsymbol{L} \cdot \boldsymbol{L} \rangle = \left\langle \sum_{j=1}^{n} \boldsymbol{b}_j \cdot \sum_{k=1}^{n} \boldsymbol{b}_k \right\rangle$$

$$= \left\langle \sum_{j,k=1}^{n} \boldsymbol{b}_j \cdot \boldsymbol{b}_k \right\rangle = \sum_{j=1}^{n} |\boldsymbol{b}_j|^2 = nb^2 \qquad (10.49)$$

Keep in mind that $\boldsymbol{b}_j \cdot \boldsymbol{b}_k = b^2 \cos \Theta_{jk}$, where the angle is that between the two bond vectors, measured from the jth to the kth. In the double sum, the angle between the jth and the kth bond will assume some value and for every such angle, one exactly π larger (or smaller) will also occur for some other values of j and k. These two cosines have opposite signs but identical absolute values and the two exactly cancel in the sum. Only when j and k have the same value does the coefficient b^2 survive, for then the angle is zero and $\cos(0) = 1$. It is worth mentioning that the vector products $\boldsymbol{b}_j \cdot \boldsymbol{b}_k$ and $\boldsymbol{b}_k \cdot \boldsymbol{b}_j$ are equal, since the angles θ_{jk} and θ_{kj} differ by 2π, not π.

The simple Gaussian nature of the final result for $P(\boldsymbol{L})$ is a consequence of the central limit theorem, which applied to this case states that in the limit of very large n, $P(\boldsymbol{L})$ must be Gaussian, independent of the fluctuations to which the individual $p(\boldsymbol{b}_j)$ are subject. If n is not very large, the Gaussian statistics are upset, and the mean square end to end distance will become larger. For real industrial polymers, it is often useful to define a characteristic ratio C_n

$$C_n = \frac{\langle \boldsymbol{L}^2 \rangle}{nb^2} \qquad (10.50)$$

The difficulty arises in measuring the mean square end to end distance, for which there are no practical laboratory methods available. In special cases, $\langle \boldsymbol{L}^2 \rangle$ can be measured. Neutron scattering experiments were conducted with polymethylene chains end-capped with methyl-d_3 groups, and an experimental measure of $\langle \boldsymbol{L}^2 \rangle$ could be extracted from the scattering data. For finite chains, the characteristic ratio departs from unity, and for Gaussian chains the ratio is 1 for all values of n. When it can be made available, then, the characteristic ratio partially characterizes real polymer chains. Very stiff chains and helices may be expected not to display Gaussian character since the δ-function model is upset. However, if these chains are made extremely long, breaks will occur, and if a sufficiently large number of breaks occur, Gaussian character may be approached. Typically the length of the "bond vector" will itself be a statisrical entity and very much longer than any ordinary chemical bond. This makes no difference to the essential chain behavior, as the length of the bond vector was never precisely defined or used in the foregoing.

If in ordinary chains the bond vector is taken as the length of a chemical bond along the chain backbone and free rotation about the bond is allowed but the bond angke is not allowed to vary, the result is to introduce a calculable numerical factor into $\langle L^2 \rangle = mnb^2$, where m is the factor. Obviously, b may be made slightly larger, and nb^2 is recovered with the new value for b. The algebraic expression for the numerical factor m will be addressed in paragraphs below. Then it is useful to divide the chain into n equal segments of length b, which satisfy $\langle \boldsymbol{L}^2 \rangle = nb^2$ and for the completely stretched out chain, $L_{max} = nb$. These segments are useful theoretical constructs, and are called Kuhn[2] statistical segments.

10.4. Conformational Entropy

It is clear that there are a very large number of conformational states that any very long chain can explore, and thus that there is considerable associated entropy. In this section, this entropy is explored. First, it is necessary to define a standard state, and for this purpose the fully stretched chain of length nb is taken as the state of zero entropy. The probability for the freely jointed chain is given by equation (10.47) above. $P(\boldsymbol{L})$ is the ratio of the number $\Omega(\boldsymbol{L})$ of ways the chain can realize end to end vector $\boldsymbol{L}$ to the total number Ω_T of ways the chain can realize all of the possible end to end vectors. There is exactly one conformation available to the fully stretched chain of length $L_s = nb$, that is, there is only one way of realizing the fully stretched state. Then

$$P(\boldsymbol{L}_s) = \frac{1}{\Omega_T} \tag{10.51}$$

Equation (10.51) may be written

$$P(\boldsymbol{L_s}) = \left(\frac{3}{2\pi n b^2}\right)^{3/2} e^{-3L_s^2/2nb^2} \tag{10.52}$$

Substituting $L_s^2 = n^2 b^2$ leads immediately to

$$\frac{1}{\Omega_T} = \left(\frac{3}{2\pi n b^2}\right)^{3/2} e^{-3n/2} \tag{10.53}$$

Since $P(\boldsymbol{L})$ can also be expressed as Ω/Ω_T, the number of ways of realizing a particular end to end distance $\boldsymbol{L}$ is $\Omega = \Omega_T P(\boldsymbol{L})$, or

$$\Omega = e^{\frac{3}{2}(n - L^2/nb^2)} \tag{10.54}$$

In the microcanonical ensemble, the entropy $S = k_B \ln \Omega$. The entropy for the freely jointed chain becomes

$$S(L) = \frac{3}{2} n k_B - \frac{3}{2} k_B \frac{L^2}{nb^2} \tag{10.55}$$

Note that the entropy drops to zero for the fully stretched chain for which $L = L_s = nb$. Note that the entropy decreases from a maximum when the end to end distance is zero, until it finally vanishes when the end to end distance can no longer increase because the chain is fully stretched. The decrease is, furthermore, strictly parabolic in L. The slope $\frac{\partial S}{\partial L}$ is related to the configurationally driven force. In thermodynamics, the combined first and second law equation for a variation in the energy is

$$dE = TdS - pdV + \mu dN + fdL \tag{10.56}$$

A change in the entropy is given by

$$dS = \frac{dE}{T} + \frac{p}{T} dV - \frac{\mu}{T} dN - \frac{f}{T} dL \tag{10.57}$$

At constant E, V and N, the end to end force f is given by

$$f = -T \left(\frac{\partial S}{\partial L}\right)_{E,V,N} \tag{10.58}$$

Evidently, the conformational force f varies linearly with L from a minimum of zero at $L = 0$ to a maximum at $L = nb$.

A polymer chain in the melt or in solution will explore all available conformations. As a result, fluctuations in the mean squared end to end distance will occur. Clearly, the average fluctuation is given by

$$\langle \Delta L \rangle^2 = \langle \langle L^2 \rangle - L^2 \rangle^2 = \langle L^4 \rangle - \langle L^2 \rangle^2 \tag{10.59}$$

From this equation the standard deviation follows. Development is left as an exercise for the reader.

10.5. Gyration Radius

The gyration radius is routinely accessible experimentally in elastic light scattering measurements. There are no experimental measures of the end to end distance, other than the special circumstance of the neutron scattering experiment mentioned earlier, hardly a routine measurement. For this reason, the radius of gyration is a more important characteristic of a real polymer chain than the end to end distance.

In classical mechanics, the radius of gyration s of a set of mass points arrayed in space is defined by the masses m_j of the points and the distances d_j of the mass points from the center of gravity of the array. The defining relation is reminiscent of the weight average molecular weight, and is

$$s^2 = \frac{\sum_j m_j d_j^2}{\sum_j m_j} \tag{10.60}$$

This definition is useful in designing sports equipment, such as baseball bats or tennis raquets, or in designing tools, such as machetes. In polymers, it allows characterization of the physical form assumed by proteins, viruses, and other polymers in solution. In the freely jointed chain, the polymer is composed of n Kuhn statistical segments, each of which has the same mass m. Then the gyration radius is given by

$$s^2 = \frac{\sum_{j=1}^{n} m d_j^2}{\sum_{j=1}^{n} m} = \frac{1}{n} \sum_{j=1}^{n} d_j^2 \tag{10.61}$$

The position of the centroid is defined by the vector sum

$$\sum_{j=1}^{n} m_j d_j = 0 \tag{10.62}$$

This results from taking moments about the centroid, and is the balance condition, that is, the sum of all the moments that would generate a clockwise torque is just balanced by the sun of all the moments that would generate an opposing torque.

Zimm and Stockmayer (3) were able to show in a straightforward way that the average gyration radius could be expressed exactly as

$$\langle s^2 \rangle = \frac{1}{2n^2} \sum_{i=1}^{n} \sum_{j=1}^{n} r_{ij}^2 \tag{10.63}$$

where r_{ij} is a vector from the ith to the jth Kuhn statistical segment in a freely jointed chain. When the sums are evaluated, the mean squared gyration radius of the freely jointed chain is found to be approximately

$$\langle s^2 \rangle \cong \frac{nb^2}{6} \tag{10.64}$$

Although Equation (10.63) is exact, Equation (10.64) is only an excellent approximation that is better the longer the chain becomes. It is worth pointing out that since the mean squared end to end distance is nb^2 for a freely jointed chain of n segments, the mean squared gyration radius of such a chain is just $1/6$ of the mean squared end to end distance.

$$\langle s^2 \rangle \cong \frac{\langle L^2 \rangle}{6} \tag{10.65}$$

The two results, $\langle L^2 \rangle = nb^2$ and $\langle s^2 \rangle \cong \frac{nb^2}{6}$, hold only in the limit of the very long freely jointed chain, and serve as the polymer chain dynamics analog of the ideal gas law. These relations are used in much the same way as limiting laws to which theories designed to accommodate other contributions, such as excluded volume, must converge in the limit of vanishing contributions.

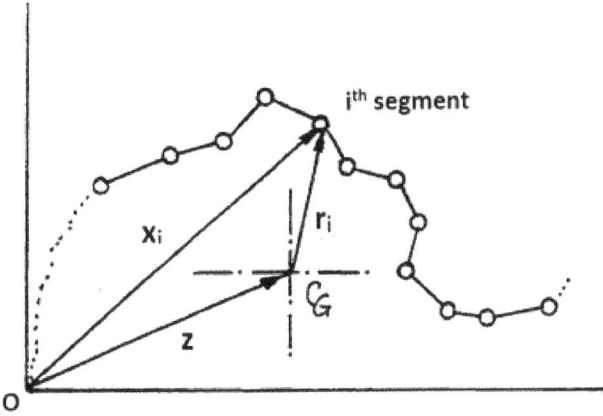

Fig. 10.3. Section of a linear freely jointed chain. The Kuhn statistical segments are represented by the small circles. The z vector is drawn from the origin of coordinates to the center of gravity of the chain while it is momentarily in the conformation shown.

To derive the result given in equation (10.64), follow Zimm and Stockmayer. Consider the diagram in Figure 10.3.

The calculation begins with the vector sum $x_i = z + r_i$. Evidently

$$r_i = x_i - z \qquad (10.66)$$

Since the masses are all the same, the condition that determines the centroid, namely, $\sum_{i=1}^{n} m_i r_i = 0$, becomes simply

$$\sum_{i=1}^{n} r_i = 0 \qquad (10.67)$$

Upon summing equation (10.66) there results

$$z = \frac{1}{n} \sum_{i=1}^{n} x_i \qquad (10.68)$$

According to the definition of the gyration radius, $s^2 = \frac{1}{n} \sum_{i=1}^{n} r_i^2$, and r_i^2 is the dot product of r_i with itself. Then, using Equation (10.66)

$$r_i^2 = x_i^2 + z^2 - 2z \cdot x_i \qquad (10.69)$$

Summing equation (10.69) over i,

$$\sum_{i=1}^{n} r_i^2 = \sum_{i=1}^{n} x_i^2 + nz^2 - 2z \cdot \sum_{i=1}^{n} x_i \qquad (10.70)$$

From equation (10.68), $\sum_{i=1}^{n} x_i = nz$, and equation (10.70) becomes

$$\sum_{i=1}^{n} r_i^2 = \sum_{i=1}^{n} x_i^2 - nz^2 \qquad (10.71)$$

Again from equation (10.68),

$$z^2 = \frac{1}{n^2} \sum_{i=1}^{n} \sum_{j=1}^{n} x_i \cdot x_j \qquad (10.72)$$

Making this substitution in equation (10.71) yields

$$\frac{1}{n} \sum_{i=1}^{n} r_i^2 = \frac{1}{n} \sum_{i=1}^{n} x_i^2 - \frac{1}{n^2} \sum_{i=1}^{n} \sum_{j=1}^{n} x_i \cdot x_j \qquad (10.73)$$

To progress, define $x_{ij} = x_i - x_j$. Note that this is a vector definition. Then it is easy to show that

$$x_i \cdot x_j = \frac{1}{2}(x_i^2 + x_j^2 - x_{ij}^2) \qquad (10.74)$$

When this result is substituted into equation (10.73) and the sums are evaluated, the result is

$$\frac{1}{n} \sum_{i=1}^{n} r_i^2 = \frac{1}{2n^2} \sum_{i=1}^{n} \sum_{j=1}^{n} x_{ij}^2 \qquad (10.75)$$

Let $r_{ij} = r_i - r_j$ and mote from Figure 10.4 that then the vector $x_{ij} = r_{ij}$ identically, as these vectors are both vectors from the ith to the jth segment. On taking the average over all possible conformations of the chain, the mean square gyration radius is

$$\langle s^2 \rangle = \frac{1}{n} \sum_{i=1}^{n} \langle r_i^2 \rangle = \frac{1}{2n^2} \sum_{i=1}^{n} \sum_{j=1}^{n} \langle r_{ij}^2 \rangle \qquad (10.76)$$

This result, due to Zimm and Stockmayer, is exact.

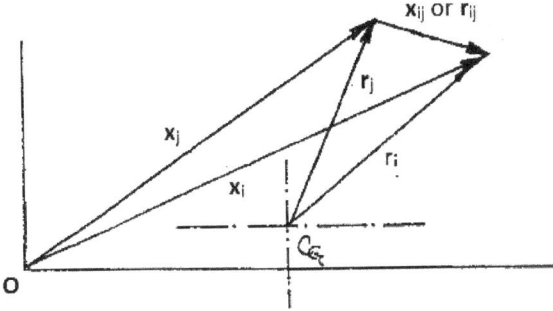

Fig. 10.4. Vectors to segments i and j.

Evaluation of the mean squared gyration radius for the very long freely jointed chain may be done as follows. First, consider the nature of the distance r_{ij}. The two defining segments are normally very far apart since the chain is very long. If the chain were cut at the ith and j^{th} segments the remaining chain would also be very long and of length $|i - j|$, because segments are numbered sequentially. The mean squared end to end distance of this chain would be

$$\langle r_{ij}^2 \rangle = |i - j|b^2 \tag{10.77}$$

The mean squared gyration radius becomes

$$\langle s^2 \rangle = \frac{b^2}{2n^2} \sum_{i=1}^{n} \sum_{j=1}^{n} |i - j| \tag{10.78}$$

In evaluating this, it is useful to note that $|i - j| = |j - i|$. The sum may be written

$$\sum_{j=1}^{n} |i - j| = \sum_{j=1}^{i} |i - j| + \sum_{j=i+1}^{n} |j - i| \tag{10.79}$$

The first sum has i terms $|i - 1| + |i - 2| + \cdots + |i - i|$, while the second sum has $n - i$ terms $|i + 1 - i| + |i + 2 - i| + \cdots + |n - i|$. In the first sum, i occurs i times for a total of i^2, and the first sum can be written $|i^2 - (1 + 2 + \cdots + i)|$. The sum of the first i integers is well known, and is $i(i + 1)/2$. The i's cancel in the second sum, and the sum is $1 + 2 + \cdots + n - i = (n - i)(n - i + 1)/2$. Then the sum in equation (10.79) becomes

$$\sum_{j=1}^{n} |j - i| = i^2 - \frac{i(i + 1)}{2} + \frac{(n - i)(n - i + 1)}{2} \tag{10.80}$$

This reduces to $i^2 - i(n + 1) + \frac{1}{2}n(n + 1)$. The sum over j is now evaluated, and the sum over i remains to be taken. To do this, another well known sum is needed, namely

$$\sum_{j=1}^{n} j^2 = \frac{1}{6}n(n + 1)(2n + 1) \tag{10.81}$$

The mean squared gyration radius becomes

$$\langle s^2 \rangle = \frac{b^2}{2n^2} \sum_{i=1}^{n} \left[i^2 - i(n + 1) + \frac{1}{2}n(n + 1) \right] \tag{10.82}$$

Evaluating all the sums produces

$$\langle s^2 \rangle = \frac{b^2}{2n^2} \left[\frac{1}{6}n(n + 1)(2n + 1) - \frac{1}{2}n(n + 1)(n + 1) + n\frac{1}{2}n(n + 1) \right] \tag{10.83}$$

After some algebra this simplifies to

$$\langle s^2 \rangle = \frac{b^2}{6} \left(n - \frac{1}{n} \right) \tag{10.84}$$

At very large n, the reciprocal $1/n$ becomes very small and the mean squared gyration radius of the very long freely jointed chain becomes essentially just $nb^2/6$. Since the mean squared end to end distance of such a chain is nb^2, the mean squared gyration radius of the chain is approximately one sixth of its end to end distance. A review of the derivation will show that several approximations were made, all of them very good ones. Short segment to segment distances r_{ij} were ignored as occurring only rarely, and $1/n$ is taken very small compared to n. The analysis does not apply to short chains.

The gyration radius provides a dependable measure of average polymer chain conformation in solution. To illustrate this fact, the

Table 10.1. Gyration radius calculated for different hypothetical polymer conformations.

Form	$\langle s^2 \rangle^{1/2}$, Angstrom
Solid sphere	41
Flexible chain	118
Solid cylinders	541
Diameter 20 A	
Length 1874 A	
Diameter 15 A	962
Length 3332 A	
Diameter 10 A	2164
Length 7495 A	
Disk	137
Diameter 387 A	
Length 5 A	

gyration radius can be calculated for an hypothetical polymer consisting of monomer units of molecular weight 150 g mol^{-1}, molecular weight 500,000 g mol^{-1}, and an average density of 1.41 g cm^{-3}. The results for several different average conformations are given in Table 10.1, which was constructed assuming that the number n of Kuhn statistical segments was given by the ratio of the molecular weights of polymer and monomer.

Some typical experimental results for light scattering measurements of the gyration radius of several biopolymers are compared in Table 10.2 with a gyration radius calculated assuming the biopolymer is uniformly distributed in a sphere with its packing density and molecular weight.

Tobacco mosaic virus is known from electron microscopic results to be a thin rod about 3000 A long. The mean squared gyration radius of a cylinder of radius r and length L is

$$\langle s^2 \rangle = \frac{r^2}{2} + \frac{L^2}{12} \tag{10.85}$$

Neglecting the small contribution of the $r^2/2$ term, the length of a rod of tobacco mosaic virus calculated from the light scattering

Table 10.2.   Comparison of observed gyration radii with radii calculated for an hypothetical sphere.

Biopolymer	Mol. wt.	$\rho, g\,cm^{-3}$	$\langle s^2 \rangle_{calc}^{1/2}, A$	$\langle s^2 \rangle_{obs}^{1/2}, A$
Albumin	66,000	1.33	18	30
Catalase	225,000	1.37	26	40
Myosin	493,000	1.37	34	470
DNA	4×10^6	1.8	63	1170
Bushy stunt virus	10.6×10^6	1.35	96	120
Tobacco mosaic virus	39×10^6	1.33	149	924

Data from C. Tanford, "Physical Chemistry of Macromolecules", Wiley, New York, 1961. Advances since 1960 show Myosin, for examples is a Y shaped protein with sturdy stem and each arm of the Y terminated by globular masses.[4]

result for the mean squared gyration radius is about 3200 A, in good agreement with electron microscopy. By the same token, DNA would appear to be long and thin rods, but other work shows that DNA is bent at a large number of places along the double helical strand, and thus is more rope-like than rod-like.

10.6. Angle Averages and Inclusions

Qualitatively, in linear chain polymers it is expected that restricting the bond angles between segments to specific values rather than allowing all possible angles should increase the mean square end to end distance and gyration radius. In this section, the effect of constraining the bond angle to a single value θ is first investigated, and then the effect of a hindering potential energy on the torsional angle is briefly discussed. In the first model, free rotation corresponds to an unrestricted torsion angle, and is allowed about all the bonds, as sketched in Figure 10.5.

Note that in the freely jointed chain, the bond angle θ may assume any value whatsoever, and that as a result the average of the cosine of the bond angle vanishes. Further, the average $\langle b_i \cdot b_j \rangle$ vanishes for all i not equal to j, and the mean squared end to end distance for the freely jointed chain is nb^2. But now the bond angle cannot assume arbitrary values, and is restricted to a single value, requiring

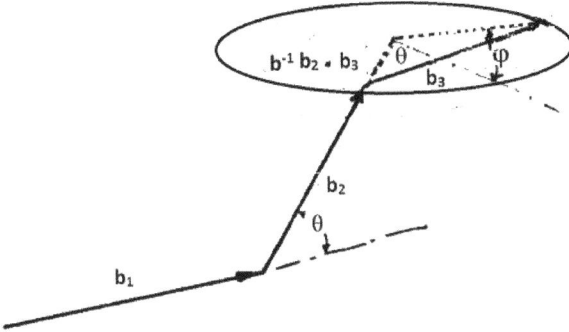

Fig. 10.5.   Bond vectors with fixed bond angle θ and unrestricted internal rotation angle φ.

$b_i \cdot b_{i+1} = b^2 \cos \Theta$ for all i. Study of Figure 10.5 shows that as the angle φ assumes all possible values between 0 and 2π, the vector b_3 traces out a cone of apex angle θ, and that $b_1 \cdot b_2 = b_2 \cdot b_3 = b^2 \cos \Theta$. As a result of the free rotation, the average of the component of vector b_3 that is perpendicular to vector b_2 vanishes. The component of b_3 that is parallel to b_2 survives averaging, and is $b \cos \theta$. In the figure, $b \cos \theta$ is the height of the cone swept out by b_3. Effectively, in the averaging process, the vector b_3 can be replaced by one parallel to b_2 of magnitude $b \cos \theta$. A unit vector in the b_2 direction is b_2/b, allowing the effective vector to be written as $b \cos \theta\ b_2/b$. Then the average $\langle b_1 \cdot b_3 \rangle$ is effectively

$$\langle b_1 \cdot b_3 \rangle = \langle b_1 \cdot b \cos \Theta (b_2/b) \rangle = \cos \Theta \langle b_1 \cdot b_2 \rangle \qquad (10.86)$$

But $b_1 \cdot b_2 = b^2 \cos \Theta$, and therefore

$$b_1 \cdot b_3 = b^2 \cos^2 \Theta \qquad (10.87)$$

Extension of this argument shows that in general

$$\langle b_i \cdot b_{i+k} \rangle = b^2 \cos^k \Theta \qquad (10.88)$$

The sum that has to be evaluated for the mean square end to end distance is

$$\langle L^2 \rangle = \sum_{i=1}^{n} \langle b_i \cdot b_i \rangle + 2 \sum_{i<j} \langle b_i \cdot b_j \rangle \qquad (10.89)$$

The first sum is nb^2. The key to evaluating the second sum is to list its terms in order. It can be written

$$\sum_{i<j=2}^{n} \langle \boldsymbol{b}_i \cdot \boldsymbol{b}_j \rangle = \sum_{i=1}^{n-1} \langle \boldsymbol{b}_i \cdot \boldsymbol{b}_{i+1} \rangle + \sum_{i=1}^{n-2} \langle \boldsymbol{b}_i \cdot \boldsymbol{b}_{i+2} \rangle + \sum_{i=1}^{n-3} \langle \boldsymbol{b}_i \cdot \boldsymbol{b}_{i+3} \rangle + \cdots$$

$$(10.90)$$

There are $n-1$ identical terms $b^2 \cos \theta$ in the first sum, $n-2$ identical terms $b^2 \cos^2 \theta$ in the second, $n-3$ identical terms $b^2 \cos^3 \theta$, and so on until the sum terminates. The mean squared end to end distance is then

$$\langle L^2 \rangle = nb^2 + 2(n-1)b^2 \cos \Theta + 2(n-2)b^2 \cos^2 \Theta$$
$$+ 2(n-3)b^2 \cos^3 \Theta + \cdots \qquad (10.91)$$

The angle θ is actually the supplement of the conventional bond angle, and is always in the first quadrant. Thus, $\cos \theta$ is always positive and less than unity. The convergence criterion for the sum of the series for an infinite chain is $\cos^2 \theta < 1$, which is clearly met except in the exceptional case in which $\theta = \pi/2$. Dividing Equation (10.91) by nb^2 forms the characteristic ratio C_n, and collecting like terms

$$C_n = 1 + 2 \cos \Theta + 2 \cos^2 \Theta + 2 \cos^3 \Theta + \cdots$$
$$- \frac{2}{n} (\cos \Theta + 2 \cos^2 \Theta + 3 \cos^3 \Theta + \cdots) \qquad (10.92)$$

In a very long chain, the negative terms contribute only trivially. The positive terms can be summed exactly. The characteristic ratio for very long chains can be written

$$C_n = 2(1 + \cos \Theta + \cos^2 \Theta + \cos^3 \Theta + \cdots + \cos^n \Theta) - 1 \qquad (10.93)$$

The sum in parentheses is very nearly the geometric series, and the function to which it converges for the infinite chain is $1/(1 - \cos \Theta)$. The convergence criterion $\cos^2 \Theta < 1$ is obviously met for all reasonable cases of bond angles in real polymers, as $\theta = \pi/2$ is of infrequent natural occurrence. Then very nearly, for finite but very long chains,

$$C_n = \frac{2}{1 - \cos\Theta} - 1 = \frac{1 + \cos\Theta}{1 - \cos\Theta} \tag{10.93}$$

The mean squared end to end distance and gyration radius becomes

$$\langle L^2 \rangle = nb^2 \frac{1 + \cos\Theta}{1 - \cos\Theta} \tag{10.94}$$

Evidently, if the segment length is taken to be σ, defined as

$$\sigma^2 = b^2 \frac{1 + \cos\Theta}{1 - \cos\Theta} \tag{10.95}$$

the usual expressions for the infinite chain, $\langle L^2 \rangle = n\sigma^2$ and $\langle s^2 \rangle = \frac{1}{6}n\sigma^2$ are recovered.

It should be emphasized that the approximate Equation (10.93) can be summed exactly. To do so, several finite sums are needed. It is helpful to expand the short finite sum

$$\sum_{i<j=2}^{5} \langle \boldsymbol{b}_i \cdot \boldsymbol{b}_j \rangle = \langle \boldsymbol{b}_1 \cdot \boldsymbol{b}_2 \rangle + \langle \boldsymbol{b}_2 \cdot \boldsymbol{b}_3 \rangle + \langle \boldsymbol{b}_3 \cdot \boldsymbol{b}_4 \rangle + \langle \boldsymbol{b}_4 \cdot \boldsymbol{b}_5 \rangle$$

$$+ \langle \boldsymbol{b}_1 \cdot \boldsymbol{b}_3 \rangle + \langle \boldsymbol{b}_2 \cdot \boldsymbol{b}_4 \rangle + \langle \boldsymbol{b}_3 \cdot \boldsymbol{b}_5 \rangle$$

$$+ \langle \boldsymbol{b}_1 \cdot \boldsymbol{b}_4 \rangle + \langle \boldsymbol{b}_2 \cdot \boldsymbol{b}_5 \rangle + \langle \boldsymbol{b}_1 \cdot \boldsymbol{b}_5 \rangle \tag{10.96}$$

Generalizing from this, it should be clear that this sum can be written as a sum of sums

$$\sum_{i<j=2}^{n} \langle \boldsymbol{b}_i \cdot \boldsymbol{b}_j \rangle = \sum_{i=1}^{n-1} \langle \boldsymbol{b}_i \cdot \boldsymbol{b}_{i+1} \rangle + \sum_{i=1}^{n-2} \langle \boldsymbol{b}_i \cdot \boldsymbol{b}_{i+2} \rangle + \cdots + \langle \boldsymbol{b}_i \cdot \boldsymbol{b}_n \rangle$$

$$\tag{10.97}$$

There are $n - 1$ identical terms $b^2 \cos\theta$ in the first sum, $n - 2$ identical terms $b^2 \cos^2\theta$ in the second, and so on until the last sum, which contains only one term. That term is $b^2 \cos^n\theta$. This allows condensation of the sum in equation (10.97) to two sums. First, write

$$\sum_{i<j=2}^{n} \langle \boldsymbol{b}_i \cdot \boldsymbol{b}_j \rangle = (n-1)b^2 \cos\Theta + (n-2)b^2 \cos^2\Theta + \cdots + b^2 \cos^n\Theta$$

$$\tag{10.98}$$

The coefficient of the last term is $n - (n - 1)$. This allows the sum to be rewritten as

$$\sum_{i<j=2}^{n} \langle \boldsymbol{b}_i \cdot \boldsymbol{b}_j \rangle = nb^2 \sum_{k=1}^{n} \cos^k \Theta - (n-1)b^2 \sum_{k=1}^{n-1} k \cos^k \Theta \qquad (10.99)$$

The first sum is easy to evaluate, and is

$$S_0 = \sum_{k=1}^{n} \cos^k \Theta = \frac{\cos \Theta - \cos^{n+1} \Theta}{1 - \cos \Theta} \qquad (10.100)$$

The second sum is related, and is given by the derivative taken with respect to $\cos \theta$, with a minor adjustment. The result is, effectively, that the derivative is taken with respect to the natural logarithm of the cosine of the bond angle.

$$S_1 = \sum_{k=1}^{n-1} k \cos^k \Theta = \cos \Theta \frac{\partial S_0}{\partial \cos \Theta} \qquad (10.101)$$

After a little algebra, S_1 evaluates as

$$S_1 = \frac{(\cos \Theta - n \cos^n \Theta + (n-1) \cos^{n+1} \Theta)}{(1 - \cos \Theta)^2} \qquad (10.102)$$

The mean squared end to end distance for a finite chain is exactly

$$\frac{\langle L^2 \rangle}{nb^2} = 1 + 2S_0 - \frac{2}{n} S_1 \qquad (10.103)$$

After some further algebra, the final result for finite chains emerges

$$\frac{\langle L^2 \rangle}{nb^2} = \frac{1 + \cos \Theta}{1 - \cos \Theta} - \frac{2}{n} \left(\frac{\cos \Theta - \cos^{n+1} \Theta}{(1 - \cos \Theta)^2} \right). \qquad (10.104)$$

For very long chains, this expression reverts to equation (10.94).

The foregoing analysis depended on the assumption of free rotation about segment vectors, while holding the angle formed by adjacent vectors to a single value. If the rotation is hindered, rather than free, the chain dimensions will be further swollen. In chains such as polymethene, the hindering potential is symmetric about the ante position. This configuration was called the trans position in early works. By convention, the torsion angle is measured from the ante configuration as the zero of angle. For polymethene and similar chains, the potential energy $V(\varphi)$ is symmetric about $\varphi = 0$. However, the potential energy exhibits two minima at approximately $\varphi = \pm 2\pi/3$. The energy difference between the ante and gauche configurations is about 460 cal mol^{-1} in the short chain hydrocarbon n-pentane.[5] The minima are separated by barriers, the highest in the cis position, which are of unknown magnitude. The torsional angles assumed by the chain segments will be of different probabilities for the ante and gauche conformers, and the averages of functions of the torsion angle will therefore be temperature dependent, as the probability is of the form $p(\varphi) = Ae^{-V(\varphi)/RT}$, where A is a normalization constant, R is the universal gas constant and T is the absolute temperature. The average of the cosine of the torsion angle is

$$\langle \cos \varphi \rangle = \frac{\int_0^{2\pi} \cos \varphi e^{-V(\varphi)/RT} d\varphi}{\int_0^{2\pi} e^{-V(\varphi)/RT} d\varphi} \tag{10.105}$$

In general, the component of the vector product perpendicular to that in the plane formed by $\boldsymbol{b}_1$ and $\boldsymbol{b}_2$ in Figure 10.5 is $\sin \vartheta \cos \varphi$. The dot product of $\boldsymbol{b}_1$ with $\boldsymbol{b}_3$ is no longer restricted to the plane formed by vectors $\boldsymbol{b}_1$ and $\boldsymbol{b}_2$, but now contains a contribution from $\sin \vartheta \cos \varphi$. By elaborating on the procedure used to obtain Equation (10.94), the final result for very long chains is

$$\langle L^2 \rangle = nb^2 \frac{1 + \cos \Theta}{1 - \cos \Theta} \frac{1 + \langle \cos \varphi \rangle}{1 - \langle \cos \varphi \rangle} \tag{10.106}$$

This equation was first reported by W. J. Taylor.[6] For the freely rotating chain, $\langle \cos \varphi \rangle$ vanishes, giving Equation (10.94). For potentials for which $V(\varphi) = V(\varphi + 2\pi/m)$, $\langle \cos \varphi \rangle$ vanishes again, leading to the same result. However, when the ante configuration lies below the gauche configuration energetically, $\langle \cos \varphi \rangle$ lies in the plane formed by $\boldsymbol{b}_1$ and $\boldsymbol{b}_2$ and does not vanish. The more closely $\langle \cos \varphi \rangle$ approaches unity, the more extended the chain becomes, until in the limit of $\langle \cos \varphi \rangle = 1$, the infinite chain assumes the all ante conformation. This is the conformation favored in short segments in crystalline long chain hydrocarbons. In contrast, when the gauche conformer lies below the trans energetically, $\langle \cos \varphi \rangle$ becomes negative, and the chain contracts. In either case, average chain dimensions are a function of temperature. The effect of various values of $\langle cos\varphi \rangle$ on mean chain dimensions is discussed in detail in Taylor's 1948 paper.[7] See also Benoit.[8]

Problems

1. The average of the torsional angle φ measured along the backcone of a long polymethylene chain endcapped with methyl groups can be written approximately as

$$\langle \cos \varphi \rangle = \frac{\cos(0)e^{-\varepsilon_0/k_BT} + 2\cos\left(\frac{2\pi}{3}\right)e^{-\varepsilon_1/k_BT}}{e^{-\varepsilon_0/k_BT} + 2e^{-\varepsilon_1/k_BT}}$$

   since each carbon–carbon bond has three torsional states, one ante and two gauche. The energy difference $\Delta\varepsilon = \varepsilon_1 - \varepsilon_0 \cong 460$ cm^{-1} is all that can be measured, since the zero of energy is undefined. The CCC bond angle is tetrahedral, and the bond vector magnitude b is taken as the CC distance. Plot the characteristic ratio (the ratio of mean squared end to end distance for the polymer to that of a freely jointed Gaussian chain) as a function of temperature from $T = 0$ to $T = 1000$ K, assuming the polymer has no ceiling temperature (i.e., does not dissociate) and does not pyrolyze or otherwise react or decompose.

2. The probability

$$P(\boldsymbol{L}) = \left(\frac{3}{2\pi nb^2}\right)^{3/2} e^{-\frac{3L^2}{2nb^2}}$$

   is the probability of occurrence of a single end to end vector $\boldsymbol{L}$. Show that the maximum in the spherical distribution function $4\pi L^2 P(\boldsymbol{L})$ occurs at a value L_{max} that is not the same as the square root of the mean squared end to end distance.

3. The radius of gyration may alternatively be expressed

$$\langle S^2 \rangle = \frac{\int d^2 \rho dV}{\int \rho dV}$$

where ρ is the density of a continuous solid and V its volume, while d is the distance of the volume element from the center of mass of the solid.

   a. Show that the mean squared gyration radius of a sphere of radius r is $3r^2/5$.

   b. Show that the mean squared gyration radius of a right circular cylinder of radius r and length L is $\frac{r^2}{2} + \frac{L^2}{12}$.

4. Use the end to end distance probability given in problem 2 above to calculate the mean square end to end distance $\langle L^2 \rangle$ and the higher moments $\langle L^4 \rangle$, $\langle L^6 \rangle$, $\langle L^8 \rangle$. Use the general set-up

$$\langle f(L) \rangle = \int_0^\infty f(L) P(\mathbf{L}) 4\pi L^2 dL$$

given $\int_{-\infty}^\infty e^{-\alpha^2 y^2} dy = \frac{\sqrt{\pi}}{\alpha}$, and the fact that differentiation to all orders with respect to α is allowed.

5. A fluctuation in the mean squared end to end distance is $\Delta L^2 = L^2 - \langle L^2 \rangle$, and the mean squared fluctuation in the end to end distance is

$$\langle (\Delta L^2)^2 \rangle = \langle L^4 \rangle - \langle L^2 \rangle^2 = \sigma_{L^2}^2$$

For a Gaussian chain, obtain the mean relative fluctuation $\frac{\sigma_{L^2}^2}{\langle L^2 \rangle^2}$.

References

1. K. Solc and W. H. Stockmayer, J. Chem. Phys. 54, 2756 (1971); K. Solc, J. Chem. Phys. **55**, 335 (1971).
2. H. Kuhn, J. Chem. Phys. **15**, 843 (1947).
3. B. H. Zimm and W. H. Stockmayer, J. Chem. Phys. **17**, 1301 (1949).
4. D. B. Stone, D. K. Schneider, Robert A. Mendelson, Biophys. J. **69**, 767 (1995).
5. A. J. LaPlante, H. D. Stidham, G. A. Guirgis, H. W. Dukes, J. Mol. Struct. **1023**, 170–175 (2012).
6. W. J. Taylor, J. Chem. Phys. **15**, 412 (1947).
7. W. J. Taylor, J. Chem. Phys. **16**, 257 (1948).
8. H. Benoit, J. Chim. Phys. **44**, 18 (1947).

Chapter 11

More Polymers

11.1. The Transfer Matrix, T

The problem of deriving the dependence of the mean squared dimensions of a long linear polymer chain on the bond angle and on the average of the cosine of the torsional angle can be efficiently attacked by the method of the transfer matrix. This method was introduced in another context by Kramers and Wannier[1] and expanded by Montroll[2] and by Lasettre and Howe.[3]

To define the transfer matrix for this problem, first consider the transformation of a vector v directed along the z axis to a new orientation defined by the angles θ and φ shown in Figure 11.1.

The components of the vector v' are given in Equations (1.1).

$$v'_x = |v'| \sin \Theta \cos \varphi$$
$$v'_y = |v'| \sin \Theta \sin \varphi$$
$$v'_z = |v'| \cos \Theta \tag{11.1}$$

Define $v = |v'|$, in order to achieve a less cumbersome notation. If the vector v' initially lay along the z axis, the components would be

$$v_x = 0$$
$$v_y = 0$$
$$v_z = v \tag{11.2}$$

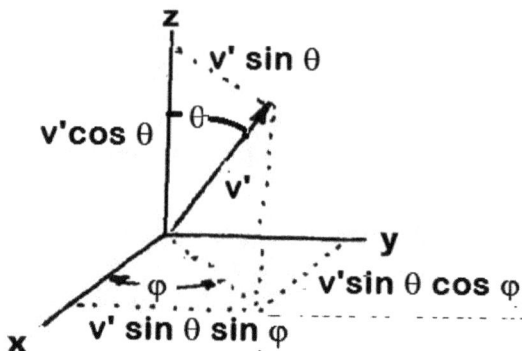

Fig. 11.1.   Components of vector v'.

Let a matrix that defines the transformation of vector v into vector v' be T. The transformation would then be written as

$$v' = Tv \qquad (11.3)$$

Expanding the vectors and the T matrix, in this case

$$\begin{pmatrix} v \sin\Theta \cos\varphi \\ v \sin\Theta \sin\varphi \\ v \cos\Theta \end{pmatrix} = \begin{pmatrix} T_{11} & T_{12} & T_{13} \\ T_{21} & T_{22} & T_{23} \\ T_{31} & T_{32} & T_{33} \end{pmatrix} \begin{pmatrix} 0 \\ 0 \\ v \end{pmatrix} \qquad (11.4)$$

When a matrix with elements A_{ij} is multiplied into a compatible matrix B_{jk}, the ik^{th} element of the product matrix AB is $AB_{ik} = \sum_j A_{ij} B_{jk}$. It is evident that $T_{13} = \sin\theta \cos\varphi$, $T_{23} = \sin\theta \sin\varphi$ and $T_{33} = \cos\theta$ is required for Equation (11.4) to hold. The other elements are as yet undefined.

To define the remaining elements, first consider a rotation by an angle θ about the z axis. There are two ways of looking at this. Either the coordinate axes x and y can remain fixed and the vector v be rotated, or the vector can be held fixed in space and the axes rotated by the angle. Figure 11.2 shows the latter consideration. The vector v has x and y components v_x and v_y in the xy plane. In the x' and y' coordinate system of the transformed vector v', the components are v'_x and v'_y. Inspection of the figure will show that these are related

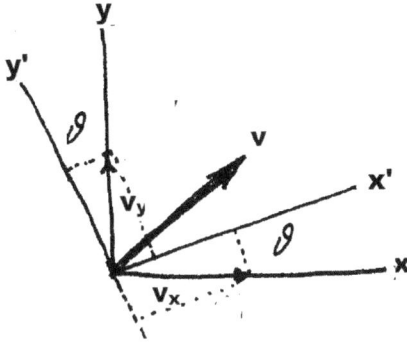

Fig. 11.2. Rotation about the z axis, holding vector v fixed and rotating the coordinate system from x, y, z to x', y', z.

as shown in Equations (11.5a,b).

$$v'_x = v_x \cos \Theta + v_y \sin \Theta$$

$$v'_y = -v_x \sin \Theta + v_y \cos \Theta$$

$$v'_z = v_z \tag{11.5}$$

The rotation is represented by a matrix R, and the rotated vector by $v' = Rv$, where R is the matrix

$$R = \begin{pmatrix} \cos \Theta & \sin \Theta & 0 \\ -\sin \Theta & \cos \Theta & 0 \\ 0 & 0 & 1 \end{pmatrix} \tag{11.6}$$

With these preliminaries, the stage is now set for solution of the problem of the mean square dimensions of long linear chains by the transfer matrix method. The reason for introducing this method on a problem already solved by direct means is to provide a more powerful mathematical framework and show that it in fact leads to correct results. More powerful methods are important, for these allow problems that do not yield to direct attack to be successfully resolved.

In Figure 11.3, a linear hydrocarbon chain with a variety of bond and torsion angles is sketched, mainly for the purpose of defining the

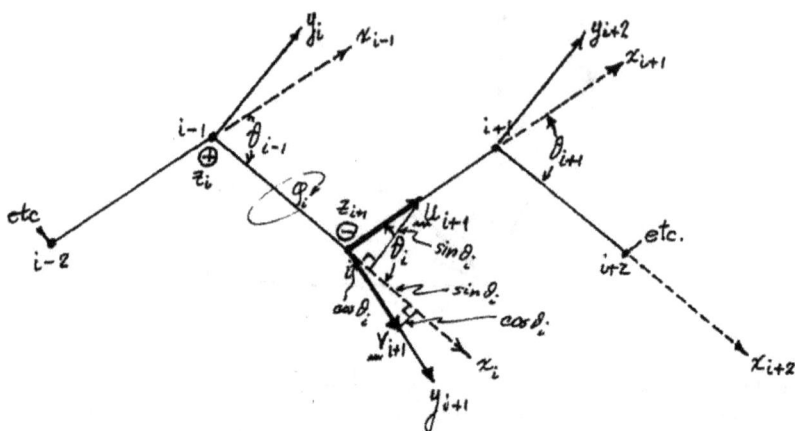

Fig. 11.3.   Parameters of a linear chain composed of masses at $i-2, i-1, i, i+1,$ $i+2$ and so on. The origin of coordinates for i is at $i-1$, where the z_i axis is out of the plane of the paper. The origin of coordinates for $i+1$ is at i. The z_{i+1} axis is into the plane of the paper, and so on, the direction of the z_j axes alternating at successive coordinate origins. All coordinate systems are Cartesian.

various quantities that will be needed in the analysis. This figure will reward some study.

Now consider a unit vector u_{i+1} directed along the x_{i+1} axis, originating at the ith mass point, and inquire, what are its components in the ith coordinate system. A moment's reflection will show that the components are

$$u_{i+1}(x_i) = \cos \Theta_i$$

$$u_{i+1}(y_i) = \sin \Theta_i \cos \varphi_i$$

$$u_{i+1}(z_i) = \sin \Theta_i \sin \varphi_i \qquad (11.7)$$

Consult Figure 11.4 to visualize the contributors to the components.

The components of a unit vector v_{i+1} directed along the positive y_{i+1} direction in the ith coordinate system are given in Equation (11.8).

$$v_{i+1}(x_i) = \sin \Theta_i$$

$$v_{i+1}(y_i) = -\cos \Theta_i \cos \varphi_i$$

$$v_{i+1}(z_i) = -\cos \Theta_i \sin \varphi_i \qquad (11.8)$$

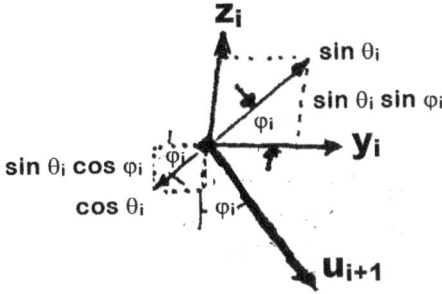

Fig. 11.4. Components of the unit vector u_{i+1} in the i^{yj} coordinate system. The x_i axis is out of the plane of the paper, and the angle θ_i is between u_{i+1} and the positive x_1 axis.

The vectors u_{i+1} and v_{i+1} define a plane in which the x_i axis lies. The y_i and z_i axes in general do not lie in this plane, though in principle either may in special cases. Finally, a unit vector w_{i+1}, directed along the positive z_{i+1} axis, has components given in Equation (11.9). Note that w_{i+1} is directed into the plane of the figure. The vector w_{i+1} lies in the y_i, z_i plane.

$$w_{i+1}\left(x_i\right) = 0$$
$$w_{i+1}\left(y_i\right) = \sin \varphi_i$$
$$w_{i+1}\left(z_i\right) = \cos \varphi_i \qquad (11.9)$$

Any vector v with components v_x, v_y and v_z in the $i + 1$ coordinate system then has components in the ith coordinate system v'_x, v'_y and v'_z given by Equation (11.10).

$$v'_x = v_x \cos \Theta_i + v_y \sin \Theta_i$$
$$v'_y = v_x \sin \Theta_i \cos \varphi_i + v_y \left(- \cos \Theta_i \cos \varphi_i\right) + v_z \sin \varphi_i$$
$$v'_z = v_x \sin \Theta_i \sin \varphi_i + v_y \left(- \cos \Theta_i \sin \varphi_i\right) + v_z \left(- \cos \varphi_i\right)$$
$$(11.10)$$

In matrix notation, this result can be stated

$$v' = T_i v \qquad (11.11)$$

The matrix T_i transforms a vector in the $i+1$ system to one in the i system. Explicitly, the matrix is

$$T_i = \begin{pmatrix} \cos \Theta_i & \sin \Theta_i & 0 \\ \sin \Theta_i \cos \varphi_i & -\cos \Theta_i \cos \varphi_i & \sin \varphi_i \\ \sin \Theta_i \sin \varphi_i & -\cos \Theta_i \sin \varphi_i & -\cos \varphi_i \end{pmatrix} \qquad (11.12)$$

The matrix T_i is orthogonal. An orthogonal matrix is one that has the special property that the transpose can be used as either the left or the right inverse. Orthogonal matrices satisfy the relation $E = T_i T_i^T = T_i^T T_i$, where the superscript T denotes the operation transpose and E is the identity matrix. The operation transpose is an exchange of rows and columns, effectively constituting a reflection of entries about the main diagonal. The identity matrix is of order 3, as is T_i, and is

$$E = \begin{pmatrix} 1 & 0 & 0 \\ 0 & 1 & 0 \\ 0 & 0 & 1 \end{pmatrix} \qquad (11.13)$$

The identity matrix has the property that if A is a compatible matrix, then $A = EA = AE$. In the foregoing, the vectors v and v' are expressed as column matrices, sometimes called column vectors or simply vectors.

$$v = \begin{bmatrix} v_x \\ v_y \\ v_z \end{bmatrix} \qquad (11.14)$$

The transpose of a column vector is a row vector. The specific entries in T_i depend on the way the i and $i+1$ coordinate systems are defined.

The scalar product of two bond vectors b_i and b_j can be expressed in terms of the matrices T_i. The trick is to express the bond vector b_j successively in coordinates for bond vectors $j-1$, $j-2$, and so on, until bond vector i is reached. The product is then

$$b_i^T \cdot b_j = b_i^T T_i \dots T_{j-2} T_{j-1} b_j \qquad (11.15)$$

Now recall that the product of bond vectors b_i and b_{i+1} is $b^2 \cos \Theta_i$. This inner vector product can be expressed in terms of T_i as

$$b_i \cdot b_{i+1} = b^2 \begin{pmatrix} 1 & 0 & 0 \end{pmatrix} T_i \begin{pmatrix} 1 \\ 0 \\ 0 \end{pmatrix} \qquad (11.16)$$

since $\cos \theta_i$ appears alone in the entries in T_i only as the $1, 1$ element, and the unit vectors $(1\ 0\ 0)$ and its transpose select that element in the matrix multiplication. Then the inner vector product of vectors b_i and b_j can be expressed as

$$b_i \cdot b_j = b^2 \begin{pmatrix} 1 & 0 & 0 \end{pmatrix} T_i \cdots T_{j-1} \begin{pmatrix} 1 \\ 0 \\ 0 \end{pmatrix} \qquad (11.17)$$

Since the unit vectors act to select the 1,1 element of the product of all the T matrices, this may be rewritten

$$b_i \cdot b_j = b^2 \left(T_i \cdots T_{j-1} \right)_{11} \qquad (11.18)$$

The average over all possible statistical states is required. If in general a potential energy $E(\{b\})$ acts at the various bonds, the required average is

$$T_i \cdots T_{j-1} = \frac{\int T_i \cdots T_{j-1} e^{-E(\{b\})/RT} d\,\{b\}}{\int e^{-E(\{b\})/RT} d\,\{b\}} \qquad (11.19)$$

Such averages are the most general, and often the full generality is not required. When the potential energy at one bond is independent of that at another,

$$\langle T_i \cdots T_{j-1} \rangle = \prod_{k=i}^{j-1} \langle T_k \rangle \qquad (11.20)$$

Though this happy circumstance is of rare occurrence in real molecules, it is nonetheless quite useful in leading to an approximate solution to some problems otherwise amenable to this treatment. For example, if the bond angle θ is fixed and the torsion angle φ is subject to a symmetric potential, requiring $E(\varphi) = E(-\varphi)$, then $\langle \sin \varphi_i \rangle = 0$ and the average transfer matrix becomes

$$\langle T_i \rangle = \begin{pmatrix} \cos \Theta_i & \sin \Theta_i & 0 \\ \sin \Theta_i \langle \cos \varphi_i \rangle & -\cos \Theta_i \langle \cos \varphi_i \rangle & 0 \\ 0 & 0 & -\langle \cos \varphi_i \rangle \end{pmatrix} \qquad (11.21)$$

If rotation about bonds is free, $E(\varphi_i) = 0$, and $\langle \cos \varphi_i \rangle = 0$. The transfer matrix is even simpler.

$$\langle T_i \rangle = \begin{pmatrix} \cos \Theta & \sin \Theta & 0 \\ 0 & 0 & 0 \\ 0 & 0 & 0 \end{pmatrix} \qquad (11.22)$$

if all the bond angles are the same. Now apply these considerations to evaluating the mean squared end to end distance. In any one conformation, the end to end distance L is $\sum_{i=1}^n b_i$. The mean squared end to end distance is given by Equation (10.89), which is quoted again here for clarity.

$$\langle L^2 \rangle = \sum_{i=1}^n \langle b_i \cdot b_i \rangle + 2 \sum_{i<j\leq n} \langle b_i \cdot b_j \rangle \qquad (11.23)$$

The first sum is obviously nb^2. The second sum may be expressed in terms of the transfer matrices.

$$2 \sum_{i<j\leq n} \langle b_i \cdot b_j \rangle = 2 \sum_{i<j\leq n} b^T \langle T \rangle^{j-i} b \qquad (11.24)$$

where b^T is the row vector (b 0 0), and b is a bond distance, all of which are the same. The notation $\langle T \rangle^{j-i} = \langle T_i \rangle \cdots \langle T_{j-i+1} \rangle \langle T_{j-i} \rangle$ since in this approximation all the T_i are the same. The b vectors are not involved in the sum, and may be factored from it. The sum itself may be rewritten

$$\sum_{i<j\leq n} \langle T \rangle^{j-i} = \sum_{k=1}^{n-1} (n-k) \langle T \rangle^k \qquad (11.25)$$

and this allows the mean square end to end distance to be expressed

$$\langle L^2 \rangle = nb^2 + 2b^T \left[\sum_{k=1}^{n-1} (n-k)\langle T \rangle^k \right] b \qquad (11.26)$$

There are two sums involved in Equation (11.26), both finite but very large. The leading sum is

$$\sum_{k=1}^{n-1} \langle T \rangle^k = \langle T \rangle + \langle T \rangle^2 + \cdots + \langle T \rangle^{n-1} = (\langle T \rangle - \langle T \rangle^n)(E - \langle T \rangle)^{-1}$$
$$(11.27)$$

Keep in mind that the $\langle T \rangle$ are matrices, not algebraic quantities, and E is the 3×3 identity matrix. The multiplication indicated is a matrix multiplication. The other sum evaluates as follows. Consider first the finite sum

$$\sum_{k=1}^{n-1} kx^k = x + 2x^2 + 3x^3 + \cdots + (n-1)x^{n-1} \qquad (11.28)$$

This sum evaluates in closed algebraic form by differentiating the finite sum in Equation (11.27) substituting x for the averaged transfer matrix $\langle T \rangle$. The result of the differentiation is

$$1 + 2x + 3x^2 + \cdots + (n-1)x^{n-2} = \frac{1 - nx^{n-1} + (n-1)x^n}{(1-x)^2}$$
$$(11.29)$$

This sum still has to be multiplied by x to bring it to the required form. Then the matrix equivalent of the sum is

$$\sum_{k=1}^{n-1} k\langle T \rangle^k = [\langle T \rangle - n\langle T \rangle^n + (n-1)\langle T \rangle^{n+1}](E - \langle T \rangle)^{-2} \quad (11.30)$$

Substitution of the evaluated sums in Equation (11.26) gives, after a little algebra, the final result

$$\langle L^2 \rangle = nb^2 \left[(E + \langle T \rangle)(E - \langle T \rangle)^{-1} \right.$$
$$\left. - \frac{2\langle T \rangle}{n}(E - \langle T \rangle^n)(E - \langle T \rangle)^{-2} \right]_{11} \qquad (11.31)$$

since in this case only the $1, 1$ element is required. The matrix products need to be evaluated, The powers of $\langle T \rangle$ are straightforward. Consider successively the square and the cube.

$$
\begin{pmatrix} \cos\Theta & \sin\Theta & 0 \\ 0 & 0 & 0 \\ 0 & 0 & 0 \end{pmatrix} \begin{pmatrix} \cos\Theta & \sin\Theta & 0 \\ 0 & 0 & 0 \\ 0 & 0 & 0 \end{pmatrix}
$$

$$
= \begin{pmatrix} \cos^2\Theta & \sin\Theta\cos\Theta & 0 \\ 0 & 0 & 0 \\ 0 & 0 & 0 \end{pmatrix}
$$

$$
\begin{pmatrix} \cos\Theta & \sin\Theta & 0 \\ 0 & 0 & 0 \\ 0 & 0 & 0 \end{pmatrix} \begin{pmatrix} \cos^2\Theta & \sin\Theta\cos\Theta & 0 \\ 0 & 0 & 0 \\ 0 & 0 & 0 \end{pmatrix}
$$

$$
= \begin{pmatrix} \cos^3\Theta & \sin\Theta\cos^2\Theta & 0 \\ 0 & 0 & 0 \\ 0 & 0 & 0 \end{pmatrix}
$$

By induction then

$$
\langle T \rangle^k = \begin{pmatrix} \cos^k\Theta & \sin\Theta\cos^{k-1}\Theta & 0 \\ 0 & 0 & 0 \\ 0 & 0 & 0 \end{pmatrix} \tag{11.32}
$$

Evidently, the $1, 1$ element is $\cos^k\theta$. The inverse of $E - \langle T \rangle$ is a harder matter. The matrix to be inverted is

$$
E - \langle T \rangle = \begin{pmatrix} 1 - \cos\Theta & -\sin\Theta & 0 \\ 0 & 1 & 0 \\ 0 & 0 & 1 \end{pmatrix} \tag{11.33}
$$

The inverse must have the property $AA^{-1} = E$. This leads to a system of nine simultaneous equations that are solved for the nine components a_{ij}^{-1} of the inverse matrix A^{-1}. Forming the product for the case at hand and representing the inverse components as a_{ij}^{-1} for brevity produces

$$\begin{pmatrix} 1-\cos\Theta & -\sin\Theta & 0 \\ 0 & 1 & 0 \\ 0 & 0 & 1 \end{pmatrix} \begin{pmatrix} a_{11}^{-1} & a_{12}^{-1} & a_{13}^{-1} \\ a_{21}^{-1} & a_{22}^{-1} & a_{23}^{-1} \\ a_{31}^{-1} & a_{32}^{-1} & a_{33}^{-1} \end{pmatrix}$$

$$= \begin{pmatrix} (1-\cos\Theta)a_{11}^{-1}-a_{21}^{-1}\sin\Theta & a_{12}^{-1}(1-\cos\Theta)-a_{22}^{-1}\sin\Theta & a_{13}^{-1}(1-\cos\Theta)-a_{23}^{-1}\sin\Theta \\ a_{21}^{-1} & a_{22}^{-1} & a_{23}^{-1} \\ a_{31}^{-1} & a_{32}^{-1} & a_{33}^{-1} \end{pmatrix}$$

It is immediately evident that only the first row is involved, as the second and third are unchanged from the matrix $E - \langle T \rangle$. There follows

$$a_{11}^{-1} = \frac{1}{1-\cos\Theta}$$

$$a_{12}^{-1} = \frac{\sin\Theta}{1-\cos\Theta}$$

$$a_{13}^{-1} = 0 \tag{11.34}$$

The test of whether or not the inverse has been correctly calculated is to multiply together the two matrices $(E - \langle T \rangle)(E - \langle T \rangle)^{-1}$. This must reduce to E.

$$\begin{pmatrix} 1-\cos\Theta & -\sin\Theta & 0 \\ 0 & 1 & 0 \\ 0 & 0 & 1 \end{pmatrix} \begin{pmatrix} \dfrac{1}{1-\cos\Theta} & \dfrac{\sin\Theta}{1-\cos\Theta} & 0 \\ 0 & 1 & 0 \\ 0 & 0 & 1 \end{pmatrix}$$

$$= \begin{pmatrix} 1 & 0 & 0 \\ 0 & 1 & 0 \\ 0 & 0 & 1 \end{pmatrix} \tag{11.35}$$

The matrix Equation (11.30) may now be properly evaluated. Extracting the 1,1 element from the products indicated in this equation gives the familiar result

$$\langle L^2 \rangle = nb^2 \left[\frac{1 + \cos \Theta}{1 - \cos \Theta} - \frac{2 \cos \Theta}{n} \frac{(1 - \cos^n \Theta)}{(1 - \cos \Theta)^2} \right] \tag{11.36}$$

At large n this reduces to Taylor's result, given in Equation (10.95). If rotation about the bonds is hindered by identical potentials, the average $\langle \cos \varphi \rangle$ must be included in the transfer matrix. The appropriate matrix is

$$\langle T \rangle = \begin{pmatrix} \cos \Theta & \sin \Theta & 0 \\ \sin \Theta \langle \cos \varphi \rangle & - \cos \Theta \langle \cos \varphi \rangle & 0 \\ 0 & 0 & -\langle \cos \varphi \rangle \end{pmatrix} \tag{11.37}$$

Replacing the transfer matrix in Equation (11.30) and following the algebra through produces Taylor's approximate result, also obtained earlier by the Japanese worker Oka

$$\langle L^2 \rangle = nb^2 \frac{1 + \cos \Theta}{1 - \cos \Theta} \frac{1 + \langle \cos \varphi \rangle}{1 - \langle \cos \varphi \rangle} \tag{11.38}$$

11.2. Phase Transitions in Polymeric Structures

Polymeric structures are capable of a number of phase transitions that monomeric matter is not able to display. Some of these are given in Table 11.1, together with a few that are also undergone by monomeric matter.

Table 11.1.   Classes of polymeric phase transitions.

1. Helix to random coil transition
2. Linear polymer threading a small hole in a membrane
3. Adsorption onto a surface
4. Collapse transition, DNA and protein folding
5. Polymer-polymer and polymer-solvent liquids
6. Liquid crystals and plastic crystals
7. Glasses and Gels
8. Crystallization
9. Equilibrium polymerization in dilute solution
10. Catch all category: block co-polymers, branching, micelles, membranes, vesicles, soaps, etc.

The first five listed are specific to isolated linear chains, and are not possible in monomeric matter. The remaining five require multiple chains. Table 1.1 is based on an historical description of statistical mechanics applied to polymer chains by DiMarzio.[4] To illustrate one approach to the single chain phase transition, the first listed problem, the helix to random coil transition, is treated here. The original work was reported by Gibbs and DiMarzio,[5] who used a method different from that given here.

The early literature concerning the helix to random coil transition is reviewed in a paper by Reiss, McQuarrie, McTague and Cohen.[6] The helix random coil transition has long been a principal problem of chain dynamics, and the approach used here[7] is considerably different from that used by the first solution of the problem. An experimental realization suitable for use by careful undergraduate physical chemistry laboratories uses poly-γ-benzyl-l-glutamate dissolved in a mixed solvent of dichloracetic acid and 1,2-dichloroethane or, better, 1,3-dichloropropane, which gives more reproducible results. The experiment is described in detail in the laboratory manual by Garland, Nibler and Shoemaker.[8]

It is assumed that the chain has already at least one helical region. The details concerning the cause of helix formation are ignored. Assign a number φ_k to each segment in the linear chain. Let this number be 0 if the segment is in a random coil, and let it be 1 if it is in a helix. For purposes of the theory, the chain configuration is completely specified by giving all of the φ_k in sequence, e.g. $\varphi_1 \, \varphi_2 \, \varphi_3 \, \varphi_4 \, \varphi_5 \, \varphi_6 \, \varphi_7 \, \varphi_8 \, \varphi_9 \cdots \varphi_N = \{\varphi\}$ for a chain of N segments. The potential energy $V\{\varphi\}$ depends on the chain configuration. The probabilities P are given as $P(\varphi) \sim e^{-V\{\varphi\}/kT}$, where the potential consists of two parts

$$V\{\varphi\} = \sum_{i=1}^{N} v(\varphi_i) + \sum_{i=1}^{N-1} v(\varphi_i \varphi_{i+1}) \tag{11.39}$$

for N very large. This form of the potential allows the state of the nearest neighbor segment (φ_{i+1}) to contribute to the potential energy, but not the next nearest neighbor (φ_{i+2}). With this selection, the probability will contain factors of the form $e^{-v(\varphi_i)/k_B T}$

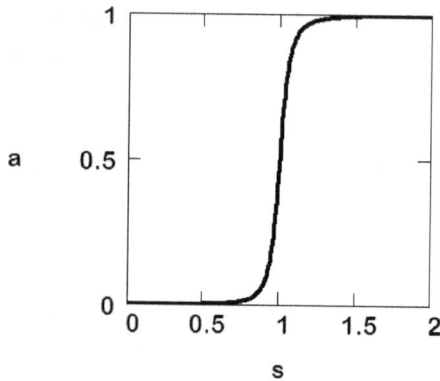

Fig. 11.5.   The fraction of segments in helical form as a function of s.

and $e^{-v(\varphi_i,\varphi_{i+1})/k_B T}$. Now, let a be the fraction of segments in helical form. When regarded as a function of temperature or pressure, it is expected that a will display a typical titration curve, with the sharpness of the transition defined essentially by the configurational entropy decrease that accompanies the formation of the first loop of the helix. The larger the decrease, the more difficult it is to form a turn of the helix, but the sharper the transition becomes. Once a helix turn is formed, however, it acts as a nucleus to stabilize the formation of more turns of the same helix. The expected form of the titration curve is shown in Figure 11.5.

Define $e^{-v(0)/k_B T} = 1$ and $e^{-v(1)/k_B T} = s$. Then the potential $v(0) = 0$ and $v(1) = -k_B T \ln(s)$. If $s > 1$, the helical form is favored, while if $s < 1$, random coil is favored. Note that $v(1)$ is the energy change that accompanies a transition from random coil to helix form. There are four possibilities for assigning values to $v(\varphi_i \, \varphi_{i+1})$. These are

$v(\varphi_i\varphi_{i+1})$	$v(0,0) = 0$	no connectivity effect for a random coil to random coil transition
	$v(1,1) = 0$	no connectivity effect for a helix to helix transition
	$v(1,0) = 0$	no connectivity effect for a helix to random coil transition
	$v(0,1) = -k_B T \ln(\sigma)$	connectivity effect for a random coil to helix transition

When σ is very small (0.0001, say) the random coil to helix transition is relatively disfavored. A segment cannot be in a helix all by itself. There must be some number of segments already laid in helical form. Thus, $v(0,1)$ requires some previous segments to already be in helical form. When σ is large, the random coil to helix transition is favored, leading to longer and fewer helices. When s is very small, connectivity effects are overcome, leading to very few and very short helices. The residue of a few short helices at very small $s \cong 0$ (corresponding to very large temperature) is analogous to the survival of a few holes, or imperfections, in crystals at low temperatures. In both cases, the entropy decrease associated with disappearance of the last helix, or the last hole, is insurmountably large.

With these definitions, the next task is to calculate the partition function, Z. Since the partition function is a sum over all possible states, it is

$$Z = \sum_{\varphi_1=0}^{1} \sum_{\varphi_2=0}^{1} \sum_{\varphi_3=0}^{1} \cdots \sum_{\varphi_N=0}^{1} e^{-\sum_{i=1}^{N}[v(\varphi_i)+v(\varphi_i\varphi_{i+1})]/k_BT}$$

(11.40)

In Equation (11.40), the sum in the argument of the exponential has to be interpreted as running from 1 to N for $v(\varphi_i)$ and from 1 to $N-1$ for $v(\varphi_i, \varphi_{i+1})$. Some study of the multiple sum suggests forming the quantity $U(\varphi_i, \varphi_{i+1})$, defined as

$$U(\varphi_i, \varphi_{i+1}) = e^{-[v(\varphi_i)+v(\varphi_i,\varphi_{i+1})]/k_BT}$$

(11.41)

There is a formal resemblance to the Ising lattice problem, suggesting that the formalism developed for the solution of that problem might be applicable here. One of these methods is the transfer matrix. Given the definition of $U(\varphi_i, \varphi_{i+1})$ in Equation (11.41), and defining $U(\varphi_N, \varphi_{N+1})$ to be unity, the partition function can be expressed

$$Z = \sum_{\varphi_1=0}^{1} \sum_{\varphi_2=0}^{1} \cdots \sum_{\varphi_N=0}^{1} \prod_{i=1}^{N} U(\varphi_i, \varphi_{i+1})$$

(11.42)

Now consider part of the sums, The sums are all finite (each contains only two terms), and the order in which the sums are written does not alter the value of the partition function. Thus, the first three sums can be written

$$\sum_{\varphi_1=0}^{1} \sum_{\varphi_3=0}^{1} \sum_{\varphi_2=0}^{1} U(\varphi_1, \varphi_2) U(\varphi_2, \varphi_3)$$

$$= \sum_{\varphi_1=0}^{1} \sum_{\varphi_3=0}^{1} \left(\sum_{\varphi_2=0}^{1} U(\varphi_1, \varphi_2) U(\varphi_2, \varphi_3) \right) \qquad (11.43)$$

These matrices are all the same, and the quantity in large parentheses is the matrix product of a U matrix with itself, leading to the result

$$\sum_{\varphi_1=0}^{1} \sum_{\varphi_3=0}^{1} \sum_{\varphi_2=0}^{1} U(\varphi_1, \varphi_2) U(\varphi_2, \varphi_3) = \sum_{\varphi_1=0}^{1} \sum_{\varphi_3=0}^{1} U^2(\varphi_1, \varphi_3)$$

$$(11.44)$$

The process is repeated until at length the last matrix is reached. The result is

$$Z = \sum_{\varphi_1=0}^{1} \sum_{\varphi_N=0}^{1} U^N(\varphi_1, \varphi_N) \qquad (11.45)$$

This simplification suggests that the detailed nature of the elements of the U matrices be considered. Evidently, from the definitions introduced above,

$$U(0,0) = e^{-[v(o)+v(0,0)]/k_B T} = e^0 = 1$$
$$U(0,1) = e^{-[v(0)+v(0,1)]/k_B T} = e^{\ln \sigma} = \sigma$$
$$U(1,0) = e^{-[v(1)+v(1,0)]/k_B T} = e^{\ln s} = s$$
$$U(1,1) = e^{-[v(1)+v(1,1)]/k_B T} = e^{\ln s} = s \qquad (11.46)$$

The entries in the U matrix refer to the various possible transitions, such as from random coil to helix (represented by $U_{01} = \sigma$). The interpretation of the other entries is

$$\begin{pmatrix} U_{00} & U_{01} \\ U_{10} & U_{11} \end{pmatrix} = \begin{pmatrix} random\ to\ random & random\ to\ helix \\ helix\ to\ random & helix\ to\ helix \end{pmatrix} = \begin{pmatrix} 1 & \sigma \\ s & s \end{pmatrix} \tag{11.47}$$

Rigorously, the partition function is given by Equation (11.45). However, statistically it does not matter what happens to a single segment in a chain of N segments, when N is very large. The task of performing the sum indicated in Equation (11.45) can be eased by insisting that the first and the last segment be in the same state, that is, that both are in random coil, or both are in helices, rejecting the 01 and 10 transitions for these segments. With this simplification, the partition function becomes

$$Z \cong \sum_{\varphi=0}^{1} U^N(\varphi, \varphi) = Tr(U^N) \tag{11.48}$$

The trace is invariant under a similarity transformation. Such a transformation diagonalizes the matrix U, displaying its spectrum along the main diagonal. Suppose these roots are Λ_1 and Λ_2. The corresponding U^N is

$$U^N = \begin{pmatrix} \Lambda_1^N & 0 \\ 0 & \Lambda_2^N \end{pmatrix} \tag{11.49}$$

The trace of this matrix is $\Lambda_1^N + \Lambda_2^N$. One of these roots is, in general, larger than the other. Let the larger root be Λ_1. Factoring the larger root from the trace yields

$$Z \cong \Lambda_1^N \left(1 + \left(\frac{\Lambda_2}{\Lambda_1} \right)^N \right) \tag{11.50}$$

Since Λ_1 is greater than Λ_2, the ratio Λ_2/Λ_1 is necessarily less than one. But N is large, and raising a number less than one to a very large power produces a negligibly small, number. In this case, the partition function is essentially the larger root raised to a very large power,

$$Z \cong \Lambda_1^N \tag{11.51}$$

To extract the roots of the matrix U^N, form the determinant

$$\begin{vmatrix} 1 - \Lambda & \sigma \\ s & s - \Lambda \end{vmatrix} = (1 - \Lambda)(s - \Lambda) - s\sigma = 0 \tag{11.52}$$

This expands to an equation quadratic in Λ, namely $\Lambda^2 - (1 + s)$ $\Lambda + s(1 - \sigma) = 0$ which has the larger root $\Lambda_1 = \frac{1}{2}[(1 + s) + \sqrt{(1 + s)^2 - 4s(1 - \sigma)}]$. The smaller root uses a minus sign before the radical. The partition function is then explicitly

$$Z \cong \frac{1}{2^N} (1 + s)^N \left[1 + \sqrt{1 - \frac{4s(1 - \sigma)}{(1 + s)^2}} \right]^N \tag{11.53}$$

When the connectivity effect is so small that a segment in a random coil is never converted to one in a helix, the parameter σ essentially vanishes. If D_1 and D_2 are taken as the roots when σ is zero, the determinant that determines the roots D_1 and D_2 is

$$\begin{vmatrix} 1 - D_1 & 0 \\ s & s - D_2 \end{vmatrix} = (1 - D_1)(s - D_2) = 0 \tag{11.54}$$

Thus, when σ vanishes, $D_1 = 1$ and $D_2 = s$.

The reader will remember that $e^{-v(1)/k_B T} = s$ and $e^{-v(0)/k_B T} = 1$. In order to form averages, the probability $P = C e^{-V\{\varphi\}/k_B T}$ is required, where $V\{\varphi\}$ is defined by Equation (11.39). Suppose that in one configuration of the chain, there are exactly k segments in helical conformation. In this case, the probability will contain k identical factors s, and $N - k$ factors 1, allowing the probability for this specification of k to be written

$$P \sim s^k e^{-v(\varphi_i, \varphi_{i+1})/k_B T} \tag{11.55}$$

while the partition function in these terms is

$$Z = \sum s^k e^{-v(\varphi_i, \varphi_{i+1})/k_B T} \tag{11.56}$$

The unspecified sum is over all possible values of k. Now, the average number of segments in helical form is given by

$$\langle k \rangle = \sum kP \tag{11.57}$$

Evidently,

$$\langle k \rangle = \frac{\sum k s^k e^{-v(\varphi_i, \varphi_{i+1})/k_B T}}{\sum s^k e^{-v(\varphi_i, \varphi_{i+1})/k_B T}} \tag{11.58}$$

Consider now the derivative of $\ln(Z)$ with respect to s. A few moments of calculation will show convincingly that

$$\langle k \rangle = \frac{\partial \ln (Z)}{\partial \ln(s)} \tag{11.59}$$

since $\frac{\partial \ln(Z)}{\partial \ln(s)} = s \frac{\partial \ln(Z)}{\partial s} = s \frac{\sum k s^{k-1} e^{-v(\varphi_i, \varphi_{i+1})/k_B T}}{\sum s^k e^{-v(\varphi_i, \varphi_{i+1})/k_B T}}$. Almost sublime in its simplicity, Equation (11.59) is an important relation that allows calculation of the average number of helical segments, given the explicit functional form of the partition function, obtained for this problem in Equation (11.53).

Suppose now that in one configuration there is exactly one helix. Then the probability will contain a factor σ, while if there are two helices in another configuration, the probability will contain a factor σ^2, and so on. If in general a configuration contains n helices, the probability for that configuration will contain a factor σ^n, and in general the probability is proportional to $s^k \sigma^n$. The partition function is given by

$$Z = \sum_{k,n} s^k \sigma^n \tag{11.60}$$

and the average number of helices is calculated from

$$\langle n \rangle = \frac{\partial \ln(Z)}{\partial \ln(\sigma)} \tag{11.61}$$

as the algebra is the same as that used to demonstrate Equation (11.59). With these fundamental relations and the approximate partition function of Equation (11.53), specific values for the mean number of segments in helical form and the mean number of helices may be calculated. Given the mean number of segments in helical form, the fraction of segments $a = \langle k \rangle / N$ can be calculated. The results of the differentiations are

$$\langle k \rangle = N \left[\frac{s}{1+s} - \frac{2s\,(1-s)\,(1-\sigma)}{(1+s)^3\,(1+\sqrt{1-F})\,(\sqrt{1-F})} \right] \qquad (11.62)$$

and

$$\langle n \rangle = N \frac{2s\sigma}{(1+s)^2(1+\sqrt{1-F})(\sqrt{1-F})} \qquad (11.63)$$

where

$$F = \frac{4s\,(1-\sigma)}{(1+s)^2} \qquad (11.64)$$

The logarithm of the partition function is

$$\ln(Z) = N \ln\,(1+s) + N \ln \left(1 + \sqrt{1-F} \right) - N \ln 2 \qquad (11.65)$$

Equation (11.65) is needed in the calculation of the configurational entropy of the condensing chain. Since the fundamental definition of entropy is

$$S = -k_B \sum_{k,n} P\,(k,n) \ln P\,(k,n) \qquad (11.66)$$

where k_B is Boltzmann's constant and the probability and partition function may be considered as

$$P\,(k,n) = \frac{s^k \sigma^n}{Z} \qquad (11.67)$$

$$Z = \sum_{k,n} s^k \sigma^n \qquad (11.68)$$

Substituting these two equations into Equation (11.66) yields

$$S = -k_B \left[\langle k \rangle \ln s + \langle n \rangle \ln \sigma - \ln Z \right] \qquad (11.69)$$

The mean number of helices relative to the number of segments $\langle n \rangle / N$, is given as a function of s in Figure 11.6. The configurational

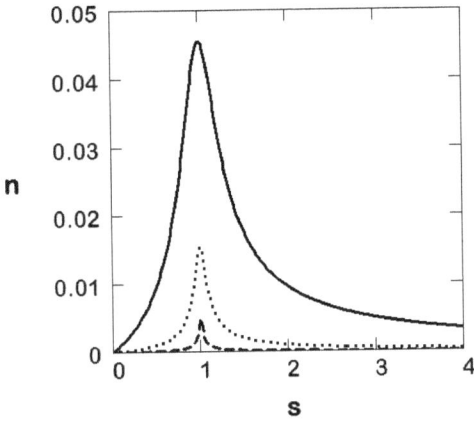

Fig. 11.6. The mean number of helices $n = \langle n \rangle / N$ as a function of s for three values of the parameter σ. Solid line, $\sigma = 0.01$. Dotted line, $\sigma = 0.001$. Dashed line, $\sigma = 0.0001$.

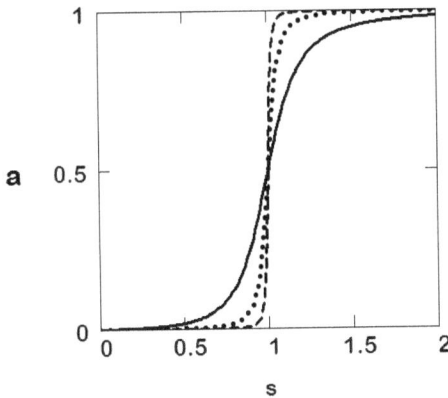

Fig. 11.7. The fraction of segments in helical form $a = \frac{\langle k \rangle}{N}$ as a function of s for three values of the parameter σ. Solid line, $\sigma = 0.01$. Dotted line, $\sigma = 0.001$. Dashed line, $\sigma = 0.0001$.

entropy of the chain relative to the number of segments and Boltzmann's constant (S/Nk_B) is displayed in Figure 11.7. Note that $\langle n \rangle / N$ approaches zero at small s and $1/N$ at large s. There are no helices at $s = 0$, and there is only one very long helix at large s.

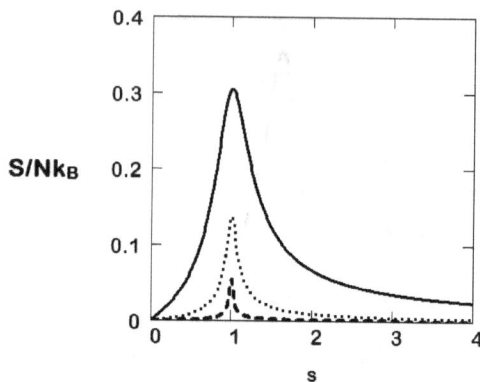

Fig. 11.8.   Configurational entropy for the chain relative to Nk_B as a function of s for three values of the parameter σ. Solid line, $\sigma = 0.01$. Dotted line, $\sigma = 0.001$. Dashed line, $\sigma = 0.0001$.

11.3. Linear Polymer Threading a Small Hole
in a Membrane

The problem of a linear polymer threading a hole in a membrane was first successfully treated by DiMarzio and Mandell.[9] This introductory solution has been followed by an active literature, and well over 80 papers have since appeared considering some aspect of this fundamental problem. The introduction presented here will consider only the case of a Gaussian chain penetrating the membrane, but neither anchored to it nor adsorbed on it, and will ignore the refinement occasioned by the fact that the number of conformations available to the parts of the chain on either side of the hole must be less than the number available to a free chain of the same length, since the interior segment passing through the hole is constrained by the hole and is thus not free. In contrast, both ends of the chain are completely free.

Let there be n segments in the chain, and consider the volume occupied by the polymer as divided into regions 1 and 2 by the membrane, which itself stretches indefinitely in either of two orthogonal directions to the extremities of the volume, itself indefinitely large. The problem considered is sketched in Figure 11.9. The hole is so small that only one segment at a time can pass through it.

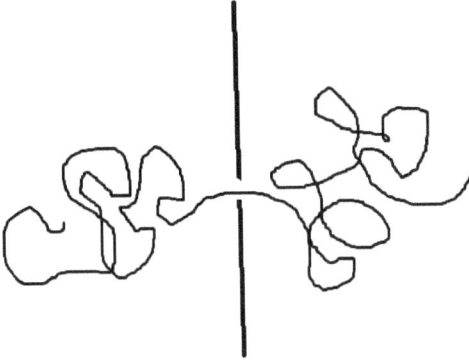

Fig. 11.9. Gaussian polymer threading a hole in a membrane. Regions 1 and 2 are on either side of the membrane.

Now consider the probability. If a segment is in region 1, the probability will contain a factor $e^{-v(1)/k_B T}$, where $v(1) = -k_B T \ln(x_1)$, and x_1 is the probability the segment is in region 1. If a second segment is in region 1, the probability will contain a factor x_1^2, and so on, until the total number, k, of segments in region 1 is achieved. Similarly, if a segment is in region 2, the probability will contain a factor $e^{-v(2)/k_B T}$, where $v(2) = -k_B T \ln(x_2)$, and x_2 is the probability the segment is in region 2. The same logic used for region 1 applies in region 2, and the probability will contain a factor x_2^m, where m is the number of segments in region 2. The partition function $Z(n)$ is then

$$Z(n) = \sum_{k=0}^{n-1} \sum_{m=0}^{n-1} x_1^k x_2^m \qquad (11.70)$$

subject to the restriction that $k + m = n - 1$. The sums must start with k or m equal to zero, in order to ensure the chain completely translocates across the membrane, and ends up entirely in region 1 or region 2, depending on which way it is going. The sums evaluate easily, once the algebraic trick is realized. Start by writing the trivial case $n = 2$. The sum is $x_1 + x_2$. Obviously $x_1 + x_2 = \frac{x_1^2 - x_2^2}{x_1 - x_2}$. Now consider $n = 3$. For this case, the partition sum is

$$x_1^2 + x_1 x_2 + x_2^2 = \frac{x_1^3 - x_2^3}{x_1 - x_2} \qquad (11.71)$$

which is easy to establish using synthetic division. Proceeding by induction, then, the partition function becomes

$$Z(n) = \frac{x_1^n - x_2^n}{x_1 - x_2} \tag{11.72}$$

The equivalent of Equation (11.59) gives the average number of segments in region 1, $\langle k \rangle$, as

$$\langle k \rangle = \frac{\partial \ln Z}{\partial \ln x_1} \tag{11.73}$$

Carrying out the differentiation gives the preliminary result

$$\langle k \rangle = \frac{n x_1^n}{x_1^n - x_2^n} - \frac{x_1}{x_1 - x_2} \tag{11.74}$$

Now set the ratio $r = \frac{x_2}{x_1}$. The average fraction of segments in region 1 can be expressed

$$f(r) = \frac{\langle k \rangle}{n} = \frac{1}{1 - r^n} - \frac{1}{n(1 - r)} \tag{11.75}$$

This fraction is shown as a function of r for values of $n = 20$, 50 and 100 in Figure 11.10. The longer the chain, the sharper the transition. Note that the fraction in Equation (11.75) does not diverge as r approaches 1. It is easy to show that the limiting value as r approaches 1 indefinitely closely is $\frac{1}{2}(1 - \frac{1}{n})$. The longer the chain, the more closely $1/2$ is approached as r approaches 1.

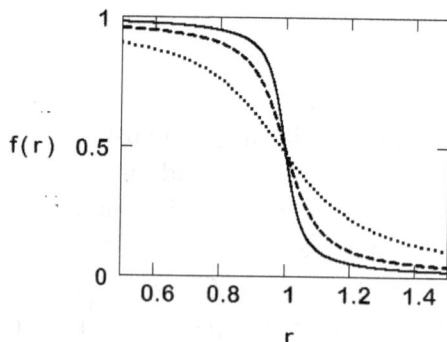

Fig. 11.10.   Fraction of segments in region 1 as a function of the ratio r. Dotted line, $n = 20$. Dashed line, $n = 50$. Solid line, $n = 100$.

The solution to the problem is an exact solution of an approximate problem statement. Its position in the theory of a linear polymer threading a hole is analogous to the position of the ideal gas law in the theory of real gases. The solution approaches a first order phase transition arbitrarily closely as the length of the Gaussian chain grows boundlessly. A first order phase transition has discontinuous thermodynamic variables and their derivatives at the phase transition. Refinements include consideration of the effect of constraining an interior segment of the chain to the position of the hole, constraining the end or ends of the polymer by adsorption onto the membrane near the hole, inclusion of the effects of the volume excluded to the chain segments by the impenetrability of the chain by itself, inclusion of the effects of osmotic swelling of the chain by the solvent in which it is suspended, and consideration of the problem of threading of two or more successive holes in membranes located a distance d apart. The effect of introducing the length of the hole (for a membrane of finite thickness) into the problem is analogous.

In this chapter the transfer matrix was introduced by solving a problem solved in the last chapter, then used to treat the first of the single chain polymer phase transition problems, the helix to random coil transition. Finally, another phase transition of a single polymer chain was treated in first approximation, the threading of a hole in a membrane.

References

1. H. A. Kramers and G. H. Wannier, Phys. Rev. **60**, 252 (1941); **60**, 263 (1941).
2. E. W. Montroll, J. Chem. Phys. **9**, 706 (1941).
3. E. N. Lassettre and J. P. Howe, J. Chem. Phys. **9**, 747 (1941); **9**, 801 (1941).
4. E. A. DiMarzio, J. Polym. Sci. Part B Polym. Phys. **37**, 3 (1999).
5. J. H. Gibbs and E. A. Dimarzio, J. Chem. Phys. **30**, 271 (1959).
6. H. Reiss, D. A. McQuarrie, J. P. McTague and E. R. Cohen, J. Chem. Phys. **44**, 4567 (1966).
7. M. Fixman, class notes.
8. C. W. Garland, J. W. Nibler and D. P. Shoemaker, *Experiments in Physical Chemistry-Seventh Edition*, McGraw Hill, New York, 2003, p. 228.
9. E. A. DiMarzio and A. J. Mandell, J. Chem. Phys. **107**, 5510 (1997).

Chapter 12

Excluded Volume
of Polymer Chains

12.1. Introduction

The segments of a freely jointed chain can be regarded as recording the steps in a completely random walk. In this walk, the chain may cross and even superpose segments, something that cannot occur in a real polymer chain. The result is that even in the most dilute solution, a real polymer chain cannot penetrate itself, but must avoid crossing at segment contacts. The effect is to swell the chain to larger dimensions than those of the equivalent freely jointed chain. While this may be accommodated by simply increasing the length of the statistical segment, the question immediately arises, by how much? The swelling of the chain is real, and can be detected by any of a variety of physical measurements. For example, light scattering or low angle X-ray scattering experiments lead to an experimental determination of the mean square gyration radius. The osmotic second virial coefficient is affected by excluded volume, and the friction constant measured by sedimentation provides another measure. The kinetic theory of gases shows how the viscosity can be used as a molecular micrometer, yielding mean square molecular diameters of small molecules such as nitrogen or carbon dioxide, and the intrinsic viscosity of polymer solutions provides a similar measure of the size of polymer molecules in solution. An exact theory of polymer solution viscosity is probably as impossible to give as an exact theory of the excluded volume effect. Nonetheless, the measurements are deceptively easy to make and viscosity is often used as a means of characterizing polymers. The main theoretical details are as follows.

In 1906, Einstein showed that for hard spheres suspended in a solvent of viscosity η_0, the viscosity η of the mixture could be expressed[1]

$$\eta = \eta_0 \left(1 + \frac{5}{2}\varphi \right) \tag{12.1}$$

where φ is the volume fraction occupied by hard spheres. In solution work, the intrinsic viscosity is defined as

$$[\eta] = \lim_{c \to 0} \frac{\eta - \eta_0}{c} \tag{12.2}$$

It is not hard to show that

$$\eta \propto \frac{L^3}{M} \tag{12.3}$$

where $L^3 = \langle L^2 \rangle^{3/2}$ and M is the molecular weight of the polymer. If there are n monomers of molecular weight M_0 each, $M = nM_0$. For $\langle L^2 \rangle = nb^2$, where b is the mean segment length, then

$$[\eta] \propto \frac{n^{3/2}b^3}{nM_0} \propto n^{1/2} \tag{12.4}$$

Thus, in the absence of excluded volume effects, it would be expected that

$$[\eta] \propto M^{1/2} \tag{12.5}$$

In osmometry, the second virial coefficient may vanish under certain conditions, and when this happens, the solvent is spoken of as the theta solvent. In the theta solvent, the intrinsic viscosity is indeed proportional to the square root of the molecular weight of the polymer. Otherwise, the intrinsic viscosity varies at a somewhat larger power than $1/2$, usually about 0.3 or so greater in a good solvent. One of the objectives of a theory of excluded volume is to provide a theoretical basis for powers higher than 0.5 in the dependence of the intrinsic viscosity on molecular weight.

The theory for excluded volume starts from that for the freely jointed chain, which it takes as a limiting law that incorporates the bulk of the chain behavior. The excluded volume is treated as a small perturbation. As a bonus, it provides an exact series expansion for both $\langle L^2 \rangle / \langle L_0^2 \rangle$ and $\langle R^2 \rangle / \langle R_0^2 \rangle$, where $\langle L_0^2 \rangle = nb^2$, is the mean square end to end distance of the freely jointed chain of the same number n of segments as $\langle L^2 \rangle$, and R represents the gyration radius similarly defined. The coefficients of the expansion parameter z (defined later) are known exactly for the linear and quadratic terms. The results are

$$\langle L^2 \rangle = \langle L_0^2 \rangle \left[1 + \frac{4}{3}z + \left(\frac{28\pi}{27} - \frac{16}{3} \right) z^2 + \cdots \right] \tag{12.6}$$

$$\langle R^2 \rangle = \langle R_0^2 \rangle \left[1 + \frac{134}{105}z + \left(\frac{1247}{1296}\pi - \frac{536}{105} \right) z^2 + \cdots \right] \tag{12.7}$$

In the following, the coefficient of the linear term in the end to end distance is derived. Derivation of the quadratic term is non-trivial, and will not be undertaken in these introductory notes. In later paragraphs, the gyration radius is treated in fair approximation, but the exact ratio $134/105$ is left in the literature.

12.2. End to End Distance

For the freely jointed chain composed of bond vectors $\boldsymbol{b}_i$ each of length b, the probability is

$$P\{b\}d\{b\} = \prod_{i=1}^{n} p(\boldsymbol{b}_i)d(\boldsymbol{b}_i) \tag{12.8}$$

where $\{b\}$ represents the entire set of vectors $\boldsymbol{b}_i$, and

$$p(\boldsymbol{b}_i)d(\boldsymbol{b}_i) \propto e^{-3b_i^2/2b^2} \tag{12.9}$$

The approach is to regard the probability for the Gaussian chain as

$$P\{b\} \propto e^{-V_0\{b\}/k_B T} \tag{12.10}$$

In this equation, the potential energy V_0 for chain conformation is

$$V_0\{b\} = \frac{3k_BT}{2b^2} \sum_{i=1}^{n} b_i^2 \qquad (12.11)$$

So far, this is just a way of rewriting the Gaussian probability. When excluded volume effects are to be included, a term is simply added to the potential to account for the intersegment repulsive force. Thus

$$V = V_0 + V_{ex} \qquad (12.12)$$

The probability including excluded volume effects is then proportional to the exponential e^{-V/k_BT}.

The nature of V_{ex} is impossible to describe exactly, but it is possible to say some useful qualitative things about it. First, interactions are expected to be pairwise, and the potential energy should rise rapidly to indefinitely large values as segments approach one another. At large distances between interacting segments, the potential energy should drop to vanishing. However, owing to the very large number of possible chain configurations, it is not possible to remark usefully on the form of the potential energy at intermediate distances.

Let the pairwise interaction between segments i and j be denoted v_{ij}, where $v_{ij} = v(r_{ij})$ and r_{ij} is the distance between segments i and j. The potential energy for excluded volume V_{ex} is then

$$V_{ex} = \sum_{i<j} v_{ij} \qquad (12.13)$$

Each v_{ij} has repulsive behavior in the short range, vanishing at long ranges, and of indeterminate character at intermediate ranges. Including the excluded volume potential, the mean squared end to end distance is

$$\langle L^2 \rangle = \frac{\int L^2 e^{-\frac{V\{b\}+V_{ex}}{k_BT}} d\{b\}}{\int e^{-\frac{V\{b\}+V_{ex}}{k_BT}} d\{b\}} \qquad (12.14)$$

This has never been evaluated for the general case. Several useful approximations suggest themselves.

1. Perturbation theory, in the limit of sufficiently small V_{ex}.

2. Statistical approximation of all possible chain conformations, leading to a smoothed segment density.

The first approach will be sketched here, following Fixman.[2] Suppose V_{ex} is everywhere small, without assuming any particular v_{ij} is small. Indeed, at small r_{ij}, v_{ij} must become large, otherwise there would be no need for the theory. The development that follows is familiar from the Ursell-Mayer development of the imperfect gas theory, but does not follow the same evolution. The crux of the matter is to define f_{ij} as

$$f_{ij} = 1 - e^{-v(r_{ij})/k_B T} \tag{12.15}$$

for then $1 - f_{ij} = e^{-v(r_{ij})/k_B T}$, and

$$e^{-V_{ex}/k_B T} = e^{-\sum_{i<j} v(r_{ij})/k_B T} = \prod_{i<j}(1 - f_{ij})$$

$$= 1 - \sum_{i<j} f_{ij} + \sum_{i<j}\sum_{k<l} f_{ij} f_{kl} + \cdots \tag{12.16}$$

Before proceeding, it is time to think about what is being addressed. How does it differ from the Ursell-Mayer gas problem? A polymer chain in solution has, within the confines of the chain volume, a segment density, equivalent to the gas density in volume V. The difference is that the gas density may increase to liquid proportions, while the polymer, remaining in solution, has only a relatively small segment density range, near 1 percent or so. In these circumstances, the chain behavior will dominantly be that of a Gaussian free flight chain, modified slightly by an occasional intersegment interaction. Indeed, for most chain conformations, two segment-segment interactions at different places on the chain will be relatively rare, and contribute little to the average chain properties. Thus, it should suffice to accommodate at most one intersegment interaction in the excluded volume contribution to chain properties. To the extent that this is so, simply truncate the series expansion at the linear term and write

$$e^{-V_0/k_B T} \cong 1 - \sum_{i<j} f_{ij} \tag{12.17}$$

Of course, if denser segment concentrations are encountered, so that ternary or double binary encounters become appreciable, Mayer's method of counting could be invoked. However, the polymer probably would not dissolve if segment density became so large.

To progress, it is convenient to define two different probabilities. Let

$F\{b\}$ be the probability distribution function for the chain with excluded volume.

$P\{b\}$ be the probability distribution function for the Gaussian chain.

Thus

$$F(b) = \frac{e^{-V_{ex}/k_B T} P\{b\}}{Z} \qquad (12.18)$$

where

$$Z = \int e^{-V_{ex}/k_B T} P\{b\} d\{b\} \qquad (12.19)$$

The normalizing constant Z evaluates as

$$Z = \int \prod_{i<j} (1 - f_{ij}) P\{b\} d\{b\} \cong 1 - \sum_{i<j} \int f_{ij} P\{b\} d\{b\}$$

$$\cong 1 - \sum_{i<j} f_{ij} p(r_{ij}) dr_{ij} \qquad (12.20)$$

Here

$$p(r_{ij}) = \left(\frac{3}{2\pi |i - j| b^2} \right)^{3/2} e^{-3r_{ij}^2 / 2|i-j| b^2} \qquad (12.21)$$

under the restriction that $|i - j| \gg 1$, applicable to very long chains. In the final result, only large $|i - j|$ is important, so for simplicity the restriction is introduced here. Note that $p(r_{ij})$ is Gaussian. This means $p(r_{ij})$ drops off rapidly with increasing r_{ij}. Generally, the mean square intersegment distance r_{ij} is expected to be

$$r_{ij}^2 = |i - j|b^2 \tag{12.22}$$

This states that on average not many intersegment interactions are expected. For the few that do occur, $r_{ij}^2 \ll |i - j|b^2$. For these few, define

$$[P(r_{ij})]_{r_{ij} \to 0} \equiv P(O_{ij}) \tag{12.23}$$

In terms of this infrequent probability the normalization constant Z becomes

$$Z \cong 1 - \sum_{i<j} \int f_{ij} dr_{ij} P(O_{ij}) \cong 1 - X \sum_{i<j} P(O_{ij}) \tag{12.24}$$

where

$$X = \int f_{ij} dr_{ij} = \int (1 - e^{-v(r_{ij})/k_B T}) dr_{ij} \tag{12.25}$$

For example, if the segments were all hard spheres of radius s, the distance of closest approach D would be $2s$ and the potential $v(r_{ij})$ would be infinite from zero to D and zero from D to infinity. The coefficient X becomes in this extreme case

$$X = \int_0^{2\pi} \int_0^{\pi} \int_0^{\infty} (1 - e^{-v(r_{ij})/k_B T}) r_{ij}^2 \sin\theta \, dr_{ij} d\theta d\varphi$$

$$= 4\pi \int_0^D (1 - e^{-\infty}) r^2 dr = 4\pi D^3/3 \tag{12.26}$$

which is 8 times the volume of one hard sphere. The normalization constant Z is

$$Z = 1 - X \sum_{i<j} P(O_{ij}) \tag{12.27}$$

and the probability F is

$$F\{b\} \cong \frac{e^{-V_{ex}/k_B T} P\{b\}}{1 - X \sum_{i<j} P(O_{ij})} \cong \frac{e^{-\sum_{i<j} v(r_{ij})/k_B T} P\{b\}}{1 - X \sum_{i<j} P(O_{ij})}$$

$$\cong \frac{(1 - \sum_{i<j} f_{ij}) P\{b\}}{1 - X \sum_{i<j} P(O_{ij})} \tag{12.28}$$

The function f_{ij} differs substantially from zero only near the origin for real polymer chains with real repulsive forces. Indeed, f_{ij} is expected to decay rapidly as $\mathbf{r}_{ij}$ increases from zero, and any sufficiently sharply peaked function may fairly successfully model it. Accordingly, f_{ij} is approximated as

$$f_{ij} = X\delta(\mathbf{r}_{ij}) \tag{12.29}$$

where $\delta(\mathbf{r}_{ij})$ is a three dimensional Dirac delta function, or any other function similar but sufficiently sharply peaked at $\mathbf{r}_{ij} = 0$. Then the probability F becomes

$$F\{\mathbf{b}\} = \frac{(1 - X\sum_{i<j}\delta(\mathbf{r}_{ij}) + \cdots)P\{\mathbf{b}\}}{1 - X\sum_{i<j}P(O_{ij}) + \cdots} \tag{12.30}$$

This is the fundamental probability distribution function for a polymer chain in solution swollen by excluded volume effects. In principle, it may be obtained correct to any desired order. It is thus rigorous to the order calculated. It is also incomplete, owing to the increasingly arduous task of bookkeeping in higher orders of approximation. Probably the main use of the theory is to serve as a check or guide for more approximate theories when it is applied to calculation of the mean squared end to end distance or the gyration radius.

The computational detail required to secure the excluded volume expansions of the end to end distance $\langle L^2 \rangle$ and the gyration radius $\langle R^2 \rangle$ is illustrated below for the linear term in the expansion of $\langle L^2 \rangle$. The calculation starts with the probability, which as written above refers to each bond vector $\mathbf{b}_i$ in a particular configuration. A probability with the end to end vector alone constrained to a single state is more useful. Since $\mathbf{L} = \sum_i \mathbf{b}_i$ this is

$$F(\mathbf{L}) = \int F\{\mathbf{b}\}\delta(\mathbf{L} - \sum_i \mathbf{b}_i)d\{\mathbf{b}\} \tag{12.31}$$

Substituting $F\{\mathbf{b}\}$ from equation (12.30) gives

$$F(\mathbf{L}) = \frac{\int \delta(\mathbf{L} - \sum_i \mathbf{b}_i)\left[1 - X\sum_{i<j}\delta(\mathbf{r}_{ij}) + \cdots\right]P\{\mathbf{b}\}d\{\mathbf{b}\}}{1 - X\sum_{i<j}P(O_{ij}) + \cdots}$$

$$\tag{12.32}$$

which is more compactly written

$$F(\boldsymbol{L}) = \frac{P(\boldsymbol{L}) - X\sum_{i<j} P(\boldsymbol{L}, O_{ij}) + \cdots}{1 - X\sum_{i<j} P(O_{ij}) + \cdots} \qquad (12.33)$$

where $P(\boldsymbol{L}, O_{ij})$ is a conditional probability: it is the probability that the ith and j^{th} segment are in contact while the end to end distance is constrained to $\boldsymbol{L}$. That is, the numerator refers to $\boldsymbol{L}$ fixed while the denominator refers to $\boldsymbol{L}$ free. Both have segments i and j in contact. Since the term $X\sum_{i<j} P(O_{ij})$ is small, the denominator can be expanded in geometric series.

$$\frac{1}{1-x} = 1 + x + x^2 + \cdots \qquad (12.34)$$

Then $F(\boldsymbol{L})$ becomes

$$F(\boldsymbol{L}) \cong \left[P(\boldsymbol{L}) - X\sum_{i<j} P(\boldsymbol{L}, O_{ij}) + \cdots \right]\left[1 + X\sum_{i<j} P(O_{ij}) + \cdots \right] \qquad (12.35)$$

and multiplying this out to the linear terms

$$F(\boldsymbol{L}) = P(\boldsymbol{L}) + X\left[\sum_{i<j}(P(\boldsymbol{L})P(O_{ij}) - P(\boldsymbol{L}, O_{ij})) \right] + \cdots \qquad (12.36)$$

This shows that the dominant behavior is collected in the Gaussian $P(\boldsymbol{L})$, with the excluded volume effect a small additive correction term. Note that $F(\boldsymbol{L})$ is normalized, and the average of the squared end to end distance is calculated as

$$\langle L^2 \rangle = \int L^2 F(\boldsymbol{L}) d\boldsymbol{L} \qquad (12.37)$$

Define the Gaussian mean squared end to end distance as

$$\langle L_0^2 \rangle = \int L^2 P(\boldsymbol{L}) d\boldsymbol{L} = nb^2 \qquad (12.38)$$

Then

$$\langle L^2 \rangle = \langle L_0^2 \rangle + X\sum_{i<j} \int L^2 \left[P(\boldsymbol{L})P(O_{ij}) - P(\boldsymbol{L}, O_{ij}) \right] d\boldsymbol{L} \qquad (12.39)$$

After some reflection, it appears that

$$P(\boldsymbol{L})P(O_{ij}) = P_{n-|i-j|}(\boldsymbol{L})P(O_{ij}) \tag{12.40}$$

This is not very obvious, until it is realized that the conditional probabilities are really bivariate Gaussians. If the loop that runs from i to j is conceptually cut off, then exactly the same end to end probability will occur, but for fewer segments. As n is very large, $n - |i - j|$ is still also very large and the statistics are unaltered by the removal. With this alteration, the mean square end to end distance in the chain with excluded volume becomes

$$\langle L^2 \rangle = \langle L_0^2 \rangle + X \sum_{i<j} \int L^2 \left[P(\boldsymbol{L})P(O_{ij}) - P_{n-|i-j|}(\boldsymbol{L})P(O_{ij}) \right] d\boldsymbol{L} + \cdots$$
$$\tag{12.41}$$

Evaluating the integrals,

$$\langle L^2 \rangle = \langle L_0^2 \rangle + X \sum_{i<j} \left[nb^2 - (n - |i - j|)b^2 \right] P(O_{ij}) + \cdots \tag{12.42}$$

Since n cancels this becomes

$$\langle L^2 \rangle = \langle L_0^2 \rangle + Xb^2 \sum_{i<j} |i - j| P(O_{ij}) + \cdots \tag{12.43}$$

By definition

$$P(O_{ij}) = [P(r_{ij})]_{r_{ij} \to 0} \equiv \left(\frac{3}{2\pi b^2 |i - j|} \right)^{3/2} \left[e^{-\frac{3r_{ij}^2}{2|i-j|b^2}} \right]_{r_{ij} \to 0}$$

$$= \left(\frac{3}{2\pi b^2} \right)^{3/2} \frac{1}{|i - j|^{3/2}} \tag{12.44}$$

Then

$$\langle L^2 \rangle = \langle L_0^2 \rangle + Xb^2 \sum_{i<j} |i - j| \left(\frac{3}{2\pi b^2} \right)^{3/2} \frac{1}{|i - j|^{3/2}} \tag{12.45}$$

and immediately

$$\langle L^2 \rangle = \langle L_0^2 \rangle + Xb^2 \left(\frac{3}{2\pi b^2} \right)^{3/2} \sum_{i<j} |i - j|^{-1/2} \tag{12.46}$$

The sum may be evaluated by the Euler-MacLaurin sum formula. The details are

$$\int_0^n dj \int_0^j (i-j)^{-\frac{1}{2}} di = \frac{4}{3} n^{\frac{3}{2}} \tag{12.47}$$

The result is then

$$\langle L^2 \rangle = \langle L_0^2 \rangle + nb^2 X \left(\frac{3}{2\pi b^2} \right)^{3/2} \frac{4}{3} n^{1/2} + \cdots \tag{12.48}$$

In terms of the expansion parameter z this may be expressed as

$$\langle L^2 \rangle = \langle L_0^2 \rangle \left(1 + \frac{4}{3} z + \cdots \right) \tag{12.49}$$

where the expansion parameter is defined as

$$z = n^{1/2} X \left(\frac{3}{2\pi b^2} \right)^{3/2} \tag{12.50}$$

This expansion parameter vanishes in the theta solvent, but may be as large as 10 to 20 for large polymers in good solvents, It is essentially the volume excluded by segment impenetrability modified by a few numerical factors. The coefficient of the z^2 term may be obtained by similar means. For counting details, see Fixman.[3] The theory is not normally carried beyond the quadratic term, owing partly to the great algebraic labor involved, but largely to the slow convergence of the series. The extra labor would not produce a truly useful result. As long as segment density remains low, the theory serves as a check on other approximate theories, since for low segment density it is exact as far as it goes. Flory and Fisk[4] have published a smoothed density approximation for the excluded volume effect. Fixman[5,6] and Stidham and Fixman[7] have provided an approximate treatment of excluded volume in dilute polymer solution using algebra isomorphic with that of boson operators, which is discussed next.

12.3. Raising and Lowering Operators for Hermite Orthogonal Functions

The Hermite orthogonal functions $\varphi_n(x)$ can be made the basis of an approximation scheme for accommodating the effects of excluded volume in high polymers in dilute solution. These functions are sometimes represented by the bra and ket notation introduced by Dirac in quantum mechanics. For example, the ket $|n\rangle$ may be used to represent $\varphi_n(x)$ where

$$\varphi_n(x) = \frac{H_n(x)e^{-x^2/2}}{\sqrt{\pi^{1/2}n!2^n}} \tag{12.51}$$

The Hermite polynomials $H_n(x)$ may be defined by the recursion relations

$$\frac{dH_n(x)}{dx} = 2nH_{n-1}(x) \tag{12.52}$$

$$2xH_n(x) = H_{n+1}(x) + 2nH_{n-1}(x) \tag{12.53}$$

Equation (12.53) is given, together with a number of other recursion relations for other functions, in Abramowitz and Stegun, Chapter 22.[11] The lowering operator results from considering the derivative

$$\frac{d|n\rangle}{dx} = \frac{d\varphi_n(x)}{dx} = \frac{\left\{\left[\frac{dH_n(x)}{dx}\right]e^{-x^2} - xH_n(x)e^{-x^2/1}\right\}}{\sqrt{\pi^{1/2}n!2^n}} \tag{12.54}$$

Applying the recursion relations shows

$$\frac{d|n\rangle}{dx} = (2n)^{1/2}\varphi_{n-1}(x) - x\varphi_n(x) \tag{12.55}$$

Substituting the definition of the ket gives

$$\frac{d|n\rangle}{dx} = (2n)^{1/2}|n-1\rangle - x|n\rangle \tag{12.56}$$

The ket simplification of the notation suggests the definition

$$b \equiv 2^{-1/2}\left(\frac{d}{dx} + x\right) \tag{12.57}$$

which has the property

$$b|n\rangle = n^{1/2}|n-1\rangle \tag{12.58}$$

The operator b lowers the index n by one unit, leaving a footprint of $n^{1/2}$. Further, subtracting $x\varphi_n(x)$ from both sides of the derivative $\frac{d\varphi_n(x)}{dx}$ and using the second recursion relation provides the result

$$\frac{d\varphi_n(x)}{dx} - x\varphi_n(x) = -2^{1/2}(n+1)^{1/2}\varphi_{n+1}(x) \tag{12.59}$$

suggesting the definition

$$b^+ \equiv 2^{-1/2}\left(x - \frac{d}{dx}\right) \tag{12.60}$$

for then

$$b^+|n\rangle = (n+1)^{1/2}|n+1\rangle \tag{12.61}$$

The raising operator b^+ and the lowering operator b may then be used to give an operator representation of x and d/dx as

$$x = 2^{-1/2}(b + b^+) \tag{12.62}$$

$$\frac{d}{dx} = 2^{-1/2}(b - b^+) \tag{12.63}$$

The raising and lowering operators obey the commutation relations

$$(b, b^+) = bb^+ - b^+b = 1 \tag{12.64}$$

$$(1, b) = 0 \tag{12.65}$$

$$(1, b^+) = 0 \tag{12.66}$$

where 1 is the identity operator. Multiplication by unity leaves every function unaltered, whether pre or post multiplication is involved.

These commutation relations form the multiplication table of the unique three dimensional Lie algebra with operation the commutation isomorphic with that generated by the boson operators of quantum mechanics. Under commutation, the operators b, b^+ and 1 satisfy the Jacobi identity and therefore form the basis for the algebra. It has become common to refer to the operators b and b^+

as boson operators, even though these operators have nothing to do with Bose-Einstein statistics. The application is restricted to the Hermite orthogonal functions, and to functions capable of expansion in terms of the Hermite orthogonal functions. Note that the boson operators derive their commutation relations from the fundamental commutator

$$\left(\frac{d}{dx}, x\right) = 1 \tag{12.67}$$

These boson operators may be applied to an arbitrary function, provided the function may be expanded in a series of Hermite orthogonal functions, Thus, $bf(x)$ has a non-trivial meaning. It is required that

$$f(x) = \sum_n f_n \varphi_n(x) \tag{12.68}$$

so that

$$bf(x) = \sum_n f_n n^{1/2} \varphi_{n-1}(x) \tag{12.69}$$

The leading term in this series is $f_2 2^{1/2} \varphi_1(x)$, since $\varphi_0(x) = 0$.
    The Hermite orthogonal functions are orthonormal, that is

$$\langle n|m \rangle = \int_{-\infty}^{\infty} \varphi_n(x) \varphi_m(x) dx = \delta_{n,m} \tag{12.70}$$

which permits a matrix representation of an arbitrary operator A.

$$\langle n|A|m \rangle = \int_{-\infty}^{\infty} \varphi_n(x) A \varphi_m(x) dx = A_{nm} \tag{12.71}$$

It may happen that A_{nm} is not equal to A_{mn}. If the operator A is a function of coordinates alone, then

$$A\varphi_m(x) = \varphi_m(x)A \tag{12.72}$$

and $A_{mn} = A_{nm}$. However, if the operator A contains derivatives, this will not be so. Consider the matrix elements of the derivative

d/dx.

$$\left\langle n \left| \frac{d}{dx} \right| m \right\rangle = \langle n | 2^{-1/2}(b - b^+) | m \rangle \qquad (12.73)$$

$$\left\langle n \left| \frac{d}{dx} \right| m \right\rangle = 2^{-1/2}(\langle n | b | m \rangle - \langle n | b^+ | m \rangle) \qquad (12.74)$$

$$\left\langle n \left| \frac{d}{dx} \right| m \right\rangle = 2^{-1/2}(\langle n | m^{1/2} | m - 1 \rangle - \langle n | (m+1)^{1/2} | m + 1 \rangle)$$
$$(12.75)$$

$$\left\langle n \left| \frac{d}{dx} \right| m \right\rangle = 2^{-1/2}(m^{1/2} \delta_{n,m-1} - (m+1)^{1/2} \delta_{n,m+1}) \qquad (12.76)$$

and the non-vanishing matrix elements of the differential operator d/dx are

$$\langle n | \frac{d}{dx} | n + 1 \rangle = 2^{-1/2}(n+1)^{1/2} \qquad (12.77)$$

$$\langle n | \frac{d}{dx} | n - 1 \rangle = 2^{-1/2} n^{1/2} \qquad (12.78)$$

When the operator A is a function of coordinates alone, the full notation for the matrix elements is somewhat complicated. Specifically,

$$A_{nm} = \int_{-\infty}^{\infty} \frac{e^{-x^2} H_n(x) H_m(x)}{\sqrt{2^n n! \pi^{1/2} 2^m m! \pi^{1/2}}} A \, dx \qquad (12.79)$$

In the interest of simplifying the notation, define several things for use in the case that the operator A is a function of coordinates alone. First, set

$$N_n = \frac{1}{\sqrt{2^n n!}} \qquad (12.80)$$

Second, define

$$\psi_\alpha = \pi^{-1/2} e^{-x^2} \qquad (12.81)$$

Finally

$$\phi_n = N_n H_n(x) \qquad (12.82)$$

With these definitions, the matrix elements of A are

$$A_{nm} = \int_{-\infty}^{\infty} \psi_\alpha \phi_n A \phi_m dx \tag{12.83}$$

However, one must be careful not to apply this formalism to operators containing derivatives, as incomplete and incorrect results will be obtained. For example, application to the derivative d/dx results in

$$\left(\frac{d}{dx}\right)_{nm} = (2m)^{1/2}\delta_{n,m-1}$$

which is incomplete and incorrect.

The boson representation allows facile evaluation of the matrix elements of x and its smaller powers. Thus

$$\langle n|x|m \rangle = 2^{-1/2}\langle n|b + b^+|m \rangle \tag{12.84}$$

$$\langle n|x|m \rangle = 2^{-1/2}(\langle n|m^{1/2}|m-1 \rangle + \langle n|(m+1)^{1/2}|m+1 \rangle) \tag{12.85}$$

$$\langle n|x|m \rangle = 2^{-1/2}(m^{1/2}\delta_{n,m-1} + (m+1)^{1/2}\delta_{n,m+1}) \tag{12.86}$$

An alternate way of expressing these results is

$$\langle n|x|n+1 \rangle = 2^{-1/2}(n+1)^{1/2}$$

$$\langle n|x|n-1 \rangle = 2^{-1/2}n^{1/2}$$

Calculation of the matrix elements of powers of x are facilitated by use of the commutation relation $(b, b^+) = bb^+ - b^+b = 1$. The object of the application is to bring all the raising operators as far to the left as these will go. To illustrate, the calculation for x^2 is as follows:

$$\langle n|x^2|m \rangle = 2^{-1}\langle n|(b + b^+)(b + b^+)|m \rangle \tag{12.87}$$

and the essential algebra is

$$(b + b^+)(b + b^+) = bb + bb^+ + b^+b + b^+b^+ \tag{12.88}$$

Applying the commutator once removes bb^+ from the operator and results in

$$(b + b^+)(b + b^+) = bb + (b^+b + 1) + b^+b + b^+b^+ \tag{12.89}$$

Collecting like terms produces

$$(b + b^+)(b + b^+) = bb + 2b^+b + b^+b^+ + 1 \qquad (12.90)$$

The three matrix elements are

$$\langle n|bb|m\rangle = m^{1/2}(m-1)^{1/2}\delta_{n,m-2}$$
$$\langle n|2b^+b + 1|m\rangle = (2m+1)\delta_{n,m}$$
$$\langle n|b^+b^+|m\rangle = (m+1)^{1/2}(m+2)^{1/2}\delta_{n,m+2}$$

The two indices on the delta function must be equal, as otherwise the matrix element vanishes. In the first of these, that means n must equal $m-2$, or $m = n+2$. The matrix element is then

$$\langle n|bb|n+2\rangle = (n+1)^{1/2}(n+2)^{1/2} \qquad (12.91)$$

Similarly in the second, $m = n$ and the matrix element is

$$\langle n|2b^+b + 1|n\rangle = 2n + 1 \qquad (12.92)$$

In the third case, $m = n - 2$ and this matrix element is

$$\langle n|b^+b^+|n-2\rangle = (n-1)^{1/2}n^{1/2} \qquad (12.93)$$

Division of these matrix elements by 2 provides the non-vanishing matrix elements of x^2.

It should be emphasized that in all these relations, x is a dimensionless variable (it appears in the exponent of the exponential in defining the Hermite orthogonal functions). If x is to have dimensions, then the argument of the exponential should be written $\alpha^2 x^2$ and that of the Hermite polynomials as αx, where the parameter α has dimensions that are inverse to those of x, and appears as a constant. The non-vanishing matrix elements of x^3 and x^4 are not hard to work out by using the raising and lowering operators and are tabulated in the book by Wilson, Decius and Cross[9] in Appendix III. The fifth and sixth powers were published by Foldes and Sandorfy[10] in an appendix to their paper.

One further comment on the Hermite polynomials needs to be made. These functions obey a particular differential equation

$$\frac{d}{dx}\left(e^{-x^2}\frac{dH_n(x)}{dx}\right) + \lambda e^{-x^2}H_n(x) = 0 \qquad (12.94)$$

where the eigenvalues $\lambda = 0, 2, 4, \ldots$ See Courant and Hilbert, Methods of Mathematical Physics, Vol. I, p. 328. This information will appear from time to time in the following.

12.4. Equations of Motion

First, consider the model, then the forces acting on a polymer chain in solution. A polymer chain is modeled here by a string of beads, each representing one Kuhn statistical segment. The beads maintain connectivity but are otherwise randomly scattered in the solvent, itself composed of sufficiently small molecules that the solution can be regarded as an isotropic fluid.

If the beads were disconnected and dispersed randomly in the solvent, the probability of finding a bead in an element of volume in the solvent would be the concentration $\psi_i(x)$ in ideal solution. The index i refers to the kind of bead, and the variable x to the position in the volume of the solution. The chemical potential is

$$\mu_i(x) = \mu_i^0 + k_B T \ln \psi_i(x) \qquad (12.95)$$

In non-ideal solution, $\psi_i(x)$ is the probability, which includes all the factors that make the solution non-ideal. The gradient of the chemical potential is

$$\frac{\partial \mu_i(x)}{\partial x} = k_B T \frac{\partial \ln \psi_i(x)}{\partial x} \qquad (12.96)$$

along a single space direction x. If three spatial directions are involved, which is the more usual case, the position of a bead may be defined by a position vector $\boldsymbol{r}$, and the chemical potential $\mu_i = \mu_i(\boldsymbol{r})$. The chemical potential is a potential energy per molecule, and a force on the molecule is described by the negative gradient of the potential.

This is the entropic force, and is of the form

$$f_r = -\nabla \mu_i(r) = -k_B T \nabla \ln \psi_i(r) \tag{12.97}$$

where ∇ is the gradient operator, often written as $i\frac{\partial}{\partial x} + j\frac{\partial}{\partial y} + k\frac{\partial}{\partial z}$ with i, j and k unit vectors along the x, y and z directions respectively. It has other expressions in other coordinate systems. For example, in spherical coordinates r, θ and φ (θ is the angle made by the radius vector r from the origin and the z axis, and φ is the angle made by the projection of the radius vector onto the xy plane, measured from the x axis) the gradient of a function f is

$$\nabla f = \frac{\partial f}{\partial r} u_r + \frac{\partial f}{\partial \theta} \frac{u_\theta}{r} + \frac{\partial f}{\partial \varphi} \frac{u_\varphi}{r \sin \theta} \tag{12.98}$$

where u_i are unit vectors directed towards an increase in the coordinate i. The entropic force derives from the manifold collisions solvent molecules make with a segment over a time indefinitely large in comparison with the average time between collisions. As the force is determined by concentration gradients, it acts on different beads differently but in such a way as to diminish the gradient, striving towards a completely random distribution of beads in essentially isotropic surroundings, Initial concentration gradients are diminished to vanishing. The force is essentially Brownian, and its use implies that forces of other origins are also average forces.

The total force acting on the ith bead located at position r consists of three contributors. First is the entropic force discussed above. Second, as a bead moves through solvent with a velocity v_i it encounters viscous drag proportional to the velocity at low velocities relative to that of the solvent. Third, the bead will move under the influence of an externally imposed field, such as gravitational or centrifugal, represented in a potential energy V, The effects of excluded volume are included with these fields. An equation accommodating all three forces is

$$m_i \frac{dv_i}{dt} = -k_B T \nabla \ln \psi_i(r) - \beta v_i - \nabla V \tag{12.99}$$

The constant β governs the viscous drag and is known as the friction constant. When V is non-zero, $\psi_i(r)$ departs from its ideal interpretation as the concentration, and may be written

$$\psi_i(r) = Ae^{-V(r)} \tag{12.100}$$

where A is a normalization constant The function $\psi_i(r)$ is still quite rigorously a probability. Since probability is conserved, it is governed by a continuity equation

$$\frac{\partial \psi_i(r)}{\partial t} = -\nabla \cdot \boldsymbol{v}_i \psi_i(r) \tag{12.101}$$

where $\boldsymbol{v}_i$ is the mean vector velocity of the ith segment or bead, and $\boldsymbol{v}_i \psi_i(r)$ is a flux of dimensions molecules cm^{-2} sec^{-1}. Together with an equation for the velocity $\boldsymbol{v}_i$, these equations form the basis for the theory of excluded volume considered in this chapter. In the following, several simplified cases will be considered, then these equations will be applied to polymer chains.

Example 1

Consider the case of no potential V, such as might be experienced on a space station, but there is a small concentration gradient. The equation of motion is

$$m_i \frac{dv_i}{dt} = -k_B T \nabla \ln \psi_i(r) - \beta v_i \tag{12.102}$$

Assume $\beta = 6\pi\eta a$. Calculate a typical value of the ratio β/m by estimating the viscosity η of water at 20 C as about 1×10^{-2} poise (units mass $\times$ length^{-1} $\times$ time^{-1}) and a proton diffusing through such a continuous medium would have a diameter a of about 1 Angstrom (10^{-8} cm), and a mass of about 1 amu. In cgs units

$$\frac{\beta}{m} = \frac{6\pi\eta a}{m} = \frac{20 \times 10^{-2} \times 10^{-8}}{1} \times 6 \times 10^{23} = 10^{15} sec^{-1}$$

This is the order of magnitude of the highest frequency expected. Polymer segments tend to be 100 times more massive and 10 times larger than a proton, which would suggest a frequency of the order of 10^{14} sec^{-1}. The worst case imaginable would lead to a frequency near 10^{13} sec^{-1}. In any reasonable case, the ratio β/m is a very large frequency. This means

$$m\frac{d\boldsymbol{v}_i}{dt} \ll \beta \boldsymbol{v}_i$$

that is, on the average the velocities do not change quickly. The low frequency response of a segment is required, and the acceleration term may be ignored, allowing the approximation

$$0 \cong -k_B T \nabla \ln \psi_i(r) - \beta \boldsymbol{v}_i \qquad (12.103)$$

Then

$$\boldsymbol{v}_i \psi_i(r) \cong \frac{k_B T}{\beta} \nabla \psi_i(r) \qquad (12.104)$$

Substituting this into the continuity equation gives the second order differential equation

$$\frac{\partial \psi_i(r)}{\partial t} = \frac{k_B T}{\beta} \nabla^2 \psi_i(r) \qquad (12.105)$$

This is the familiar form of Fick's second law, with the diffusion constant D given by

$$D = \frac{k_B T}{\beta} \qquad (12.106)$$

Example 2

In this case, the concentration $\psi_i(r)$ is constant but the potential V is not constant. The force f is

$$\boldsymbol{f} = -\nabla V \qquad (12.107)$$

The potential might be due to gravity, in which case the field is uniform over the sample, or it might be due to centrifugation, in which case the field depends on the distance from the axis of rotation of the centrifuge, or ... Now, βv_i still vastly exceeds the mass acceleration term $m_i \frac{dv_i}{dt}$, and the approximation

$$0 \cong -\beta v_i + f \tag{12.108}$$

or $f \cong \beta v_i$. The velocity of these segments depends on the applied force, and this is the same for every segment in a uniform gravitational field. In a centrifuge, segments at greater distances from the axis of rotation experience greater force and over time a concentration gradient will evolve.

12.5. General Theory

All of the foregoing was concerned with the behavior of one segment or bead in the absence of all of the others in a polymer chain. In a polymer chain composed of n segments represented by beads, the equation of continuity is

$$\frac{\partial \psi}{\partial t} = -\sum_{i=1}^{n} \nabla_i \cdot (v_i \psi) \tag{12.109}$$

The equation of motion of any segment i is essentially the same as for a single bead given above, modified to take account of intersegment interaction. Let the segment velocity be the vector v_i and the velocity of the solvent at the segment, unmodified by the presence of other segments, be the vector v_i'. The relative velocity of the segment with respect to the solvent is the vector $v_i - v_i'$. The equation of motion is

$$m \frac{\partial v_i}{\partial t} = -\beta(v_i - v_i') - k_B T \nabla_i \ln \psi - \nabla_i V \tag{12.110}$$

Again, as $\beta/m \approx 10^{13}$ sec^{-1}, just as above for a single bead, $m \frac{\partial v_i}{\partial t} \cong 0$, and

$$\beta(v_i - v_i') \cong -k_B T \nabla_i \ln \psi - \nabla_i V \tag{12.111}$$

The solvent velocity is disturbed by the motion of the segment and by that of its neighbors. Kirkwood and Riseman[12] showed that hydrodynamic interaction may be approximated by introducing the Oseen tensor T_{ij}, This tensor essentially gives the ith velocity perturbation due to the jth segment. Evidently $T_{ii} = 0$ and when i and j are different

$$T_{ij} = \frac{1}{8\pi\eta_0 r_{ij}^2}(r_{ij}^2 1 + r_{ij}r_{ij}) \tag{12.112}$$

The solvent velocity at the segment is approximately

$$v_i' \cong v_i^0 + \sum_j T_{ij} \cdot \beta(v_j - v_j') \tag{12.113}$$

Here, $r_{ij} = r_i - r_j$ is a vector from the ith to the jth segment, η_0 is solvent viscosity amd v_i^0 is solvent velocity in the absence of segments, Combining these ideas, the segment velocity is approximately

$$v_i = v_i^0 + \sum_j (\beta^{-1}\delta_{ij} 1 + T_{ij}) \cdot (-k_B T \nabla_j \ln \psi - \nabla_j V) \tag{12.114}$$

To simplify notation, let the velocity and differential operators be represented by column vectors with entries $v_{i\alpha}$ and $\nabla_{i\alpha}$ with $\alpha = x, y, z$. If there are M segments in solution, the velocity and differential operator vectors are of dimensionality $3M \times 1$. Further, let T^H be a $3M \times 3M$ matrix with elements the components of T_{ij}. Define an operator D as

$$D \equiv \beta^{-1} 1 + T^H \tag{12.115}$$

The velocity at segment i can be more compactly written

$$v = v^0 + D(-k_B T \ln \psi - \nabla V) \tag{12.116}$$

The equation of continuity is in these simplified terms

$$\frac{\partial \psi}{\partial t} + \nabla^T(v\psi) = 0 \tag{12.117}$$

where the superscript T denotes the operation transpose (converting a column vector to a row vector). Combining this with the velocity gives the time evolution of the probability that describes the distribution

$$\frac{\partial \psi}{\partial t} + \nabla^T \left[v^0 \psi - D(k_B T \nabla \psi + \psi \nabla V) \right] = 0 \qquad (12.118)$$

Let U be an intersegment potential energy and W be an external potential energy. If these are the only potentials, $V = U + W$ and equation (12.118) becomes

$$\frac{\partial \psi}{\partial t} + \nabla^T \left[v^0 \psi - D(k_B T \nabla \psi + \psi \nabla (U + W)) \right] = 0 \qquad (12.119)$$

This is the fundamental equation of the theory.

If the chains were freely jointed and there were neither flows nor fluxes the segment distribution would be Gaussian. The distribution function ψ^α in the absence of distortions is

$$\psi^\alpha \sim \exp(-S^\alpha / k_B T) \qquad (12.120)$$

where

$$S^\alpha = \frac{3 k_B T}{2 b^2} \sum_{i=1}^{N} |r_i - r_{i+1}|^2 \qquad (12.121)$$

where b is the length of a Kuhn statistical segment. To take into account all distorting influences, write

$$\psi = \psi^\alpha \rho \qquad (12.122)$$

Here ρ is a correction factor for the probability that collects all the distorting influences. Substituting this into the equation of continuity gives, after some algebra

$$\frac{\partial \rho}{\partial t} + L\rho = 0 \qquad (12.123)$$

The operator L is somewhat complicated, and is conveniently broken into two parts by defining $L = L_a + L_b$, where

$$L_a = -k_B T [\nabla^T - (\nabla^T S^\alpha / k_B T)] D \{ \nabla + [\nabla(U - S^\alpha + W)/k_B T] \} \qquad (12.124)$$

$$L_b = (\nabla^T - \nabla^T S^\alpha / k_B T) v_0 \qquad (12.125)$$

In the equilibrium condition, $\frac{\partial \rho}{\partial t} = 0, L_b = 0$ since $v_0 = 0$, and the equilibrium ρ, which includes the excluded volume contribution, satisfies the equation

$$\{\nabla + [\nabla(U - S^\alpha + W)/k_B T]\}\rho = 0 \tag{12.126}$$

which will obviously be the case if

$$\rho \sim exp\left[-(U - S^\alpha + W)/k_B T\right] \tag{12.127}$$

This solution holds independently of the approximation introduced in the operator D to take account of the effects of hydrodynamic interaction. The equation can be applied to the end to end distance or the gyration radius with equal confidence. However, when movement is involved, as in sedimentation or viscosity measurements, this will not be so.

The next problem is how best to express the equilibrium ρ. Consider a special form of the operator L_a, designed to keep ρ near the solution to $L_a\rho = 0$ when L_a is simplified by deletion of all the potential effects, and when hydrodynamic interaction is omitted. This suggests expanding ρ in a series of eigenfunctions of the simplified operator A^α

$$A^\alpha \equiv -\frac{k_B T}{\beta}\left[\nabla^T - (\nabla^T S^\alpha/k_B T)\right]\nabla \tag{12.128}$$

which may be written more simply

$$A^\alpha = -\left[\frac{k_B T}{\beta}\nabla^T + \beta^{-1}(\nabla^T S^\alpha)\right]\nabla \tag{12.129}$$

Since the hydrodynamic interaction has been removed by setting $T^H = 0$ and $D = \beta^{-1}\mathbf{1}$, the operator $\mathbf{A}^\alpha$ is called the free-draining operator. Its eigenfunctions satisfy the equation

$$A^\alpha \varphi_n = \lambda_n^0 \varphi_n \tag{12.130}$$

where n is a suitable index, and λ_n^0 is an eigenvalue of the operator. That the eigenfunctions are Hermite polynomials may be shown in the following way. Premultiply by $-\beta/k_B T$ and take $\lambda_n^0 = k_B T \lambda_n/\beta$. Writing the operator out in full shows details of the operator

$$\left[\nabla^T - \frac{1}{k_B T}(\nabla^T S^\alpha)\right] \nabla \varphi_n = -\lambda_n \varphi_n \qquad (12.131)$$

Premultiply by $\psi^\alpha = \text{constant} \times e^{-S^\alpha/k_B T}$ and commute ψ^α through ∇^T to obtain

$$\nabla^T \psi^\alpha = -\frac{\psi^\alpha}{k_B T} \nabla^T S^\alpha \qquad (12.132)$$

Hence

$$\psi^\alpha \left[\nabla^T - \frac{1}{k_B T}(\nabla^T S^\alpha)\right] \nabla = \left[\psi^\alpha \nabla^T + (\nabla^T \psi^\alpha)\right] \nabla \qquad (12.133)$$

and, as

$$\nabla^T \left[\psi^\alpha \nabla \phi_n\right] = \left[\psi^\alpha \nabla^T + (\nabla^T \psi^\alpha)\right] \nabla \phi_n \qquad (12.134)$$

the eigenvalue equation may be written

$$\nabla^T \left[\psi^\alpha \nabla \phi_n\right] + \lambda_n \psi^\alpha \phi_n = 0 \qquad (12.135)$$

If only one dimension were involved, both differential operators ∇^T and ∇ would be simple derivatives, while ψ^α would be a Gaussian in one dimension such as $\psi^\alpha = \text{constant } e^{-x^2}$. Apart from the normalization constant for ψ^α, which may be regarded as included in the eigenvalue λ_n the equation becomes in one dimension

$$\frac{\partial}{\partial x} e^{-x^2} \frac{\partial \phi_n}{\partial x} + \lambda_n e^{-x^2} \phi_n = 0 \qquad (12.136)$$

Comparison with the equation defining the Hermite polynomials given earlier shows that the $\phi_n = H_n(x)$, the Hermite polynomials of order n. When more than one dimension is involved, the functional dependence is not so obvious. The variables are extricably intertwined, and when a familiar linear transformation is made, the eigenvalue equation separates into a sum of equations each of the form of the equation defining the Hermite polynomials. The result is a set of basis functions that are products of Hermite polynomials, which may then be used to expand the equilibrium ρ.

12.6. Normal Coordinates

In a very long polymer chain, the statistics are dominated by the Gaussian. Excluded volume and hydrodynamic interactions are small perturbations on this basic behavior. The probability distribution function P is given by

$$P(\boldsymbol{r}_i - \boldsymbol{r}_{i+1}) = \left(\frac{3}{2\pi b^2}\right)^{3/2} e^{-3(\boldsymbol{r}_i - \boldsymbol{r}_{i+1})^2/2b^2} \qquad (12.137)$$

where $b = |\boldsymbol{r}_i - \boldsymbol{r}_{i+1}|$ is the intersegment distance for adjacent segments. In the absence of perturbing influences, the probability for the whole Gaussian chain is

$$\psi^\alpha = \prod_{i=1}^{N-1} P(\boldsymbol{r}_i - \boldsymbol{r}_{i+1}) \qquad (12.138)$$

$$\psi^\alpha = \left(\frac{3}{2\pi b^2}\right)^{3N/2} e^{-3\sum_{i=1}^{N-1}(\boldsymbol{r}_i - \boldsymbol{r}_{i+1})^2/2b^2} \qquad (12.139)$$

$$\psi^\alpha = Ce^{-S^\alpha/k_B T} \qquad (12.140)$$

The function S^α is called the spring potential. The sum and product run only to $N-1$ since there is no vector $\boldsymbol{r}_{N+1}$, that is, there are $N-1$ bonds in a chain of N beads. As it stands, S^α contains cross terms of the form $-2\boldsymbol{r}_i \cdot \boldsymbol{r}_{i+1}$, entwining the variables. By introducing the proper linear transformation, generating new functions that are themselves functions of the $\boldsymbol{r}_i$, S^α may be brought to diagonal form, containing only squares of the new coordinates. The transformation is familiar, as it was encountered in earlier work on the vibrations of crystals. The general theory is somewhat opaque, and in the interest of displaying the main principles, consider a polymer of three segments. The spring potential is

$$S_3^\alpha = \frac{3k_B T}{2b^2}\left[(\boldsymbol{r}_1 - \boldsymbol{r}_2)^2 + (\boldsymbol{r}_2 - \boldsymbol{r}_3)^2\right] \qquad (12.141)$$

The following substitutions will bring S_3^α to diagonal form.

$$q_0 = \frac{1}{\sqrt{3}}(\boldsymbol{r}_1 + \boldsymbol{r}_2 + \boldsymbol{r}_3)$$

$$q_1 = \frac{1}{\sqrt{2}}(\mathbf{r}_1 - \mathbf{r}_3)$$

$$q_2 = \frac{1}{\sqrt{6}}(\mathbf{r}_1 - 2\mathbf{r}_2 + \mathbf{r}_3)$$

To make these substitutions, these equations must be solved for the $\mathbf{r}_i$. The result is

$$\mathbf{r}_1 = \frac{1}{\sqrt{3}}q_1 + \frac{1}{\sqrt{2}}q_2 + \frac{1}{\sqrt{6}}q_3$$

$$\mathbf{r}_2 = \frac{1}{\sqrt{3}}q_1 - \frac{2}{\sqrt{6}}q_3$$

$$\mathbf{r}_3 = \frac{1}{\sqrt{3}}q_1 - \frac{1}{\sqrt{2}}q_2 + \frac{1}{\sqrt{6}}q_3 \tag{12.142}$$

Straightforward algebra shows

$$(\mathbf{r}_1 - \mathbf{r}_2)^2 + (\mathbf{r}_2 - \mathbf{r}_3)^2 = 0q_1^2 + 1q_2^2 + 3q_3^2$$

There are three eigenvalues, 0, 1 and $\sqrt{3}$.

If the trimer is formed into a ring, S^α contains $(\mathbf{r}_1 - \mathbf{r}_2)^2 + (\mathbf{r}_2 - \mathbf{r}_3)^2 + (\mathbf{r}_1 - \mathbf{r}_3)^2$, and since $\mathbf{r}_1 - \mathbf{r}_3 = \sqrt{2}q_2$, the same transformation suffices for diagonalization. The result contains a degenerate root, $\sqrt{3}$.

$$(\mathbf{r}_1 - \mathbf{r}_2)^2 + (\mathbf{r}_2 - \mathbf{r}_3)^2 + (\mathbf{r}_1 - \mathbf{r}_3)^2 = 0q_1^2 + 3q_2^2 + 3q_3^2$$

For longer polymers, the transformations can be written

$$\mathbf{r}_i = \sum_{k=0}^{N-1} Q_{ik}q_k \quad q_k = \sum_{i=1}^{N} Q_{ki}\mathbf{r}_i \tag{12.143}$$

$$Q_{ik} \cong N^{-1/2}(2 - \delta_{k0})^{1/2} \cos(ik\pi/N) \tag{12.144}$$

The transformation deserves some comment. This equation is not exact and does not give the transformation coefficients quoted above for the chain of three segments. It is set up to accommodate chains of great length. Applied to polymers, chains are never short, but are always very long. The distinction between N and $N - 1$ required of chains with only a few segments is unnecessary, as N and $N - 1$

become essentially indistinguishable for very large N. For short chains of N beads, counting the beads beginning with $i = 1$ and ending with $i = N$, while counting the eigenvalues with $k = 0$ and ending with $k = N - 1$, the transformation coefficients are

$$Q_{ik} = N^{-1/2}(2 - \delta_{k0})^{1/2} \cos((2i - 1)k\pi/2N) \qquad (12.145)$$

This works for any small number of segments. In fact, it describes Q_{ik} for all chains, From trigonometry

$$\cos(a - b) = \cos(a)\cos(b) + \sin(a)\sin(b) \qquad (12.146)$$

there follows

$$\cos\left(\frac{ik\pi}{N} - \frac{k\pi}{2N}\right) = \cos\left(\frac{ik\pi}{N}\right)\cos\left(\frac{k\pi}{2N}\right) + \sin\left(\frac{ik\pi}{N}\right)\sin\left(\frac{k\pi}{2N}\right) \qquad (12.147)$$

It will emerge later that only the first few hundred values of k contribute appreciably to the averages, and $2N$ is very much larger than $k = 500$ or so. Thus, $\frac{k\pi}{2N}$ is essentially zero. Operationally, the transformation coefficients for a long polymer chain are then

$$Q_{ik} \cong N^{-1/2}(2 - \delta_{k0})^{1/2} \cos(ik\pi/N) \qquad (12.148)$$

although this approximation will not provide correct coefficients for the unique transformation from r coordinates to normal coordinates q for short chains. The eigenvalues are exactly

$$\alpha_k^2 = \frac{6}{b^2} \sin^2\left(\frac{k\pi}{2N}\right) \quad k = 0, 1, 2 \ldots N - 1 \qquad (12.149)$$

no matter which transformation coefficient Q_{ik} is chosen. S^α becomes

$$S^\alpha = k_B T \sum_{k=0}^{N-1} \alpha_k^2 q_k^2 \qquad (12.150)$$

Three roots, or eigenvalues, are all that are allowed in the three dimensional case, and the zero root is a valid root. This root occurs in a chain of any length, and contributes nothing to the averages. For this reason the sums over any index referencing the normal coordinates, such as k or l, start with 1 and go to N even though N does not occur (the space spanned must be N dimensional, and the zero root coordinate spans one of those dimensions). If the general applicability and accuracy of the diagonalization is calculated using a computer program such as MathCad, start both indexes i and k with zero and run them to $N - 1$, using the coefficients of transformation

$$Q_{ik} = \sqrt{\frac{2 - \delta_{k0}}{N}} \cos\left(\frac{(2i + 1)k\pi}{2N}\right) \tag{12.151}$$

For a potential V

$$V = (r_1 - r_0)^2 + \cdots + (r_{N-1} - r_{N-2})^2 = r^T F r \tag{12.152}$$

where r is a $1 \times N$ matrix of the coordinates r_i and F is an $N \times N$ matrix of the coefficients of the coordinates r_i. For a 3 segment chain, F has the form

$$F = \begin{pmatrix} 1 & -1 & 0 \\ -1 & 2 & -1 \\ 0 & -1 & 1 \end{pmatrix} \tag{12.153}$$

The squared eigenvalues Λ_k of the transformation are

$$\Lambda_k^2 = 4\sin^2\left(\frac{k\pi}{2N}\right) \quad k = 0, 1, 2 \tag{12.154}$$

Let q be a column matrix $(1 \times N)$ of the normal coordinates q_k, and Q be the matrix of the transformation coefficients. All the transformation equations are then expressed compactly as

$$r = Qq \tag{12.155}$$

It is easy to show that Q is an orthonormal matrix, obeying the relation $Q^T Q = QQ^T = E$ where E is the identity matrix of the same dimensionality as Q. Then $q = Q^T r$ (the T denotes the operation transposition) and

$$V = q^T Q^T F Q q = q^T \Lambda q \tag{12.156}$$

showing that the eigenvalues are

$$\Lambda = Q^T F Q \tag{12.157}$$

This analysis differs from that for the polymer chain only in the constants (4 instead of $6/b^2$ in the squared eigenvalues)

After transforming the coordinates to normal, the operators L_a and L_b operate on them, and need to be converted to normal form. Suppose these operators have to work on a function of the normal coordinates. The function f has differentials defined by noting that

$$f = f(q_1, q_2, \ldots, q_N) \tag{12.158}$$

and as

$$df = \sum_{k=1}^{N} \frac{\partial f}{\partial q_k} dq_k \tag{12.159}$$

the derivatives of f with respect to the r_i are

$$\frac{\partial f}{\partial r_i} = \sum_{k=1}^{N} \frac{\partial f}{\partial q_k} \frac{\partial q_k}{\partial r_i} = \sum_{k=1}^{N} \frac{\partial f}{\partial q_k} Q_{ki} \tag{12.160}$$

Then, after some algebra, the operator L_a becomes

$$L_a = -k_B T \sum_{k=1}^{N} \sum_{l=1}^{N} \left(\frac{\partial}{\partial q_k} - 2\alpha_k^2 q_k \right) \cdot F_{kl} \cdot \left(\frac{\partial}{\partial q_l} + \frac{\partial V}{\partial q_l} \right) \tag{12.161}$$

where

$$F_{kl} = \sum_{i=1}^{N} \sum_{j=1}^{N} Q_{ki} Q_{lj} D_{ij} \tag{12.162}$$

The potential is still $V = (U - S^\alpha + W)/k_B T$.

L_b is not required in this application (mean square end to end distance and gyration radius, in both of which solvent velocity is zero).

If the hydrodynamic interaction is turned off to obtain the free draining operator in normal coordinates (for which $V = 0$ and $D_{ij} = \beta^{-1}\delta_{ij}$) then

$$F_{kl}^a = \beta^{-1} \sum_{i=1}^{N} \sum_{j=1}^{N} Q_{ki} Q_{lj} \delta_{ij} \tag{12.163}$$

which evaluates as

$$F_{kl}^a = \beta^{-1}\delta_{kl} \tag{12.164}$$

That is, F_{kl} has only diagonal terms. Then L_a is a sum of operators, each of which is identical in form to the form the free draining operator would assume if only one coordinate r were involved. This leads to the equation that the Hermite polynomials satisfy, one such equation for each normal coordinate. When expressed in normal coordinates, the free draining operator L_a has eigenfunctions that are products of Hermite polynomials, one for each Cartesian component of a normal coordinate $q_{k_i}, i = x, y, z$. These functions are the logical ones to select in expanding ρ, as ρ does not vary much from unity. The free draining operator captures the majority of the behavior of the polymer chain, and is a good approximation to the exact operator. Its eigenfunctions constitute good first approximations to the exact.

The basis functions obtained from the free draining operator and expressed in normal coordinates are

$$|n\rangle \equiv \phi_n = \prod_{l=1}^{N} \phi_{n(lx)}(q_{lx})\phi_{n(ly)}(q_{ly})\phi_{n(lz)}(q_{lz}) \tag{12.165}$$

where

$$\phi_{n(lx)}(q_{lx}) = N_{n(lx)}H_{n(lx)}(\alpha_l q_{lx}) \tag{12.166}$$

and

$$N_{n(lx)} = \left[2^{n(lx)}n(lx)!\right]^{-1/2} \tag{12.167}$$

The matrix representation of an operator L is

$$\langle n|L|n'\rangle = L_{nn'} = \int \psi^\alpha \phi_n L \phi_n, dq_1, dq_2 \ldots dq_N \tag{12.168}$$

The normalization constant is included in ψ^α, and the special case

$$\langle 0|0\rangle = \int \psi^\alpha dq_1 dq_2 \cdots dq_N = 1 \tag{12.169}$$

The expansion of ρ in a series of the basis functions is of the form

$$\rho = \sum_n \rho_n |n\rangle \tag{12.170}$$

where the expansion coefficients are given by

$$\rho_n = \langle \boldsymbol{n} | \rho \rangle \tag{12.171}$$

The average of any function of the normal coordinates $\mathbf{q}$ is

$$\langle Q \rangle = \int \psi^\alpha \rho Q dq_1 dq_2 \cdots dq_N \tag{12.172}$$

since the probability including the excluded volume effects is $\psi^\alpha \rho$. In the bra and ket formalism, this is written

$$\langle Q \rangle = \langle 0 | Q | \rho \rangle \tag{12.173}$$

The function Q may be the end to end distance or the gyration radius in the following.

12.7. The Boson Representation

In normal coordinates, S^α is diagonal, and the weighting function ψ^α is expressed as

$$\psi^\alpha = \prod_{k=1}^{N} \prod_{i=1}^{3} \frac{\alpha_k}{\pi^{1/2}} e^{-\alpha_k^2 q_{ki}^2} \tag{12.174}$$

Differentiation of the functions $\phi_{n(lx)}$ produces

$$\frac{\partial \phi_{n(lx)}}{\partial q_{lx}} = N_{n(lx)} \frac{\partial H_{n(lx)}(\alpha_l q_{lx})}{\partial q_{lx}} \tag{12.175}$$

Multiplying numerator and denominator by α_l produces

$$\frac{\partial \phi_{n(lx)}}{\partial q_{lx}} = \alpha_l N_{n(lx)} \frac{\partial H_{n(lx)}(\alpha_l q_{lx})}{\partial \alpha_l q_{lx}} \tag{12.176}$$

allowing use of the property of the Hermite polynomials

$$\frac{\partial \phi_{n(lx)}}{\partial q_{lx}} = \alpha_l N_{n(lx)} \left[2n(lx) H_{n(lx)-1}(\alpha_l q_{lx}) \right] \tag{12.177}$$

More compactly,

$$\frac{\partial \phi_{n(lx)}}{\partial q_{lx}} = \alpha_l (2n(lx))^{1/2} \phi_{n(lx)-1} \tag{12.178}$$

This suggests defining the annihilation operator as

$$b_{lx}|n(lx)\rangle = (n(lx))^{1/2}|n(lx) - 1\rangle \qquad (12.179)$$

The boson representation of the derivative is taken as

$$\frac{\partial}{\partial q_l} = \alpha_l 2^{1/2} b_l \qquad (12.180)$$

Note that this representation is not proportional to $b - b^+$. The operators defined here nonetheless obey the same commutation rule the operators defined in the discussion of the Hermite orthogonal functions above, and it is the algebra that is critical.

Insisting that the creation operator retain the property

$$b^+_{n(lx)}|n(lx)\rangle = (n(lx) + 1)^{1/2}|n(lx) + 1\rangle \qquad (12.181)$$

leads immediately to

$$q_l = \frac{1}{\alpha_l 2^{1/2}}(b_l + b^+_l) \qquad (12.182)$$

where

$$b_l = \sum_{i=1}^{3} u_i b_{li} \qquad b^+_l = \sum_{i=1}^{3} u_i b^+_{li} \qquad (12.183)$$

The u_i are unit vectors directed along orthogonal axes, and the b_{li} and b^+_{li} are components of the vector boson operators b and b^+. The next step is to cast the operator L_a into boson form. To this end, consider

$$\frac{\partial}{\partial q_k} - 2\alpha_k^2 q_k = \alpha_k 2^{1/2} b_k - 2\alpha_k^2 \left(\frac{1}{\alpha_k 2^{1/2}}(b_k + b^+_k) \right) \qquad (12.184)$$

or, collecting like terms,

$$\frac{\partial}{\partial q_k} - 2\alpha_k^2 q_k = -\alpha_k 2^{1/2} b^+_k \qquad (12.185)$$

The postmultiplier is treated by regarding $\frac{\partial V}{\partial q_k}$ as an operator that passes every function written on its right, allowing $f\frac{\partial V}{\partial q_k} = \frac{\partial V}{\partial q_k} f$ to be written. Then the commutator of the derivative with V has the property

$$\left(\frac{\partial}{\partial q_k}, V\right) f = f \frac{\partial V}{\partial q_k} \tag{12.186}$$

The derivative operates on everything to its right and to demonstrate this identity, evaluate the commutator working on the function f

$$\left(\frac{\partial}{\partial q_l}, V\right) f = \frac{\partial V f}{\partial q_l} - V \frac{\partial f}{\partial q_l} = \frac{\partial V}{\partial q_l} f + V \frac{\partial f}{\partial q_l} - V \frac{\partial f}{\partial q_l} = \frac{\partial V}{\partial q_l} f$$

$$= f \frac{\partial V}{\partial q_l} \tag{12.187}$$

since $f \frac{\partial V}{\partial q_l} = \frac{\partial V}{\partial q_l} f$. This allows writing the derivative $\frac{\partial V}{\partial q_l}$ as the commutator $\left(\frac{\partial}{\partial q_l}, V\right)$. However, although the function f passes through the derivative, it does not pass through the commutator, which is left unresolved if f is written to its left. This algebra allows

$$\frac{\partial}{\partial q_l} + \frac{\partial V}{\partial q_l} = \frac{\partial}{\partial q_l} + \left(\frac{\partial}{\partial q_l}, V\right) \tag{12.188}$$

Substituting the boson equivalents obtains

$$\frac{\partial}{\partial q_l} + \frac{\partial V}{\partial q_l} = \alpha_l 2^{1/2} b_l + (\alpha_l 2^{1/2} b_l, V) \tag{12.189}$$

or, more compactly,

$$\frac{\partial}{\partial q_l} + \frac{\partial V}{\partial q_l} = \alpha_l 2^{1/2} [b_l + (b_l, V)] \tag{12.190}$$

and L_a becomes

$$L_a = -k_B T \sum_{k=1}^{N} \sum_{l=1}^{N} b_k^+ (-2\alpha_k \alpha_l F_{kl})(b_l + (b_l, V)) \tag{12.191}$$

and, defining Λ_{kl} the operator is

$$L_a = \sum_{k=1}^{N} \sum_{l=1}^{N} b_k^+ \Lambda_{kl}(b_l + (b_l, V)) \tag{12.192}$$

where $\Lambda_{kl} = 2k_B T \alpha_k \alpha_l F_{kl}$ contains the friction constant and the hydrodynamic interaction.

12.8. Excluded Volume Potential

The next step is to calculate the excluded volume potential. In calculating either the end to end distance or the gyration radius in the mean, there is no external potential. For these problems, the external potential W is set equal to zero, and the internal potential U consists only of the spring potential S and the excluded volume potential E. That is,

$$U = S + E \tag{12.193}$$

The spring potential is

$$S = \alpha^2 S^\alpha = \frac{3k_B T}{2b_0^2} \sum |r_i - r_{i+1}|^2 \tag{12.194}$$

The mean segment length b is set equal to a reference segment length b_0 modified by the parameter α as $b = \alpha b_0$. The parameter α is chosen to approximate excluded volume effects by maximizing convergence of certain series that will appear later.

As $V = (U - S^\alpha + W)/k_B T$ the potential becomes

$$k_B T V = U - S^\alpha = S + E - S^\alpha = \alpha^2 S^\alpha + E - S^\alpha = E + (\alpha^2 - 1)S^\alpha \tag{12.195}$$

There remains the choice of E. Let $V(r_{ij})$ be the actual potential for forces developed between segments i and j. These forces tend to zero at larger distances and have substantial value only when the segments are close together. The situation is very similar to that encountered in gases, and in the same spirit, define a function $\chi(r_{ij})$ as

$$e^{-V(r_{ij})/k_B T} = 1 - \chi(r_{ij}) \tag{12.196}$$

The function $\chi(r_{ij})$ is expected to approach zero with great rapidity as r_{ij} increases from zero to some finite but generally large value.

This behavior may be approximated by setting

$$\chi(\boldsymbol{r}_{ij}) = \delta(\boldsymbol{r}_{ij}) \int (1 - e^{-V(\boldsymbol{r}_{ij})/k}) d\boldsymbol{r}_{ij} = \delta(\boldsymbol{r}_{ij}) X \qquad (12.197)$$

Assume interaction between segments is restricted to single pairs. The excluded volume potential can then be written as

$$E = \sum_{i>j} V(\boldsymbol{r}_{ij}) \qquad (12.198)$$

It is an identity that

$$E = -k_B T \sum_{i>j} \ln \exp\left[-V(\boldsymbol{r}_{ij})/k_B T\right] \qquad (12.199)$$

and introducing $\chi(\boldsymbol{r}_{ij})$

$$E = -k_B T \sum_{i>j} \ln\left[1 - \chi(\boldsymbol{r}_{ij})\right] \qquad (12.200)$$

Since $\chi(\boldsymbol{r}_{ij})$ is small due to the rarity of segment to segment contacts, the excluded volume potential is approximately

$$E \cong -k_B T \sum_{i>j} -\chi(\boldsymbol{r}_{ij}) \qquad (12.201)$$

or

$$E = k_B T X \sum_{i>j} \delta(\boldsymbol{r}_{ij}) \qquad (12.202)$$

The exponential $e^{-V(\boldsymbol{r}_{ij})/k_B T}$ has a small probability of substantial fluctuation from its mean value of unity, and the effect of excluded volume is small, on the average, on observable quantities.

The correction ρ to the ideal probability ψ^a is proportional to $exp[-(U - S^\alpha + W)]/k_B T$. A boson expansion for $(U - S^\alpha + W)/k_B T$ will provide a boson expansion for ρ. In the above, $U = S + E = \alpha^2 S^\alpha + E$, As $W = 0$, boson expansions for S^α and E will give

ρ the required boson expansion. The expansion for S^a is

$$S^\alpha = \frac{1}{2}k_B T \sum_l (b_l + b_l^+)(b_l + b_l^+) \qquad (12.203)$$

The excluded volume potential E requires a boson representation of the Dirac delta function.

One of the principal properties of the delta function is

$$f(\mathbf{r}_{ij}) = \int_{-\infty}^{\infty} \delta(\mathbf{r} - \mathbf{r}_{ij}) f(\mathbf{r}) d\mathbf{r} \qquad (12.204)$$

The delta function may be formally represented as

$$\delta(\mathbf{r} - \mathbf{r}_{ij}) = \frac{1}{(2\pi)^3} \int_{-\infty}^{\infty} e^{-i\mathbf{k}\cdot(\mathbf{r}-\mathbf{r}_{ij})} d\mathbf{k} \qquad (12.205)$$

Let $\mathbf{r}_i$ be measured from the center of mass. The boson representation of r_i is

$$\mathbf{r}_i = \sum_{l=1}^{N} Q_{il}(b_l + b_l^+)/2^{1/2}\alpha_l \qquad (12.206)$$

Differences $\mathbf{r}_{ij}$ have the boson representation

$$\mathbf{r}_{ij} = \mathbf{r}_i - \mathbf{r}_j = \sum_{l=1}^{N} f_l(b_l + b_l^+) \qquad (12.207)$$

where f_l is defined as

$$f_l = (Q_{il} - Q_{jl})/2^{1/2}\alpha_l \qquad (12.208)$$

In the equation for the delta function, the part of the integrand that depends on $\mathbf{r}_{ij}$ is

$$e^{i\mathbf{k}\cdot\mathbf{r}_{ij}} = e^{i\mathbf{k}\cdot\sum_l f_l(b_l+b_l^+)} \qquad (12.209)$$

which is a product of exponential operators

$$e^{i\mathbf{k}\cdot\mathbf{r}_{ij}} = \prod_l e^{i\mathbf{k}\cdot f_l b_l + i\mathbf{k}\cdot f_l b_l^+} \qquad (12.210)$$

When a and b are operators that commute with their commutator, it is a theorem that[12]

$$e^{a+b} = e^a e^b e^{-\frac{1}{2}(a,b)} \qquad (12.211)$$

Since $a + b = b + a$, and $(b, a) = -(a, b)$, exchanging a and b gives

$$e^{a+b} = e^b e^a e^{\frac{1}{2}(a,b)} \qquad (12.212)$$

This gives rise to

$$e^{i\boldsymbol{k}\cdot\boldsymbol{r}_{ij}} = \prod_l e^{if_l\boldsymbol{k}\cdot\boldsymbol{b}_l} e^{if_l\boldsymbol{k}\cdot\boldsymbol{b}_l^+} e^{\frac{1}{2}f_l^2 k^2} \qquad (12.213)$$

where $(\boldsymbol{b}_l, \boldsymbol{b}_l^+) = 1$ has been used. Now consider the exponential operators

$$e^b e^a = e^{b+a} e^{\frac{1}{2}(b,a)} = e^{a+b} e^{-\frac{1}{2}(a,b)} = e^a e^b e^{-(a,b)} \qquad (12.214)$$

Then

$$e^{i\boldsymbol{k}\cdot\boldsymbol{r}_{ij}} = \prod_l e^{if_l\boldsymbol{k}\cdot\boldsymbol{b}_l^+} e^{if_l\boldsymbol{k}\cdot\boldsymbol{b}_l} e^{-f_l^2 k^2} e^{\frac{1}{2}f_l^2 k^2} \qquad (12.215)$$

Combining the arguments of the exponentials

$$e^{i\boldsymbol{k}\cdot\boldsymbol{r}_{ij}} = \prod_l e^{if_l\boldsymbol{k}\cdot\boldsymbol{b}_l^+} e^{if_l\boldsymbol{k}\cdot\boldsymbol{b}_l} e^{-\frac{1}{2}f_l^2 k^2} \qquad (12.216)$$

$$e^{i\boldsymbol{k}\cdot\boldsymbol{r}_{ij}} = e^{-\frac{1}{2}k^2 \sum_l f_l^2} \prod_l e^{if_l\boldsymbol{k}\cdot\boldsymbol{b}_l^+} e^{if_l\boldsymbol{k}\cdot\boldsymbol{b}_l} \qquad (12.217)$$

In order to simplify this equation, introduce the ordering operator Θ. This has the property of reordering the boson operators in the series expansion of $e^{i\boldsymbol{k}\cdot\sum_l f_l(\boldsymbol{b}_l + \boldsymbol{b}_l^+)}$ in such a way that all operators $\boldsymbol{b}^+$ are written to the left of products of the operator $\boldsymbol{b}$. Then by definition of Θ

$$e^{if_l\boldsymbol{k}\cdot\boldsymbol{b}_l^+} e^{if_l\boldsymbol{k}\cdot\boldsymbol{b}_l} \equiv \Theta e^{if_l\boldsymbol{k}\cdot(\boldsymbol{b}_l + \boldsymbol{b}_l^+)} \qquad (12.218)$$

and

$$e^{i\mathbf{k}\cdot\mathbf{r}_{ij}} = e^{-c_{ij}k^2}\Theta e^{i\mathbf{k}\cdot\mathbf{r}_{ij}} \tag{12.219}$$

Define

$$c_{ij} = \frac{1}{2}\sum_l f_l^2 \tag{12.220}$$

The delta function can then be written as

$$\delta(\mathbf{r} - \mathbf{r}_{ij}) = \frac{1}{(2\pi)^3}\Theta\int e^{-c_{ij}k^2}e^{i\mathbf{k}\cdot(\mathbf{r}_{ij}-\mathbf{r})}d\mathbf{k} \tag{12.221}$$

The equation is in an especially convenient form for integration. A theorem from the theory of Fourier integrals[14]

$$e^{-x^2} = \frac{1}{\pi^{1/2}}\int_{-\infty}^{\infty} e^{-u^2+2ixu}du \tag{12.222}$$

This may be established as follows. Consider the square $(u + ix)^2 = u^2 - 2ixu - x^2$. Then $-u^2 + 2ixu = -(u - ix)^2 - x^2$, and the integral I may be rewritten as

$$I = \frac{1}{\pi^{1/2}}\int_{-\infty}^{\infty} e^{-(u-ix)^2-x^2}du \tag{12.223}$$

Factoring the x dependence from the integration over u gives

$$I = \frac{e^{-x^2}}{\pi^{1/2}}\int_{-\infty}^{\infty} e^{-(u-ix)^2}du \tag{12.224}$$

This suggests setting the complex variable $z = u - ix$, noting that as x is constant, $du = dz$. The integral becomes

$$I = \frac{e^{-x^2}}{\pi^{1/2}}\int_{-\infty}^{\infty} e^{-z^2}dz \tag{12.225}$$

But $\int_{-\infty}^{\infty} e^{-z^2}dz = \pi^{1/2}$, and therefore $I = e^{-x^2}$, which establishes the theorem.

The theorem may be used to evaluate the delta function. Set $\mathbf{u} = c_{ij}^{1/2}\mathbf{k}$ and $d\mathbf{u} = c_{ij}^{1/2}d\mathbf{k}$. Keeping in mind that there are three dimensions involved, the theorem allows writing

$$\delta(\boldsymbol{r} - \boldsymbol{r}_{ij}) = \frac{1}{(4\pi c_{ij})^{3/2}} \Theta e^{-|r-r_{ij}|^2/4c_{ij}} \tag{12.226}$$

The numerical factors c_{ij} are

$$c_{ij} = \frac{1}{4} \sum_{l=1}^{N} (Q_{li} - Q_{lj})^2 / \alpha_l^2 \tag{12.227}$$

When N is very large, $\sin^2(\frac{l\pi}{2N}) \cong (\frac{l\pi}{2N})^2$, since the infinite series $\sin(x) = x - \frac{x^3}{3!} + \cdots$ defines $\sin(x)$ in both the fields of real and complex numbers, and for sufficiently small x the higher terms in the series contribute negligibly. The eigenvalues are approximately

$$\alpha_l^2 = \frac{6}{b^2} \sin^2 \left(\frac{l\pi}{2N} \right) \cong \frac{3l^2\pi^2}{2\alpha^2 b_0^2 N^2} \tag{12.228}$$

and c_{ij} becomes

$$c_{ij} \cong \frac{1}{4} \sum_{l=1}^{N} \frac{2}{N} \left(\cos \left(\frac{li\pi}{2N} \right) - \cos \left(\frac{lj\pi}{2N} \right) \right)^2 \frac{2\alpha^2 b_0^2 N^2}{3l^2\pi^2} \tag{12.229}$$

combining terms,

$$c_{ij} \cong \frac{N b_0^2 \alpha^2}{3\pi^2} \sum_{l=1}^{N} l^{-2} \left(\cos \left(\frac{li\pi}{2N} \right) - \cos \left(\frac{lj\pi}{2N} \right) \right)^2 \tag{12.230}$$

Jolley,[8] formula 681 gives

$$\sum_{l=1}^{\infty} l^{-2} \cos(la) \cos(l\theta) = \frac{1}{4}\theta^2 + \frac{1}{4}(a - \pi)^2 - \frac{\pi^2}{12} \tag{12.231}$$

By taking N as infinity, this formula can be used to evaluate the sum in c_{ij}. The result is

$$c_{ij} \cong \frac{1}{6}\alpha^2 b_0^2 |i - j| \tag{12.232}$$

The boson representation of the excluded volume potential E may now be written

$$E = k_B T X \sum_{i>j} \delta(\mathbf{r}_{ij}) \tag{12.233}$$

$$E = k_B T X \sum_{i>j} \frac{1}{(4\pi c_{ij})^{3/2}} \Theta e^{-r_{ij}^2/4c_{ij}} \tag{12.234}$$

$$E = k_B T X \Theta \sum_{i>j} \frac{1}{(4\pi c_{ij})^{3/2}} \left(1 - \frac{r_{ij}^2}{4c_{ij}} + \cdots \right) \tag{12.235}$$

The leading term is constant and will vanish when the gradient is taken, Higher terms than the quadratic are presumed small and presumably difficult to handle. Since $\mathbf{r}_{ij} = \sum_l f_l(\mathbf{b}_l + \mathbf{b}_l^+)$, the excluded volume potential can be approximated as

$$E \cong -\frac{k_B T X}{32\pi^{3/2}} \sum_{i>j} \frac{1}{c_{ij}^{5/2}} \sum_k \sum_l f_k f_l (\mathbf{b}_k + \mathbf{b}_k^+)(\mathbf{b}_l + \mathbf{b}_l^+) \tag{12.236}$$

This has the same boson structure as S^α, and the potential V is

$$V = \frac{1}{k_B T}(U - S^\alpha) = \frac{1}{k_B T}((\alpha^2 - 1)S^\alpha + E) \tag{12.237}$$

Giving this a boson representation

$$V = \frac{1}{2} \sum_k \sum_l G_{kl}(\mathbf{b}_k + \mathbf{b}_k^+)(\mathbf{b}_l + \mathbf{b}_l^+) \tag{12.238}$$

where

$$G_{kl} = (\alpha^2 - 1)\delta_{kl} - \frac{X f_k f_l}{16\pi^{3/2}} \sum_{i<j} c_{ij}^{-5/2} \tag{12.239}$$

The diagonal elements of G_{kl} are called G_l, and are

$$G_l = \alpha^2 - 1 - \frac{X}{16\pi^{3/2}} \sum_{i<j} \frac{f_l^2}{c_{ij}^5} \tag{12.240}$$

In order to calculate numerical values for G_l, the following equations are needed

$$c_{ij} = \frac{1}{6}|i - j|b^3$$

$$b = \alpha b_0$$

$$f_l^2 = \frac{\left[\cos\left(\frac{lj\pi}{2N}\right) - \cos\left(\frac{li\pi}{2N}\right)\right]^2}{2N\alpha_l^2}$$

$$\alpha_l^2 = \frac{6}{b^2}\sin^2\left(\frac{l\pi}{2N}\right)$$

$$X \equiv \int \left(1 - e^{-V(r)/k_B T}\right)dr$$

The parameter z is often used in series expansions that approximate excluded volume forces, and is introduced here as

$$z \equiv \left(\frac{3}{2\pi b_0^2}\right)^{3/2} N^{1/2} \int \left(1 - e^{-V(r)/k_B T}\right)dr \qquad (12.241)$$

There results

$$G_l = \alpha^2 - 1 - \frac{z}{\alpha^3}g_l \qquad (12.242)$$

where

$$g_l = \frac{1}{2}N^{-3/2}\left[\sin\left(\frac{l\pi}{2N}\right)\right]^{-2}$$

$$\sum_{i<j}\left[\cos\left(\frac{lj\pi}{2N}\right) - \cos\left(\frac{li\pi}{2N}\right)\right]^2 |i - j|^{-5/2} \qquad (12.243)$$

When N is large, the g_l become essentially independent of N and are related to the Fresnel integrals. The first few values are given in Table 12.1.

The commutator (b_l, V) occurs in the operator L_a, and is evaluated by insisting that all b^+ operators be written to the left of the corresponding operator b, making use of the fact that operators b_k

Table 12.1.   Values of g_l from 1 to 13. For $l > 13$, g_l may be calculated from an approximate formula given by Fixman,

$$2g_l \cong l^{-3/2}[1.2004217][l\pi - 1.0857865 + l^{-1/2}\{0.4501585 - 0.0000013(-1)^l\}]$$

in which a term proportional to $l^{-5/2}$ is omitted as contributing trivially. For $k \neq l$, calculation shows that the typical value of g_{kl} is quite small, and these values are ignored in this treatment.

l	g_l	l	g_l
1	1.5157	7	0.6830
2	1.1705	8	0.6421
3	0.9934	9	0.6077
4	0.8782	10	0.5784
5	0.7958	11	0.5529
6	0.7329	12	0.5304
		13	0.5107

and b_k^+ commute with all operators b_l and b_l^+ when k and l are not equal. The commutator is

$$(b_l, V) = b_l V - V b_l = b_l \frac{1}{2} \sum_k G_k (b_k + b_k^+)(b_k + b_k^+)$$

$$- \frac{1}{2} \sum_k G_k (b_k + b_k^+)(b_k + b_k^+) b_l \quad (12.244)$$

$$(b_l, V) = \frac{1}{2} G_l [b_l (b_l + b_l^+)(b_l + b_l^+)$$

$$- (b_l + b_l^+)(b_l + b_l^+) b_l] \quad (12.245)$$

an expression in which all the b subscripts are the same. Dropping these for ease of algebra, the quantity in square brakets is essentially $b(b + b^+)(b + b^+) - (b + b^+)(b + b^+)b^+$. These terms expand as

$$b(bb + bb^+ + b^+b + b^+b^+) = bbb + bbb^+ + bb^+b + bb^+b^+ \quad (12.246)$$

and

$$-(bb + bb^+ + b^+b + b^+b^+)b = -bbb - bb^+b - b^+bb - b^+b^+b \quad (12.247)$$

Canceling terms that are of opposite sign, there remains $bbb^+ + bb^+b^+ - b^+bb - b^+b^+b$.

Although both negative terms have the creation operators b^+ to the left of the destructors b, the positive terms need work. The commutator $(b, b^+) = bb^+ - b^+b = 1$, hence $bb^+ = 1 + b^+b$. Applying this repeatedly, the several terms become $bbb^+ = b + bb^+b$ and $bb^+b^+ = b^+ + b^+bb^+$. Also $b^+b = bb^+ - 1$, so $-b^+bb = -bb^+b + b$ and $-b^+b^+b = -b^+bb^+ + b^+$. The sum of the four terms is then

$$bbb^+ + bb^+b^+ - b^+bb - b^+b^+b = 2b + 2b^+ \qquad (12.248)$$

This algebra shows that the commutator

$$(\boldsymbol{b}_l, V) = G_l(\boldsymbol{b}_l + \boldsymbol{b}_l^+) \qquad (12.249)$$

Substituting this result into the operator L_a

$$L_a = \sum_k \sum_l \boldsymbol{b}_k^+ \Lambda_{kl} \left[\boldsymbol{b}_l + (\boldsymbol{b}_l, V) \right] \qquad (12.250)$$

$$L_a = \sum_k \sum_l \boldsymbol{b}_k^+ \Lambda_{kl} \left[\boldsymbol{b}_l + G_l(\boldsymbol{b}_l + \boldsymbol{b}_l^+) \right] \qquad (12.251)$$

$$L_a = \sum_k \sum_l \boldsymbol{b}_k^+ \Lambda_{kl} \left[G_l \boldsymbol{b}_l^+ + (1 + G_l) \boldsymbol{b}_l \right] \qquad (12.252)$$

This calculation suggests that a table of commutators of products of b and b^+ operators might prove useful for simplifying complicated products. A few are given in Table 12.2.

Example of calculation

Since $(b, b^+) = 1$, $bb^+ = b^+b + 1$ and

$$(b, b^+b^+) = bb^+b^+ - b^+b^+b = (1 + b^+b)b^+ - b^+(bb^+ - 1)$$
$$= b^+ + b^+bb^+ - b^+bb^+ + b^+ = 2b^+$$

The average of a function of coordinates Q is taken with respect to the probability ψ, which includes excluded volume contributions.

Table 12.2.   Some commutation relations for products of the creation and destruction operators b^+ and b.

$(b, b) = 0$	$(b, b^+) = 1$	$(b^+, b^+) = 0$
$(b, b + b^+) = 1$		
	$(b + b^+, b^+) = 1$	
	$(b^+, b) = -1$	
$(b + b^+, b) = -1$	$(b^+, b + b^+) = -1$	
$(b, bb) = 0$	$(b^+, b^+ b^+) = 0$	
$(b, bb^+) = b$	$(b^+, bb^+) = -b^+$	
$(b, b^+ b) = b$	$(b^+, b^+ b) = -b^+$	
$(b, b^+ b^+) = 2b^+$	$(b^+, bb) = -2b$	
$(bb, bb) = 0$	$(b^+ b^+, b^+ b^+) = 0$	
$(bb, bb^+) = 2bb$	$(b^+ b^+, bb^+) = -2b^+ b^+$	
$(bb, b^+ b) = 2bb$	$(b^+ b^+, b^+ b) = -2b^+ b^+$	
$(bb, b^+ b^+) = 4(b^+ b + \frac{1}{2})$	$(b^+ b^+, bb) = -4(b^+ b + \frac{1}{2})$	
$(bb^+, bb) = -2bb$	$(b^+ b, bb) = -2bb$	
$(bb^+, bb^+) = 0$	$(b^+ b, bb^+) = 0$	
$(bb^+, b^+ b) = 0$	$(b^+ b, b^+ b) = 0$	
$(bb^+, b^+ b^+) = 2b^+ b^+$	$(b^+ b, b^+ b^+) = 2b^+ b^+$	

That is,

$$\langle Q \rangle = \int \psi Q d\boldsymbol{r} \tag{12.253}$$

Since $\psi = \psi^\alpha \rho$, and $H_0 = 1 \equiv |0\rangle$, averages of Q may be written in several ways, specifically

$$\langle Q \rangle = \int \psi^\alpha \rho Q d\boldsymbol{r} = \int \psi^\alpha Q \rho d\boldsymbol{r} = \int \psi^\alpha \cdot 1 \cdot Q \cdot \rho \cdot d\boldsymbol{r} = \langle 0 | Q | \rho \rangle \tag{12.254}$$

The probability correction ρ expands as

$$\rho = \sum_n \rho_n |n\rangle \tag{12.255}$$

The coefficients ρ_n are given by

$$\rho_n = \langle n | \rho \rangle = \int \psi^\alpha \cdot H_n \cdot \rho \cdot 1 \cdot d\boldsymbol{r} = \langle n | \rho | 0 \rangle \tag{12.256}$$

which implies that

$$|\rho\rangle = \rho |0\rangle \tag{12.257}$$

when ρ is given its boson representation. Averages of functions of the coordinates alone are calculated from equation (12.254). The equilibrium solution has $\rho \sim e^{-V}$, where V is given its boson representation. That requires

$$|\rho\rangle = e^{-V}|0\rangle \tag{12.258}$$

apart from a normalization constant, and upon giving V its boson representation, the expansion can proceed. The calculation is

$$V = \frac{1}{2}\sum_l G_l(b_l + b_l^+)(b_l + b_l^+)$$

$$V = \frac{1}{2}\sum_l G_l(1 + b_l b_l + 2b_l^+ b_l + b_l^+ b_l^+)$$

Then

$$e^{-V} = \prod_l e^{-\frac{1}{2}G_l(1 + b_l b_l + 2b_l^+ b_l + b_l^+ b_l^+)}$$

$$e^{-V} = e^{-\frac{1}{2}\sum_l G_l}\prod_l e^{-\frac{1}{2}G_l(b_l b_l + 2b_l^+ b_l + b_l^+ b_l^+)}$$

The probability correction operator ρ is found by operating on the vacuum state $|0\rangle$

$$\rho = e^{-V}|0\rangle$$

$$= e^{-\frac{1}{2}\sum_l G_l}\prod_l \left[1 - \frac{1}{2}G_l(b_l b_l + 2b_l^+ b_l + b_l^+ b_l^+) + \cdots\right]|0\rangle$$

Since $b_l|0\rangle = 0$, the only survivor of the operations indicated is $b_l^+ b_l^+|0\rangle$. Higher powers are of a more complicated nature, such as $(b_l b_l + 2b_l^+ b_l + b_l^+ b_l^+)^2|0\rangle$. Of course powers of $b_l^+ b_l^+|0\rangle$ will survive, but operators such as $b_l^+ b_l b_l^+ b_l^+|0\rangle$ or $b_l b_l^+ b_l^+ b_l^+|0\rangle$ will also survive. These can be converted by application of the commutator $(b_l, b_l^+) = 1$ to a form in which only powers of $b_l^+ b_l^+$ and numerical values remain after operation on the vacuum state. For example,

$$
\begin{aligned}
b_l^+ b_l b_l^+ b_l^+ &= b_l^+ (1 + b_l^+ b_l) b_l^+ = b_l^+ b_l^+ + b_l^+ b_l^+ b_l b_l^+ \\
&= b_l^+ b_l^+ + b_l^+ b_l^+ (1 + b_l^+ b_l) \\
&= b_l^+ b_l^+ + b_l^+ b_l^+ + b_l^+ b_l^+ b_l^+ b_l
\end{aligned}
$$

Applied to the vacuum state, the only survivor is $2 b_l^+ b_l^+ |0\rangle$. To achieve such results, however, the exponential in the expression for e^{-V} cannot be treated using the exponential relation $e^{a+b} = e^a e^b e^{-\frac{1}{2}(a,b)}$ because the operators do not commute with their commutators, as is shown in Table 12.2. The desired factorization is into powers of exponentials of $b_l^+ b_l^+$ of the form $|\rho\rangle = (e^{\sum_l M_l b_l^+ b_l^+})|0\rangle$ where the M_l are numbers that remain to be determined.

To determine M_l, start with the boson representation of the operator L_a. At equilibrium, the operator demands that for each l

$$
\left[G_l b_l^+ + (1 + G_l) b_l \right] |\rho\rangle = 0 \tag{12.259}
$$

Boson operators belonging to different normal modes commute. It therefore suffices to consider a single normal mode. For any one of them

$$
\left[G_l b_l^+ + (1 + G_l) b_l \right] e^{M_l b_l^+ b_l^+} |0\rangle = 0 \tag{12.260}
$$

Premultiply by the operator $e^{-M_l b_l^+ b_l^+}$ and observe that b_l^+ commutes with both $e^{-M_l b_l^+ b_l^+}$ and $e^{M_l b_l^+ b_l^+}$.

$$
e^{-M_l b_l^+ b_l^+} \left[G_l b_l^+ + (1 + G_l) b_l \right] e^{M_l b_l^+ b_l^+} |0\rangle = 0
$$

$$
\left[G_l b_l^+ + (1 + G_l) e^{-M_l b_l^+ b_l^+} b_l e^{M_l b_l^+ b_l^+} \right] |0\rangle = 0
$$

This suggests that perhaps $b_l^{M_l} = e^{-M_l b_l^+ b_l^+} b_l e^{M_l b_l^+ b_l^+}$. Assume that this is so. The derivative is

$$
\frac{d b_l^{M_l}}{d M_l} = e^{-M_l b_l^+ b_l^+} (-b_l^+ b_l^+ b_l + b_l b_l^+ b_l^+) e^{M_l b_l^+ b_l^+}
$$

Now reduce this result.

$$\frac{d\boldsymbol{b}_l^{M_l}}{dM_l} = e^{-M_l \boldsymbol{b}_l^+ \boldsymbol{b}_l^+} (\boldsymbol{b}_l, \boldsymbol{b}_l^+ \boldsymbol{b}_l^+) e^{M_l \boldsymbol{b}_l^+ \boldsymbol{b}_l^+}$$

$$\frac{d\boldsymbol{b}_l^{M_l}}{dM_l} = e^{-M_l \boldsymbol{b}_l^+ \boldsymbol{b}_l^+} 2\boldsymbol{b}_l^+ e^{M_l \boldsymbol{b}_l^+ \boldsymbol{b}_l^+} = 2\boldsymbol{b}_l^+$$

according to an entry in Table 12.2. Integrating this derivative

$$\boldsymbol{b}_l^{M_l} = 2\boldsymbol{b}_l^+ \int dM_l = 2M_l \boldsymbol{b}_l^+ + \boldsymbol{b}_l$$

The constant of integration is evaluated by setting $M_l = 0$. Then $\boldsymbol{b}_l^0 = 0 + \text{constant} = e^0 \boldsymbol{b}_l e^0 = \boldsymbol{b}_l$, where the definition $\boldsymbol{b}_l^{M_l} = e^{-M_l \boldsymbol{b}_l^+ \boldsymbol{b}_l^+} \boldsymbol{b}_l e^{M_l \boldsymbol{b}_l^+ \boldsymbol{b}_l^+}$ has been used. Making these substitutions, for any value of l

$$(G\boldsymbol{b}^+ + (1+G)\boldsymbol{b}^M)|0\rangle = 0$$

becomes, on substituting $\boldsymbol{b}^M = 2M\boldsymbol{b}^+ + \boldsymbol{b}$,

$$\left[G + (1+G)(2M\boldsymbol{b}^+ + \boldsymbol{b})\right]|0\rangle = 0$$

But $\boldsymbol{b}|0\rangle = 0$. This implies that for any value of l,

$$(G + (1+G)2M)\boldsymbol{b}^+|0\rangle = 0$$

Since $\boldsymbol{b}^+|0\rangle$ is certainly not zero, the equality may be maintained only if

$$(G + (1+G)2M) = 0$$

whence

$$M = -\frac{1}{2}\frac{G}{1+G}$$

for every l. The equilibrium ρ becomes

$$\rho = e^{-\frac{1}{2}\sum_l G_l(1+G_l)^{-1}\boldsymbol{b}_l^+ \boldsymbol{b}_l^+}|0\rangle \tag{12.261}$$

The quantity α is not yet defined. It should be chosen in such a way as to maintain ρ as near unity as possible. Clearly, if all the G_l were set equal to zero, ρ would be exactly one, and there would be no excluded volume effect. On the other hand, if any G_l were set equal to -1, that would become indefinitely large and ρ would vanish. The choice of α must be made so that all M_l remain finite. Setting the first $G_l = 0$ does this, for then

$$G_1 = 0 = \alpha^2 - 1 - \frac{z}{\alpha^3} g_1$$

and as $g_1 = 1.5157\ldots$, α is determined by the equation

$$\alpha^5 - \alpha^3 = 1.5157\ldots z$$

For other values of l there follows

$$G_l = (g_1 - g_l)\frac{z}{\alpha^3}$$

12.9. Radius of Gyration

The reader is reminded that r_i is a vector from the center of gravity of the N segment polymer chain to the ith segment of that chain. The distance from the ith segment to the jth is a vector $\boldsymbol{r}_{ij} = \boldsymbol{r}_i - \boldsymbol{r}_j$. With these definitions, the mean square gyration radius R is defined by

$$\langle R^2 \rangle = \frac{1}{N}\sum_i \langle r_i^2 \rangle = \frac{1}{2N^2}\sum_i \sum_j \langle r_{ij}^2 \rangle \tag{12.262}$$

The second written expression is convenient for giving a boson representation to the mean square gyration radius, since the boson representation of $\boldsymbol{r}_{ij}$ is

$$\boldsymbol{r}_{ij} = \sum_l f_l(\boldsymbol{b}_l + \boldsymbol{b}_l^+) \tag{12.263}$$

The square is

$$r_{ij}^2 = \sum_k \sum_l f_k f_l(\boldsymbol{b}_k + \boldsymbol{b}_k^+)(\boldsymbol{b}_l + \boldsymbol{b}_l^+) \tag{12.264}$$

which evaluates as

$$r_{ij}^2 = \sum_k \sum_l f_k f_l(\boldsymbol{b}_k \boldsymbol{b}_l + \boldsymbol{b}_k \boldsymbol{b}_l^+ + \boldsymbol{b}_k^+ \boldsymbol{b}_l + \boldsymbol{b}_k^+ \boldsymbol{b}_l^+) \tag{12.265}$$

When k and l are different, the bosons commute. However, when k and l are the same, this is no longer so, and the commutator has to be used to cast the equation in standard form. The result is

$$r_{ij}^2 = \sum_k \sum_l f_k f_l (\boldsymbol{b}_k \boldsymbol{b}_l + \boldsymbol{b}_l^+ \boldsymbol{b}_k + \boldsymbol{b}_k^+ \boldsymbol{b}_l + \boldsymbol{b}_k^+ \boldsymbol{b}_l^+ + 3\delta_{lk}) \quad (12.266)$$

The commutator only applies when l is equal to k, and the factor of three comes from the three Cartesian components of each $\boldsymbol{b}$. Since α was selected to make the G_l small, expansion of the exponential $|\rho\rangle$ in power series is allowed. Up to quadratic terms, the expansion is

$$|\rho\rangle = \left[1 - \frac{1}{2} \sum_{l'} G_{l'} (1 + G_{l'})^{-1} \boldsymbol{b}_{l'}^+ \boldsymbol{b}_{l'}^+ \right.$$

$$\left. + \frac{1}{8} \sum_{k'} \sum_{l'} G_{k'} G_{l'} (1 + G_{k'})^{-1} (1 + G_{l'})^{-1} \boldsymbol{b}_{k'}^+ \boldsymbol{b}_{k'}^+ \boldsymbol{b}_{l'}^+ \boldsymbol{b}_{l'}^+ + \cdots \right] |0\rangle$$

$$(12.267)$$

This may be broken into parts as

$$|\rho\rangle = |\rho\rangle_1 + |\rho\rangle_2 + |\rho\rangle_3 + \cdots \quad (12.268)$$

where the parts are defined as

$$|\rho\rangle_1 = |0\rangle \quad (12.269)$$

$$|\rho\rangle_2 = -\frac{1}{2} \sum_{l'} G_{l'} (1 + G_{l'})^{-1} \boldsymbol{b}_{l'}^+ \boldsymbol{b}_{l'}^+ |0\rangle \quad (12.270)$$

$$|\rho\rangle_3 = \frac{1}{8} \sum_{k'} \sum_{l'} G_{k'} G_{l'} (1 + G_{k'})^{-1} (1 + G_{l'})^{-1} \boldsymbol{b}_{k'}^+ \boldsymbol{b}_{k'}^+ \boldsymbol{b}_{l'}^+ \boldsymbol{b}_{l'}^+ |0\rangle$$

$$(12.271)$$

The formal average is then

$$\langle 0|r_{ij}^2|\rho\rangle = \langle 0|r_{ij}^2|\rho\rangle_1 + \langle 0|r_{ij}^2|\rho\rangle_2 + \langle 0|r_{ij}^2|\rho\rangle_3 + \cdots \quad (12.272)$$

As will be shown below, due to the special nature of the boson representation of r_{ij}^2, only the first two terms occur. All of the others vanish identically, and the average is exactly of the form

$$\langle 0|r_{ij}^2|\rho\rangle = \langle 0|r_{ij}^2|\rho\rangle_1 + \langle 0|r_{ij}^2|\rho\rangle_2 \qquad (12.273)$$

This relation is exact to the extent that ρ is itself exact. The validity of this equation and by extension to any power of r_{ij}, is shown by considering the contribution to the average of each term. For the first term

$$\langle 0|r_{ij}^2|\rho\rangle_1 = \langle 0|\sum_k \sum_l f_k f_l (b_k b_l + b_l^+ b_k + b_k^+ b_l + b_k^+ b_l^+ + 3_{lk})|0\rangle$$

$$(12.274)$$

Since $b_l|0\rangle = 0$, the first three terms make no contribution. Since the basis is orthogonal, $\langle 0|b^+ b^+|0\rangle = 2^{1/2}\langle 0|2\rangle = 0$. The basis is also normal, and the only survivor is

$$\langle 0|r_{ij}^2|\rho\rangle_1 = \langle 0|\sum_k \sum_l f_k f_l 3\delta_{kl}|0\rangle = 3\sum_l f_l^2 \qquad (12.275)$$

For the second term

$$\langle 0|r_{ij}^2|\rho\rangle_2 = \sum_k \sum_l \sum_{l'} -\frac{1}{2}G_{l'}(1 + G_{l'})^{-1}$$

$$\langle 0|(b_k b_l + b_k^+ b_l + b_l^+ b_k + b_l^+ b_k^+ + 3_{lk})(b_{l'}^+ b_{l'}^+)|0\rangle$$

$$(12.276)$$

The only survivor of the orthogonality of the basis is the term that involves the matrix element $\langle 0|b_l b_l b_l^+ b_l^+|0\rangle$. This is

$$\langle 0|b_l b_l b_l^+ b_l^+|0\rangle$$

$$= \langle 0|(b_{lx}b_{lx} + b_{ly}b_{ly} + b_{lz}b_{lz})(b_{lx}^+ b_{lx}^+ + b_{ly}^+ b_{ly}^+ + b_{lz}^+ b_{lz}^+)|0\rangle$$

$$(12.277)$$

This evaluates as

$$\langle 0|b_l b_l b_l^+ b_l^+|0\rangle = \langle 0|b_{lx}b_{lx}b_{lx}^+ b_{lx}^+ + b_{ly}b_{ly}b_{ly}^+ b_{ly}^+ + b_{lz}b_{lz}b_{lz}^+ b_{lz}^+|0\rangle$$

$$(12.278)$$

The sum is $2 + 2 + 2 = 6$. Then the whole second term is

$$\langle 0 | r_{ij}^2 | \rho \rangle_2 = -\frac{1}{2} \sum_l G_l (1 + G_l)^{-1} f_l^2 \cdot 6 = -3 \sum_l G_l (1 + G_l)^{-1} f_l^2$$

(12.279)

and exactly

$$\langle 0 | r_{ij}^2 | \rho \rangle = 3 \sum_l f_l^2 \left(1 - \frac{G_l}{1 + G_l} \right) = 3 \sum_l \frac{f_l^2}{1 + G_l}$$

(12.280)

The next term is $|\rho\rangle_3$. Formally, it is

$$\langle 0 | r_{ij}^2 | \rho \rangle_3 = \langle 0 | (r_{ij}^2) \frac{1}{8} \sum_{k'} \sum_{l'} G_{k'} G_{l'} (1 + G_{k'})^{-1} (1 + G_{l'})^{-1}$$

$$\times b_{k'}^+ b_{k'}^+ b_{l'}^+ b_{l'}^+ | 0 \rangle$$

(12.281)

where

$$r_{ij}^2 = \sum_k \sum_l f_k f_l (b_k b_l + b_l^+ b_k + b_k^+ b_l + b_k^+ b_l^+ + 3\delta_{lk})$$

(12.282)

The product of these two factors is the quadruple sum

$$\langle 0 | r_{ij}^2 | \rho \rangle = \frac{1}{8} \sum_k \sum_l \sum_{k'} \sum_{l'} f_k f_l G_{k'} G_{l'} (1 + G_{k'})^{-1} (1 + G_{l'})^{-1}$$

$$\times [\langle 0 | A + B + C + D + E | 0 \rangle]$$

(12.283)

where the operators A through E are

$$A = b_k b_l b_{k'}^+ b_{k'}^+ b_{l'}^+ b_{l'}^+$$

$$B = b_k^+ b_l b_{k'}^+ b_{k'}^+ b_{l'}^+ b_{l'}^+$$

$$C = b_l^+ b_k b_{k'}^+ b_{k'}^+ b_{l'}^+ b_{l'}^+$$

$$D = b_k^+ b_l^+ b_{k'}^+ b_{k'}^+ b_{l'}^+ b_{l'}^+$$

$$E = 3\delta_{lk} b_{k'}^+ b_{k'}^+ b_{l'}^+ b_{l'}^+$$

Since the basis is orthogonal, every term vanishes. There is no way the double destructor $b_k b_k$ can return the quadruple excitation resulting from the operation $b_{k'}^+ b_{k'}^+ b_{l'}^+ b_{l'}^+$ to the ground state.

In general, the average of any power of r_{ij} is now available. Consider first an odd power, r_{ij}^{2n+1}, where n is an integer. In order for $|\rho\rangle_n$ to contribute to the average, there must be as many destructors $bbbbb...$ as there are constructors $b^+b^+b^+b^+b^+...$ But when the power of r_{ij} is odd, there must be an odd number of destructors in its boson representation, while every $|\rho\rangle_x$ must contain an even number of constructors. Basis orthogonality requires that

$$\langle 0|r_{ij}^{2n+1}|\rho\rangle = 0 \qquad (12.284)$$

Even powers may contain no more terms in the series than the maximum number of destructors that r_{ij}^{2n} will allow. There must be the same number of creators in $|\rho\rangle_n$ as destructors in r_{ij}^{2n}. The first few possibilities are listed in Table 12.3.

Table 12.3.   Averages of powers of r_{ij} with excluded volume contributions.

n	$2n$	$\langle 0|r_{ij}^{2n}|\rho\rangle$
0	0	$\langle 0|r_{ij}^0|\rho\rangle_1 = \langle 0|1|0\rangle = 1$
1	2	$\langle 0|r_{ij}^2|\rho\rangle_1 + \langle 0|r_{ij}^2|\rho\rangle_2$
2	4	$\langle 0|r_{ij}^4|\rho\rangle_1 + \langle 0|r_{ij}^4|\rho\rangle_2 + \langle 0|r_{ij}^4|\rho\rangle_3$
3	6	$\langle 0|r_{ij}^6|\rho\rangle_1 + \langle 0|r_{ij}^6|\rho\rangle_2 + \langle 0|r_{ij}^6|\rho\rangle_3 + \langle 0|r_{ij}^6|\rho\rangle_4$

The mean square gyration radius including excluded volume effects is

$$\langle R^2\rangle = \frac{1}{2N^2}\sum_i\sum_j\langle 0|r_{ij}^2|\rho\rangle \qquad (12.285)$$

Substituting equation (12.280) and rearranging gives

$$\langle R^2\rangle = \frac{3}{2N^2}\sum_l(1+G_l)^{-1}\sum_i\sum_j f_l^2 \qquad (12.286)$$

where as displayed in equation (12.208) f_l^2 is a function of i and j

$$f_l^2 = \frac{(Q_{il} - Q_{jl})^2}{2\alpha_l^2} \qquad (12.287)$$

Substituting the values of Q_{il} and Q_{jl} from equation (12.148) gives a more explicit relation

$$f_l^2 = \frac{1}{N\alpha_l^2}\left[\cos^2\left(\frac{il\pi}{N}\right) + \cos^2\left(\frac{jl\pi}{N}\right) - 2\cos\left(\frac{il\pi}{N}\right)\cos\left(\frac{jl\pi}{N}\right)\right]$$

$$(12.288)$$

The sums in equation (12.286) over i and j run over all the segments, from 1 to $N+1$.

$$\sum_{i=1}^{N+1}\sum_{j=1}^{N+1} f_l^2 = \frac{1}{N\alpha_l^2}\left[(N+1)\sum_{i=1}^{N+1}\cos^2\left(\frac{il\pi}{N}\right)\right.$$

$$+ (N+1)\sum_{j=1}^{N+1}\cos^2\left(\frac{jl\pi}{N}\right)$$

$$\left. - 2\sum_{i=1}^{N+1}\cos\left(\frac{il\pi}{N}\right)\sum_{j=1}^{N+1}\cos\left(\frac{jl\pi}{N}\right)\right] \qquad (12.289)$$

To evaluate these sums, the following sums taken from Jolley.[1] Jolley 438 is

$$\sum_{k=0}^{n}\cos^2 k\theta = \frac{n+2}{2} + \frac{\cos\left[(n+1)\theta\right]\sin(n\theta)}{2\sin(\theta)} \qquad (12.290)$$

Jolley 418 is also needed.

$$\cos\theta + \cos 2\theta + \cos 3\theta + \cdots \text{ to n terms}$$

$$= \cos\left[\frac{(n+1)\theta}{2}\right]\sin\left(\frac{n\theta}{2}\right)\cos ec\left(\frac{\theta}{2}\right) \qquad (12.291)$$

The limits on the sums in Jolley and the sum in the equation to be evaluated are a little different. To use Jolley's sums, adding and subtracting suitable terms allows evaluation. For example

$$\sum_{i=1}^{N+1} \cos^2\left(\frac{il\pi}{N}\right) = \sum_{i=0}^{N} \cos^2\left(\frac{il\pi}{N}\right) + \cos^2\left[\frac{(N+1)l\pi}{N}\right] - \cos^2 0$$

(12.292)

$$\sum_{i=1}^{N+1} \cos^2\left(\frac{il\pi}{N}\right) = \frac{N+2}{2} + 0 + \cos^2\left[l\pi\left(1+\frac{1}{N}\right)\right] - 1$$

(12.293)

Since N is very large, $\cos^2[l\pi(1+\frac{1}{N})] \cong \cos^2(l\pi) = 1$, and the sum evaluates as essentially $\frac{N+2}{2}$. Also

$$\sum_{i=1}^{N+1} \cos\left(\frac{il\pi}{N}\right) = \sum_{i=1}^{N} \cos\left(\frac{il\pi}{N}\right) + \cos\left[\frac{(N+1)l\pi}{N}\right]$$

(12.294)

$$\sum_{i=1}^{N+1} \cos\left(\frac{il\pi}{N}\right) = \cos\left[\frac{(N+1)l\pi}{2N}\right] \sin\left(\frac{N}{2}\frac{l\pi}{N}\right) \cos ec\left(\frac{l\pi}{2N}\right)$$

$$+ \cos\left[l\pi\left(1+\frac{1}{N}\right)\right] \cong 0 + \cos(l\pi) \cong (-1)^l$$

(12.295)

since $\cos(\frac{l\pi}{2}) = 0$ if l is odd, and oscillates between 1 and -1 when l is even, while $\sin(\frac{l\pi}{2}) = 0$ if l is even and oscillates between 1 and -1 when l is odd. There are two such sums.

$$-2\sum_{i=1}^{N+1} \cos\left(\frac{il\pi}{N}\right) \sum_{j=1}^{N+1} \cos\left(\frac{jl\pi}{N}\right) = -2(-1)^l(-1)^l = -2$$

(12.296)

The double sum is now evaluated

$$\sum_{i=1}^{N+1}\sum_{j=1}^{N+1} f_i^2 = \frac{1}{N\alpha_l^2}\left[(N+1)\frac{N+2}{2} + (N+1)\frac{N+2}{2} - 2\right]$$

(12.297)

Collecting terms,

$$\sum_{i=1}^{N+1}\sum_{j=1}^{N+1} f_i^2 = \frac{1}{N\alpha_l^2}\left[N^2 + 3N + 2 - 2\right] = \frac{N+3}{\alpha_l^2} \qquad (12.298)$$

The mean square radius of gyration is therefore

$$\langle R^2\rangle = \frac{3}{2N^2}\sum_l (1+G_l)^{-1}\frac{N+3}{\alpha_l^2} \qquad (12.299)$$

$$\langle R^2\rangle = \frac{3(N+3)}{2N^2}\sum_l (1+G_l)^{-1}\alpha_l^{-2} \qquad (12.300)$$

Approximating $\sin(x)$ by x for x small, N large gives for small l

$$\alpha_l^2 = 6b^{-2}\sin^2\left(\frac{l\pi}{2N}\right) \cong 6b^{-2}\frac{l^2\pi^2}{4N^2} = 3b^{-2}\frac{l^2\pi^2}{2N^2} \qquad (12.301)$$

Since the index l runs from 1 to $N+1$, there is a whole region in which N is not large in comparison with l. Making this approximation effectively asserts that only small l values are responsible for the majority of the excluded volume. As N is large, $N+3$ is essentially N, and under this assertion, the radius becomes approximately

$$R^2 \cong \frac{Nb^2}{\pi^2}\sum_l l^{-2}(1+G_l)^{-1} \qquad (12.302)$$

Both G_l and α^2 are functions of the expansion parameter z through the equations

$$\alpha^2 = 1 + \frac{g_1}{\alpha^3}z \qquad (12.303)$$

$$G_l = \frac{g_1 - g_l}{\alpha^3}z \qquad (12.304)$$

An expansion of the mean square gyration radius in powers of z is not straightforward, but requires an expansion of $(1+G_l)^{-1}$, which in turn requires an expansion of $1/\alpha^3$ in powers of z. Begin with the latter. Exactly

$$\frac{1}{\alpha^3} = \left(1 + \frac{g_1}{\alpha^3}z\right)^{-3/2} \tag{12.305}$$

As z is to be small, an expansion of $(1+y)^{-3/2}$ with $y = \frac{g_1}{\alpha^3}z$ will do. For small y

$$(1+y)^{-3/2} = 1 - \frac{3}{2}y + \frac{15}{8}y^2 + \cdots \tag{12.306}$$

Applying this series expansion to $\frac{1}{\alpha^3}$ gives

$$\frac{1}{\alpha^3} = 1 - \frac{3}{2}\frac{g_1}{\alpha^3}z + \frac{15}{8}\left(\frac{g_1}{\alpha^3}z\right)^2 + \cdots \tag{12.307}$$

Iterating the substitution obtains

$$\frac{1}{\alpha^3} = 1 - \frac{3}{2}g_1 z\left(1 - \frac{g_1}{\alpha^3}z + \cdots\right) + \frac{15}{8}g_1^2 z^2\left(1 - \frac{g_1}{\alpha^3}z + \cdots\right) + \cdots \tag{12.308}$$

and multiplying these factors gives

$$\frac{1}{\alpha^3} = 1 - \frac{3}{2}g_1 z + \frac{33}{8}g_1^2 z^2 - \frac{189}{16}g_1^3 z^3 + \cdots \tag{12.309}$$

Thus

$$\alpha^2 = 1 + g_1 z\left(\frac{1}{\alpha^3}\right) = 1 + g_1 z\left(1 - \frac{3}{2}g_1 z + \frac{33}{8}g_1^2 z^2 - \frac{189}{16}g_1^3 z^3 + \cdots\right) \tag{12.310}$$

or

$$\alpha^2 = 1 + g_1 z - \frac{3}{2}g_1^2 z^2 + \frac{33}{8}g_1^3 z^3 + \cdots \tag{12.311}$$

For small G_l, $(1 + G_l)^{-1} = 1 - G_l + G_l^2 - \cdots$, where G_l is given by

$$G_l = (g_1 - g_l)z\left[1 - \frac{3}{2}g_1 z + \frac{33}{8}g_1^2 z^2 - \frac{189}{16}g_1^3 z^3 + \cdots\right] \tag{12.312}$$

which is

$$G_l = (g_1 - g_l)z - \frac{3}{2}g_1(g_1 - g_l)z^2 + \frac{33}{8}g_1^2(g_1 - g_l)z^3 + \cdots \tag{12.313}$$

Then

$$(1 + G_l)^{-1} = 1 - \left[(g_1 - g_l)z - \frac{3}{2}g_1(g_1 - g_l)z^2 \right.$$

$$\left. + \frac{33}{8}g_1^2(g_1 - g_l)z^3 + \cdots \right] + (g_1 - g_l)^2 z^2 + \cdots$$

$$(12.314)$$

and this is

$$(1 + G_l)^{-1} = 1 - (g_1 - g_l)z + (g_1 - g_l)\left(\frac{5}{2}g_1 - g_l\right)z^2 + \cdots \quad (12.315)$$

For the freely jointed chain with Gaussian statistics, the mean square gyration radius is

$$\langle R^2 \rangle_0 = \frac{Nb^2}{6} \tag{12.316}$$

The expansion parameter α_R^2 is defined as

$$\alpha_R^2 \equiv \frac{\langle R^2 \rangle}{\langle R^2 \rangle_0} \tag{12.317}$$

This ratio is

$$\alpha_R^2 = \frac{6\alpha^2}{\pi^2} \sum_l l^{-2}(1 + G_l)^{-1} \tag{12.318}$$

To terms quadratic in z, the product of α^2 and $(1 + G_l)^{-1}$ is

$$\alpha^2(1 + G_l)^{-1} = \left(1 + g_1 z - \frac{3}{2}g_1^2 z^2 + \frac{33}{8}g_1^3 z^3 + \cdots \right)$$

$$\times \left(1 - (g_1 - g_l)z + (g_1 - g_l)\left(\frac{5}{2}g_1 - g_l\right)z^2 + \cdots \right)$$

$$(12.319)$$

Collecting terms

$$\alpha^2(1 + G_l)^{-1} = 1 + g_1 z + \left[(g_1 - g_l)\left(\frac{3}{2}g_1 - g_l\right) - \frac{3}{2}g_1^2 \right]z^2 + \cdots$$

$$(12.320)$$

and α_R^2 becomes, to the quadratic term in z

$$\alpha_R^2 = \frac{6}{\pi^2} \sum_l l^{-2} \left\{ 1 + g_1 z + \left[(g_1 - g_l)\left(\frac{3}{2}g_1 - g_l\right) - \frac{3}{2}g_1^2 \right] z^2 \right\}$$

(12.321)

The leading term in this sum is the familiar Riemann zeta function of order 2.

$$\sum_l \frac{1}{l^2} = 1 + \frac{1}{4} + \frac{1}{9} + \frac{1}{16} + \frac{1}{25} + \cdots = \frac{\pi^2}{6}$$

(12.322)

This evaluation can be done using Jolley[8] equation 319. The expansion parameter is then

$$\alpha_R^2 = 1 + z\frac{6}{\pi^2}\sum_l l^{-2}g_l + z^2\frac{6}{\pi^2}\sum_l l^{-2}$$

$$\times \left[(g_1 - g_l)\left(\frac{3}{2}g_1 - g_l\right) - \frac{3}{2}g_1^2 \right] + \cdots$$

(12.323)

When the sums are carried out by a computer program, the result is

$$\alpha_R^2 = 1 + 1.276178z - 3.087z^2$$

(12.324)

Fixman[15] showed that the exact ratio to the term linear in z is

$$\alpha_R^2 = 1 + \frac{134}{105}z$$

(12.325)

Later, Yamakawa *et al.*[16] showed that the exact result to the quadratic term is

$$\alpha_R^2 = 1 + \frac{134}{105}z + \left(\frac{536}{105} - \frac{1247}{1296}\pi\right)z^2$$

(12.326)

Numerical evaluation gives

$$\alpha_R^2 = 1 + 1.276z - 2.082z^2$$

(12.327)

The boson expansion succeeds very well in the linear term, but fails in the quadratic. This disagreement may have its source in the assumption that only low values of the index l need be considered, allowing the approximation $\sin(l\pi/2N) \cong l\pi/2N$. However, the disagreement may have other sources.

When the boson expansion is applied to the mean square end to end distance, the result is

$$\alpha_L^2 = 1 + z \left\{ \frac{2}{\pi^2} \sum_l l^{-2} \left[(-1)^l - 1 \right] g_l \right\}$$

$$+ z^2 \left\{ \frac{2}{\pi^2} \sum_l l^{-2} \left[(-1)^l - 1 \right]^2 \left[(g_1 - g_l) \left(\frac{3}{2} g_1 - g_l \right) - \frac{3}{2} g_1^2 \right] \right\}$$

$$(12.328)$$

Numeric evaluation yields

$$\alpha_L^2 = 1 + 1.37693z - 3.229z^2 \qquad (12.329)$$

where $\langle L^2 \rangle_0 = Nb^2$ and $\alpha_L^2 = \langle L^2 \rangle / \langle L^2 \rangle_0$. Since the exact expansion is

$$\alpha_L^2 = 1 + \frac{4}{3} z - \left(\frac{16}{3} - \frac{28\pi}{27} \right) z^2 = 1 + 1.33333z - 2.0754z^2 \quad (12.330)$$

the agreement with the linear term is unimpressive. The gyration radius agrees in the linear term but the end to end distance agrees less well. This fact may have an explanation in that in the radius of gyration calculation, all the g_l contributed, whereas in the end to end distance calculation, the presence of the factor $[(-1)^l - 1]$ in the sum over l has the effect of removing every other g_l, causing half of these quantities to be left out of the calculation entirely.

Appendix

The diagonalization of the spring potential

$$V = (r_1 - r_0)^2 + (r_2 - r_1)^2 + \cdots$$

can be accomplished efficiently using matrix algebra. Consider the 3 dimensional case. Let R be a column matrix with entries $(r_0 \ r_1 \ r_2)$ and q be a column vector with entries $(q_0 \ q_1 \ q_2)$, and let the matrix of the transformation be Q. Designate the transpose of a matrix with a superscript T. Then

$$V = R^T F R = r_0^2 + r_1^2 + r_2^2 - r_0 r_1 - r_1 r_0 - r_1 r_2 - r_2 r_1$$

where in this three dimensional case,

$$F = \begin{pmatrix} 1 & -1 & 0 \\ -1 & 2 & -1 \\ 0 & -1 & 1 \end{pmatrix}$$

is the matrix of the coefficients of the products $r_i^2, r_i r_{i+1}$ that occur when V is expanded. R is related to q by Q.

$$R = Qq \quad q = Q^T R$$

since the transformation is orthonormal, that is

$$Q^{-1} = Q^T$$

and

$$QQ^{-1} = Q^{-1}Q = QQ^T = Q^T Q = E$$

where E is the identity matrix. The relations are general, but in the three dimensional case, E has the appearance

$$E = \begin{pmatrix} 1 & 0 & 0 \\ 0 & 1 & 0 \\ 0 & 0 & 1 \end{pmatrix}$$

Immediately

$$R^T F R = q^T Q^T F Q q = q^T \Lambda q$$

where Λ is a diagonal matrix with the eigenvalues, or roots, displayed along the main diagonal as squares, α_k^2. Since $\Lambda = Q^T F Q$, the eigenvalue eigenvector equation is obtained by premultiplying both sides with Q and making use of the identity matrix to get

$$Q\Lambda = QQ^T F Q = EFQ = FQ$$

The standard form is $FQ = Q\Lambda$. In the case of 3 dimensions,

$$
Q = \begin{pmatrix} \dfrac{1}{\sqrt{3}} & \dfrac{1}{\sqrt{2}} & \dfrac{1}{\sqrt{6}} \\[2mm] \dfrac{1}{\sqrt{3}} & 0 & -\dfrac{2}{\sqrt{6}} \\[2mm] \dfrac{1}{\sqrt{3}} & -\dfrac{1}{\sqrt{2}} & \dfrac{1}{\sqrt{6}} \end{pmatrix} \qquad Q^T = \begin{pmatrix} \dfrac{1}{\sqrt{3}} & \dfrac{1}{\sqrt{3}} & \dfrac{1}{\sqrt{3}} \\[2mm] \dfrac{1}{\sqrt{2}} & 0 & -\dfrac{1}{\sqrt{2}} \\[2mm] \dfrac{1}{\sqrt{6}} & -\dfrac{2}{\sqrt{6}} & \dfrac{1}{\sqrt{6}} \end{pmatrix}
$$

The reader should verify for himself that $QQ^T = E$, and that $Q^T FQ = \Lambda$, where the entries on the main diagonal are 0, 1 and 3.

It is worth remarking that the $k = 0$ normal coordinate always occurs as

$$
q_0 = \frac{1}{\sqrt{N}} \sum_{i=0}^{N-1} r_i
$$

regardless of the value of N. The corresponding eigenvalue is always 0. The other normal coordinates follow

$$
q_k = \sum_{i=0}^{N-1} q_{ik} r_i
$$

where

$$
q_{ik} = N^{-1/2} (2 - \delta_{k0})^{1/2} \cos \left(\frac{(2i + 1)k\pi}{2N} \right)
$$

Eigenvalues are given by $\lambda_k = 4 \sin^2 \left(\frac{k\pi}{2N} \right)$.

Problems

1. The quantity g_1 is of frequent occurrence in calculating the mean squared gyration radius. Calculate it for values of $N = 100$, 500, 1000 and 10,000. By plotting the values of g_1 against $N^{-1/2}$, show that the limiting value of g_1 for an infinitely long chain is 1.51566009... If instead N is taken as 1000, 3000, 7000 and 10,000, g_1 is 1.51566532... Why is plotting these g_l values against $1/N^{1/2}$ essentially linear?

Hint: Use the relation

$$g_1 = \frac{1}{2N^{3/2} \sin^2\left(\frac{\pi}{2N}\right)} \sum_{j=2}^{N} \sum_{i=1}^{j-1}$$

$$\times \left(\cos\left(\frac{j\pi}{2N}\right) - \cos\left(\frac{i\pi}{2N}\right)\right)^2 (|i - j|)^{-\frac{5}{2}}$$

Clearly, this will require programming a computer in some sophisticated computational program. The original work was done in FORTRAN II, and the calculation has been checked in MathCad.

2. Calculate an extension to Table 12.1 for $l = 20$. Compare g_{20} as calculated with g_{20} calculated using

$$2g_l \cong l^{-3/2}[1.2004217][l\pi - 1.0857865$$
$$+ l^{-1/2}\{0.4501585 - 0.0000013(-1)^l\}]$$

Answer: $g_{20} = 0.4150$

4. For a chain of four segments the potential is

$$V = (r_1 - r_0)^2 + (r_2 - r_1)^2 + (r_3 - r_2)^2$$

a. Show that the matrix Q' given below obeys $Q'^T Q' = E = Q'Q'^T$ where E is the 4×4 identity matrix.

$$\begin{pmatrix} 1/2 & 1/2 & 1/2 & 1/2 \\ 1/2 & 1/2 & -1/2 & -1/2 \\ 1/2 & -1/2 & -1/2 & 1/2 \\ 1/2 & -1/2 & 1/2 & -1/2 \end{pmatrix}$$

b. Show that the matrix Q' does not diagonalize the potential V.

c. Number the normal coordinates q_k from zero. These are then q_0, q_1, q_2 and q_3. The transformation coefficients are

$$q_{ik} = \sqrt{\frac{2 - \delta_{k0}}{4}} \cos\left(\frac{(2i + 1)k\pi}{8}\right)$$

Calculate the entries in the matrix Q, show that these also obey $QQ^T = Q^T Q = E$, but that these diagonalize V. Note that the sign structure of both Q' and Q are identical.

d. The eigenvalues are $\alpha_k^2 = 4\sin^2(\frac{k\pi}{8})$. Show that these eigenvalues are $0, 2 - \sqrt{2}, 2, 2 + \sqrt{2}$.

e. Arrange the non-zero eigenvalues α_k to scale on graph paper, with the horizontal axis running from 0 to 2, and one set of eigenvalues on each horizontal line, running N from 2 downward to 10. The appearance will be something like

$$
\begin{array}{l}
N \\
2 \qquad | \\
3 \quad | \quad | \\
4 \; | \quad | \quad |
\end{array}
$$

and so on. Note that the lowest non-zero eigenvalue is characteristic of N, and can be used to measure N.

References

1. A. Einstein, Ann. Physik **19**, 289 (1906).
2. M. Fixman, J. Chem. Phys. **34**, 1656 (1955).
3. M. Fixman, J, Chem. Phys. **23**, 1656 (1955).
4. P. J. Flory and S. Fisk, J. Chem. Phys. **44**, 2243 (1966).
5. M. Fixman, J. Chem. Phys. **45**, 785 (1966).
6. M. Fixman, J. Chem. Phys. **45**, 793 (1966).
7. H. D. Stidham and M. Fixman, J. Chem. Phys. **48**, 3092 (1968).
8. L. B. W. Jolley, "Summation of Series," Dover, New York, 1961: Reprinted from first publication by Chapman and Hall, London, 1925.
9. E. B. Wilson, Jr., J. C. Decius and P. Cross, "Molecular Vibrations," McGraw-Hill, New York, N. Y., 1955.
10. A. Foldes and C. Sandorfy, J. Mol. Spectrosc. **20**, 262 (1966).
11. M. Abramowitz and I. A. Stegun, Nat'l Bur. Stds. AMS 55, 1964, 5th Printing.
12. J. G. Kirkwood and J. Riseman, J. Chem. Phys. **16**, 565 (1948).
13. C. P. Slichter, "Principles of Magnetic Resonance," Ed. 3, Springer, Berlin Heidelberg, 1990, corrected 1996, Appendix A.
14. E. C. Titschmarsh, "Introduction to the Theory of Fourier Integrals," Oxford Univ. Press, London, Ed. 2, 1948, Unnumbered equation at bottom of page 77.
15. M. Fixman, J. Chem. Phys. **23**, 1656 (1955).
16. H. Yamakawa, A. Aoki and G. Tanaka, J. Chem. Phys. **45**, 1938 (1966).

Chapter 13

Imperfect Gases and Liquids

13.1. Introduction, Part I

We first consider the van der Waals equation. In earlier work, it was shown that if no forces of any kind acted between structureless masses in a gas, the translational partition function

$$Q = \frac{q^N}{N!} \tag{13.1}$$

led almost immediately to the ideal gas law, $pV = Nk_BT$. Here $q = V/\Lambda^3$ and the thermal wavelength $\Lambda = h/\sqrt{2\pi mk_BT}$ (analogous to de Broglie's hypothesized, $\lambda = h/p$ where p stands for momentum). While this relation holds exactly at infinite dilution of any real gas, at finite concentrations, both repulsive and attractive forces really do act between colliding molecules. No molecule is a point mass, and each molecule must occupy some small but finite space. If the molecules of a gas are considered as hard spheres that do not interpenetrate during collision, and no other forces act, each molecule in a colliding pair will find a certain volume of the available space excluded to its center of mass. If the distance between centers of mass of spherical molecules in collisional contact is σ, the volume excluded to one of the molecules by the other is $\frac{4}{3}\pi\sigma^3$. If there are N molecules distributed over the volume V, the total volume excluded to any one molecule is $\frac{4}{3}(N-1)\pi\sigma^3$. Each binary collision requires two molecules (assuming that ternary and higher collisions are negligibly infrequent), and the excluded volume per molecule per collision is half the amount calculated for one molecule. Neglecting unity in comparison with an N of the order of Avogadro's number, the total excluded volume is $\frac{2}{3}\pi N\sigma^3$. The free volume V_f is thus

$$V_f = V - N\frac{2}{3}\pi\sigma^3 \tag{13.2}$$

This suggests that the excluded volume may be accommodated by writing

$$q = \frac{V_f}{\Lambda^3} \tag{13.3}$$

The route to thermodynamics is unaltered, and the equation of state is

$$p\left(V - N\frac{2}{3}\pi\sigma^3\right) = Nk_BT \tag{13.4}$$

The potential energy function $u(r)$ that describes the interaction between two hard spheres r distant apart is

$$\begin{aligned} u(r) &= \infty \quad 0 < r < \sigma \\ u(r) &= 0 \quad\;\; \sigma < r < \infty \end{aligned} \tag{13.5}$$

as is shown in Figure 13.1.

Molecules have attractive forces that act between them as well as the repulsive ones that define their volume. A frequent choice for

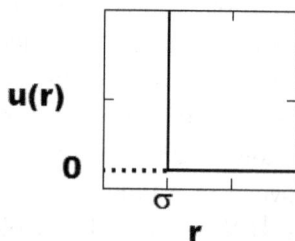

Fig. 13.1.   The potential energy function $u(r)$ for colliding hard sphere molecules. The dotted line is the r axis.

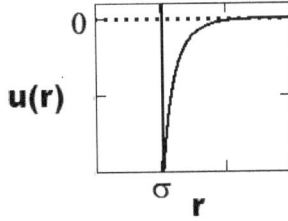

Fig. 13.2. Potential energy function $u(r)$ for hard spheres with inverse sixth power attractive forces acting between colliding molecules. The dotted line represents the r axis.

attractive forces is an inverse sixth power law such as

$$u(r) = \infty \qquad 0 < r < \sigma$$
$$u(r) = -\varepsilon \left(\frac{\sigma}{r}\right)^6 \qquad \sigma < r < \infty$$

$$(13.6)$$

This function is plotted in Figure 13.2.

Let the total potential energy due to non-colliding interaction be φ. This energy adds to the kinetic energy and causes the molecular partition function to be rewritten as

$$q = \frac{V_f e^{-\varphi/2k_B T}}{\Lambda^3} \tag{13.7}$$

The factor of two is due to the fact that it takes two molecules to make a collision. The density of molecules from 0 to σ is of course zero, since this is the region of excluded volume for hard spheres. From σ to infinity the density of molecules is essentially N/V. The contribution to φ made by molecules in a spherical shell of volume $4\pi r^2 dr$ is $-\varepsilon(\frac{\sigma}{r})^6 \frac{N}{V} 4\pi r^2 dr$. The total potential energy of interaction of all the other molecules with one molecule is

$$\varphi = -\varepsilon \int_\sigma^\infty \left(\frac{\sigma}{r}\right)^6 \frac{N}{V} 4\pi r^2 dr = -\frac{4\pi\sigma^3}{3}\varepsilon\frac{N}{V} \tag{13.8}$$

and the partition function becomes

$$q = \frac{V_f e^{+\frac{2\pi\sigma^3}{3}\varepsilon\frac{N}{Vk_B T}}}{\Lambda^3} \tag{13.9}$$

The canonical ensemble partition function for this model is

$$Q = \frac{V_f^N e^{\frac{2\pi\sigma^3}{3}\varepsilon\frac{N^2}{Vk_BT}}}{N!\Lambda^{3N}} \tag{13.10}$$

At this point it is convenient to introduce two factors, $a_\sigma = 2\pi\sigma^3\varepsilon/3$ and $b_\sigma = 2\pi\sigma^3/3$ for then $V_f = V - Nb_\sigma$ and $\frac{2\pi\sigma^3}{3}\varepsilon\frac{N^2}{Vk_BT} = a_\sigma\frac{N^2}{Vk_BT}$. These definitions allow Q to be more compactly expressed as

$$Q = \frac{(V - Nb_\sigma)^N e^{a_\sigma\frac{N^2}{Vk_BT}}}{N!\Lambda^{3N}} \tag{13.11}$$

The equation of state is obtained by the usual route

$$p = k_BT\left(\frac{\partial \ln Q}{\partial V}\right)_{T,N} \tag{13.12}$$

The logarithm of Q for this model is

$$\ln Q = N\ln(V - Nb_\sigma) + a_\sigma\frac{N^2}{Vk_BT} - \ln(N!\Lambda^{3N}) \tag{13.13}$$

and applying equation (13.12) produces

$$p = \frac{Nk_BT}{V - Nb_\sigma} - a_\sigma\frac{N^2}{V^2} \tag{13.14}$$

If N_0 is Avogadro's number, the universal gas constant $R = N_0k_B$ and the number of moles is $N/N_0 = n$. Then $Nk_BT = nRT$, $V - Nb_\sigma = V - nb$, and $a\frac{N^2}{V^2} = a\frac{n^2}{V^2}$ that is, $b = N_0b_\sigma$ and similarly $a = N_0^2 a_\sigma$. Upon rearrangement, the equation appears in standard form

$$\left(p + \frac{an^2}{V^2}\right)(V - nb) = nRT \tag{13.15}$$

This equation reverts to the ideal gas law at sufficiently low gas densities, and provides surprisingly accurate representation of the properties of real gases at densities near the ambient. However, at high densities, divergence of the theory from experiment become increasingly large until at the critical point the representation provided by the theory is only fair. The reason is that molecules are not hard spheres, but have electronic outer shells that are penetrated on collision increasingly with increasing mean molecular speeds. Both a and b depend on σ, and both van der Waals constants are altered in value as the density approaches the critical. If the constants were not altered by high density, the critical temperature T_c and pressure p_c could be calculated from the a and b constants by $T_c = 8a/27Rb$ and $p_c = a/27b^2$.

Although an inverse power of six was used for the attractive force, any power less than inverse three will do, $-4, -5, -7, -8$ and so on, all lead to the same formal result. A power of -3 produces a logarithmic divergence at the upper limit of the integration. Larger powers $(-2, -1, 0, 1 \ldots)$ of course all diverge in the integration.

The van der Waals equation provides a facile means of measuring the Boyle temperature. The Boyle temperature is defined thermodynamically by the relation

$$\left(\frac{\partial pV}{\partial p}\right)_T = 0 \tag{13.16}$$

and real gases at the Boyle temperature behave nearly ideally over a short range of temperatures around the Boyle temperature. To see this, first write the virial expansion in the form

$$\frac{pV}{nRT} = 1 + B_2(T)\frac{n}{V} + O\left(\frac{n^2}{V^2}\right) \tag{13.17}$$

The immediate object is to obtain pV as a function of p correct to terms linear in p. To do this, solve the virial equation for V approximately as

$$V \cong \frac{nRT}{p}\left(1 + B_2(T)\frac{n}{V} + \cdots\right) \tag{13.18}$$

Now substitute V on the right hand side of the equation by itself. The result is

$$V = \frac{nRT}{p}\left(1 + \frac{B_2(T)n}{\frac{nRT}{p}\left(1 + B_2(T)\frac{n}{V} + \cdots\right)} + \cdots\right) \tag{13.19}$$

Discard all terms involving higher powers of p than the first. The result is

$$\frac{pV}{nRT} = 1 + \frac{B_2(T)}{RT}p + O(p^2) \tag{13.20}$$

or

$$pV = nRT + nB_2(T)p + O(p^2) \tag{13.21}$$

Imposing the definition of the Boyle temperature, there results

$$\left(\frac{\partial pV}{\partial p}\right)_T = nB_2(T) + \cdots = 0 \tag{13.22}$$

and, to terms linear in p, the Boyle temperature is approximately defined by a vanishing second virial coefficient, Near the Boyle temperature, the virial expansion becomes

$$\frac{pV}{nRT} = 1 + O\left(\frac{n^2}{V^2}\right) \tag{13.23}$$

and nearly ideal behavior is expected. The vanishing of the second virial coefficient is used experimentally to measure the Boyle temperature. The second virial coefficient has a temperature dependence of the form

$$B_2(T) = b - \frac{a}{RT} \tag{13.24}$$

This may be shown from the hard sphere inverse sixth power potential used to obtain the van der Waals equation above. This suggests that if the second virial coefficient for a real gas is plotted against $1/T$, an approximately straight line should result. In fact, the line is reasonably straight, but curves away from linearity at high and at low temperatures, temperatures that fortunately bracket the Boyle temperature. The theory shows that the intercept on the $B_2(T)$ axis is the van der Waals constant b, and the straight line crosses the $1/T$ axis at the reciprocal Boyle temperature, while the slope of the straight line is $-a/R$. Actual plots for hydrogen show a maximum at about $400°C$, and a similar maximum for nitrogen near $2000°C$. The Boyle temperature for hydrogen is about 106 K, and for nitrogen, about $50°C$.

Other properties of the van der Waals equation are usually discussed in undergraduate textbooks of physical chemistry, of which there are many very good ones, and will not be discussed here.

13.2. Introduction to Imperfect Gases. Part II

The next topic is a deeper discussion of the second virial coefficient. The molecular partition function q for a gas of N structureless mass points confined to a rectangular box of volume $V = abc$ could be expressed approximately as

$$q = \int_0^\infty \int_0^\infty \int_0^\infty e^{-\frac{h^2}{8mk_BT}\left[\left(\frac{n_x}{a}\right)^2+\left(\frac{n_y}{b}\right)^2+\left(\frac{n_z}{c}\right)^2\right]} dn_x dn_y dn_z \quad (13.25)$$

This led directly to the ideal gas law, $pV = Nk_BT$. Classically, the Hamiltonian of one of the molecules is

$$H(p_x, x, p_y, y, p_z, z) = \varepsilon = \frac{p_x^2 + p_y^2 + p_z^2}{2m} \quad (13.26)$$

where ε is the kinetic energy of the molecule. The momenta and associated matter waves are related by de Broglie's hypothesis. For the x component

$$\lambda_{n_x} = \frac{h}{p_x} = \frac{2a}{n_x} \quad (13.27)$$

since n_x half wavelengths just fit in the a dimension of the volume. This relation allows writing

$$\frac{p_x^2}{2m} = \frac{h^2 n_x^2}{8ma^2} \quad (13.28)$$

The molecular partition function may be alternatively expressed in terms of the Hamiltonian

$$q = \int_0^a \int_0^b \int_0^c \int_0^\infty \int_0^\infty \int_0^\infty e^{-\frac{H(p_x,p_y,p_z,x,y,z)}{k_BT}} \frac{dxdydzdp_xdp_ydp_z}{h^3} \quad (13.29)$$

The canonical ensemble partition function requires multiple integrations, and the functional dependence of the Hamiltonian is best expressed vectorially. Thus, $\boldsymbol{p}$ is the set of all momenta in the gas of

N mass points and q is the set of all molecular coordinates. Similarly, dp and dq are all the momental and spatial infinitesimals. Then Q for this case is

$$Q = \frac{1}{N!h^{3N}} \int \cdots \int e^{-\frac{H(p,q)}{k_B T}} dp\,dq \qquad (13.30)$$

The Hamiltonian for the non-interacting gas of mass points involves only momenta

$$H(p,q) = \frac{1}{2m} \sum_{j=1}^{N} p_j^2 \qquad (13.31)$$

In a real gas, multiple binary, ternary, quaternary and so on interactions will occur. Most of these interactions in a fairly dilute gas will be binary. Ternary and higher order collisions occur with negligible frequency when the gas density is low enough. Let the potential energy for a binary interaction between molecules i and j be $u_{ij}(r_{ij})$, where r_{ij} is the distance between the ith and the jth molecule. In terms of the Cartesian coordinates of the molecules, the distance r_{ij} has the magnitude

$$r_{ij} = \sqrt{(x_i - x_j)^2 + (y_i - y_j)^2 + (z_i - z_j)^2} \qquad (13.32)$$

For this gas with only binary collisions, the Hamiltonian is modified to

$$H(p,q) = \frac{1}{2m} \sum_{j=1}^{3N} p_j^2 + \sum_{i<j} u_{ij}(r_{ij}) \qquad (13.33)$$

For this gas the canonical ensemble partition function is

$$Q = \frac{1}{N!h^{3N}} \int \cdots \int e^{-\frac{1}{2mk_B T} \sum_j p_j^2} e^{-\sum_{i<j} u_{ij}/k_B T} dp\,dq \qquad (13.34)$$

The integration over the momenta goes through as before, resulting in

$$Q = \frac{1}{N!} \left(\frac{2\pi m k_B T}{h^2} \right)^{3N/2} Q_\tau \qquad (13.35)$$

where Q_τ is called the configuration integral and is

$$Q_\tau = \int \cdots \int e^{-\sum_{i<j} u_{ij}/k_B T} d\mathbf{q} \tag{13.36}$$

The pressure is given in terms of this integral

$$\frac{p}{k_B T} = \left(\frac{\partial \ln Q}{\partial V}\right)_{T,N} = \left(\frac{\partial \ln Q_\tau}{\partial V}\right)_{T,N} \tag{13.37}$$

since there is no V dependence in the momentum integral. To progress, the configuration integral needs to be evaluated. To this end, note that exactly

$$\sum_{i<j=1}^{N} u_{ij} = \sum_{i<j}^{N} {}' u_{ij} + \sum_{i=1}^{N} u_{iN} \tag{13.38}$$

where the prime indicates a sum that has no contributions from u_{iN} terms. Next, define a volume element $d\tau_i = dx_i dy_i dz_i$. Then the configuration integral can be written as

$$Q_\tau = \int \cdots N-1 \cdots \int e^{-\sum_{i<j} {}' u_{ij}/k_B T} \prod_{i=1}^{N-1} d\tau_i \int e^{-\sum_{i=1}^{N} u_{iN}/k_B T} d\tau_N \tag{13.39}$$

By adding and subtracting 1 to the integrand of the last integral, there results

$$Q_\tau = \int \cdots N-1 \cdots \int e^{-\sum_{i<j}^{N-1} {}' u_{ij}/k_B T} \prod_{i=1}^{N-1} d\tau_i$$

$$\times \int \left(e^{-\sum_{i=1}^{N} u_{iN}/k_B T} + 1 - 1\right) d\tau_N$$

Doing this leaves the configuration integral unchanged. However, it allows writing the last integral as $V + w$, where

$$w = \int \left(e^{-\sum_{i=1}^{N} u_{iN}/k_B T} - 1\right) d\tau_N \tag{13.40}$$

Although the integral cannot be rigorously evaluated, it can be approximated. Hold all the molecules but the Nth fixed in place and allow the Nth to move over the available space. Appreciable contributions to w occur only for intermolecular distances less than a critical and relatively small distance $r*$, of the order of 5 or so molecular diameters. Since this happens in exactly the same way $N - 1$ times, w evaluates as

$$w = (N - 1) \int_0^{r*} (e^{-u(r)/k_B T} - 1) 4\pi r^2 dr \qquad (13.41)$$

since the function u is universal, the same for every identical molecule in the gas. Representing the integral by the Greek letter β, w becomes$(N - 1)\beta$, and the configuration integral becomes

$$Q_\tau = \int \cdots N - 1 \cdots \int e^{-\sum_{i<j}' u_{ij}/k_B T} \prod_{i=1}^{N-1} d\tau_i \, [V + (N - 1)\beta]$$

$$(13.42)$$

Nothing prevents repeating these operations, resulting in

$$Q_\tau = \int \cdots N - 2 \cdots \int e^{-\sum_{i<j}' u_{ij}/k_B T}$$

$$\times \prod_{i=1}^{N-2} d\tau_i \, [V + (N - 2)\beta] \, [V + (N - 1)\beta] \qquad (13.43)$$

Repeating this until the molecules are exhausted produces

$$Q_\tau = V(V + \beta)(V + 2\beta) \cdots [V + (N - 2)\beta] \, [V + (N - 1)\beta]$$

$$(13.44)$$

This may be more compactly expressed as

$$Q_\tau = \prod_{i=0}^{N-1} (V + i\beta)$$

Factoring V^N from the configuration integral gives

$$Q_\tau = V^N \prod_{i=0}^{N-1} \left(1 + i\frac{\beta}{V}\right) \qquad (13.45)$$

The logarithm is

$$\ln Q_\tau = N \ln V + \sum_{i=0}^{N-1} \ln\left(1 + i\frac{\beta}{V}\right) \tag{13.46}$$

Since the gas is dilute enough that ternary and higher collisions are rare, the quantity $\beta N/V$ is very small. The quantity $i\beta/V$ will be even smaller, as $i < N$ over the entire range of i. Then the approximation $\ln(1+x) = x + \cdots$ is valid over the range of i, and the configuration integral becomes

$$\ln Q_\tau \cong N \ln V + \sum_{i=0}^{N-1} i\frac{\beta}{V}$$

or

$$\ln Q_\tau \cong N \ln V + \frac{\beta}{V} \sum_{i=0}^{N-1} i \tag{13.47}$$

The sum is a well-known one and evaluates as $N(N-1)/2$, which is approximately $N^2/2$ for N of the order of Avogadro's number. The configuration integral is approximately

$$\ln Q_\tau \cong N \ln V + \frac{N^2\beta}{2V} \tag{13.48}$$

The pressure is calculated using equation (13.37) with the result

$$\frac{p}{k_B T} = \frac{N}{V} - \frac{N^2\beta}{2V^2}$$

which may be rearranged to

$$\frac{pV}{Nk_B T} = 1 - \frac{\beta}{2}\frac{N}{V} \tag{13.49}$$

Comparison with the virial expansion

$$\frac{pV}{nRT} = 1 + \frac{n}{V}B_2(T) + \cdots \tag{13.50}$$

where $n = N/N_0$ is the number of moles of gas (N_0 is Avogadro's number). This shows that the second virial coefficient is

$$B_2(T) = -\frac{N_0}{2}\beta \qquad (13.51)$$

or, more fully expressed,

$$B_2(T) = -\frac{N_0}{2}\int_0^\infty (e^{-u(r)/k_BT} - 1)4\pi r^2 dr \qquad (13.52)$$

where $r*$ has been replaced as the upper limit of the integral since distances larger than $r*$ do not contribute appreciably to the value of the integral. This is usually expressed as

$$B_2(T) = 2\pi N_0 \int_0^\infty (1 - e^{-u(r)/k_BT})r^2 dr \qquad (13.53)$$

13.3. Imperfect Gases. Part III

Now consider imperfect gases in complete detail. While the second virial coefficient serves to represent the properties of a dilute gas, the higher terms in the virial expansion are still there and as yet undefined. At higher densities the ternary, quaternary and higher terms contribute, increasingly with increasing density. Alas, the gas at liquid densities requires indefinitely many terms, and no closed form is available. The virial expansion is restricted to the gas phase and cannot be used to treat the liquid state. With this caveat, consider now the nature of the virial coefficient. The configuration integral is central to the issue, and may be expressed

$$Q_\tau = \int \cdots \int e^{-U/k_BT} \prod_{i=1}^N d\tau_i \qquad (13.54)$$

where each volume element $d\tau_i = dx_i dy_i dz_i$, and the function U is

$$U = \sum_{i<j=1} u_{ij} \qquad (13.55)$$

The sum runs over all possible pairs in the gas. Evaluation of the configuration integral follows the famous Ursell-Mayer derivation as

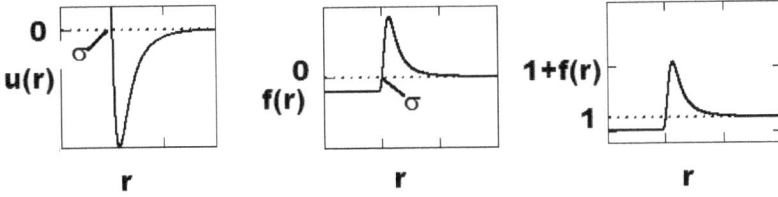

Fig. 13.3. The potential energy $u_{ij} = u(r)$ for a Lennard-Jones interaction potential, the function $f(r) = f_{ij} = e^{-u(r)/k_B T} - 1$ and the function $1 + f_{ij} = 1 + f(r)$. The hard sphere diameter σ is shown as the point where the potential curve crosses the r axis. The dotted line represents the r axis in the left and central figures, and the line $f(r) = 0$ in the rightmost figure.

described in Mayer and Mayer's book.[1,2] The key to theoretical progress is to define the function f_{ij} by writing

$$e^{-u_{ij}/k_B T} = 1 + f_{ij} \tag{13.56}$$

This definition requires

$$e^{-\sum_{all\ pairs} u_{ij}/k_B T} = \prod_{all\ pairs} (1 + f_{ij}) \tag{13.57}$$

To see how this goes, suppose the gas has only three molecules in it. Then

$$e^{-\sum_{all\ pairs} u_{ij}/k_B T} = e^{-u_{12}/k_B T} e^{-u_{13}/k_B T} e^{-u_{23}/k_B T} \tag{13.58}$$

$$e^{-\sum_{all\ pairs} u_{ij}/k_B T} = (1 + f_{12})(1 + f_{13})(1 + f_{23})$$

$$e^{-\sum_{all\ pairs} u_{ij}/k_B T} = 1 + f_{12} + f_{13} + f_{23} + f_{12}f_{13}$$
$$+ f_{12}f_{23} + f_{13}f_{23} + f_{12}f_{13}f_{23} \tag{13.59}$$

If the gas has four molecules in it, the total number of terms increases considerably. The last term in the expansion is $f_{12}f_{13}f_{14}f_{23}f_{24}f_{34}$. Although it has no application here, it is important to keep in mind that the expansion is finite and always has a terminal term involving every f_{ij}.

The basic mathematics that allows evaluation of the f_{ij} integrals is

$$\iint f_{ij} d\tau_i d\tau_j = V \int_0^\infty f(r) 4\pi r^2 dr \qquad (13.60)$$

$$\int f_{ij} d\tau_i d\tau_j = \int f_{nm} d\tau_n d\tau_m \qquad (13.61)$$

Again, unless molecules i and j are closer together than r_{ij}^* the integral effectively vanishes. Now consider the sum of the f_{ij} products for a gas of N molecules, where N is very large. Somewhere in the sum there will occur the term

$$f_{12} f_{34} f_{56} f_{67} f_{57}$$

This term will contribute appreciably to the configuration integral if simultaneously molecule 1 is near molecule 2, molecule 3 is near molecule 4 and molecules 5, 6 and 7 are near each other. This may be represented by a diagram

showing that the term contributes two binary clusters and one ternary cluster to the configuration integral. Another term that will appear in the sum is $f_{12}f_{34}f_{56}f_{67}$. and this has the diagram

showing another kind of ternary cluster. Graphs such as these were first introduced by Mayer (Mayer and Mayer, Chapter 12 and

Appendix X). More generally, the first few possible interactions and the number of each are

1 Binary

4 Ternary

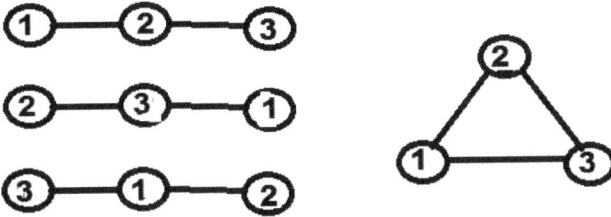

38 Quaternary, broken down as

12 linear, such as $f_{12}f_{23}f_{34}$

12 triangular with a binary tail, such as $f_{12}f_{24}f_{14}f_{23}$

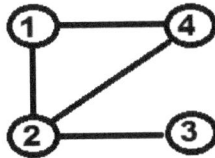

6 double triangle, such as $f_{12}f_{23}f_{34}f_{14}f_{24}$

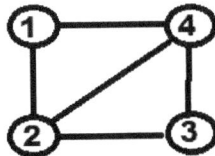

3 rectangles, such as $f_{12}f_{23}f_{34}f_{14}$

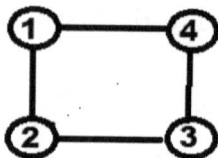

4 arrowheads, such as $f_{12}f_{13}f_{14}$

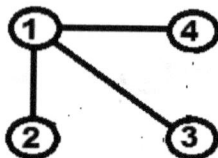

1 all interacting, $f_{12}f_{13}f_{14}f_{23}f_{24}f_{34}$

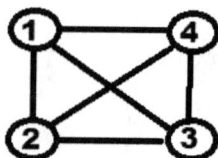

All 38 configurations contribute to the configuration integral. Note that the number of f_{ij} functions varies from 3 to 6, and all are involved in a ternary cluster. It is not possible to achieve successive approximation by considering terms with 2, then 3, then 4 f_{ij} functions in the product. The next job is to classify the terms in the expansion on the basis of the probability of occurrence. Clearly $f_{12}f_{34}f_{56}$ is all binary and will contribute appreciably to the configuration integral. On the other hand, the term $f_{12}f_{34}f_{45}$ is a product of a binary f_{12} and a ternary $f_{34}f_{45}$ cluster, and will contribute less.

The approach that clarifies the murkiness of these issues is to consider successively a gas consisting of 1, 2, 3 and then 4 molecules. The general relation will then emerge. For 1 molecule, the configuration integral is just

$$Q_\tau = \int 1 d\tau = V \qquad (13.62)$$

since there is no interaction possible. A molecule cannot collide with itself.

When N is 2, a binary collision is possible, and the configuration integral becomes

$$Q_\tau = \int (1 + f_{12}) d\tau_1 d\tau_2 = V^2 + V \int_0^\infty f(r) 4\pi r^2 dr = V^2 + V B_2$$

(13.63)

where $B_2 = \int_0^\infty f(r) 4\pi r^2 dr$. This function was called β in the elementary analysis presented in Section 13.2 above, and should not be confused with the second virial coefficient $B_2(T)$. The reason for naming it B_2 will emerge later in this analysis.

For $N = 3$, the configuration integral assumes the form

$$Q_\tau = \int (1 + f_{12} + f_{23} + f_{13} + f_{12}f_{13} + f_{12}f_{23} + f_{13}f_{23}$$

$$+ f_{12}f_{23}f_{13}) d\tau_1 d\tau_2 d\tau_3$$

(13.64)

Clearly, there are three terms of the form $B_2 V^2$. Lump the remaining terms in a term called $B_3 V$. The result is

$$Q_\tau = V^3 + 3B_2 V^2 + B_3 V$$

(13.65)

where B_3 is defined as

$$B_3 V = \int (f_{12}f_{13} + f_{12}f_{23} + f_{13}f_{23} + f_{12}f_{23}f_{13}) d\tau_1 d\tau_2 d\tau_3$$

(13.66)

Note that there are three ways three molecules can collide in pairs, 12, 23, 13, and only one way all three can form a ternary cluster 123. These are the coefficients of B_2 and B_3 in the configuration integral. Perhaps this observation holds generally, and to test it out, try a gas with four molecules. When $N = 4$, the configuration integral is expected to assume the form

$$Q_\tau = V^4 + AV^3 + BV^2 + CV$$

(13.67)

where A, B and C remain to be determined. There are 6 possible pairs, 12, 13, 14, 23, 24 and 34. In addition there are three double pairs, 12 and 34, 13 and 24, 14 and 23. There are 4 ternary clusters,

123, 124, 134, 234. This suggests the configuration integral should have the form

$$Q_\tau = V^4 + 6B_2V^3 + 3B_2^2V^2 + 4B_3V^2 + B_4V \tag{13.68}$$

After some reflection, this suggests that the lth cluster integral be defined as

$$B_l = \frac{1}{V} \int \sum \left(\prod_{p<q} f_{pq} \right) \prod_{i=1}^{l} d\tau_i \tag{13.69}$$

The sum is taken over all combinations consistent with one cluster. The product is to be consistent with the order of the cluster l. Let n_l be the number of l-ary clusters. The number of molecules in the gas up to 5 is listed in Table 13.1.

Note that V always appears raised to the sum of the n_l numbers in a specification. Also, cross terms like $B_2B_3V^2$ in $N = 5$ are beginning to make an appearance, but only when the corresponding n_l is non-zero. In fact, the term could be written

$$B_2^{n_2} B_3^{n_3} V^{n_2+n_3} = (B_2V)^{n_2} (B_3V)^{n_3}$$

apart from the factor 10, which is the number of different ways the 5 molecules can be combined into a binary and a ternary cluster. This suggests that the general term should be proportional to

$$\prod_l (B_lV)^{n_l}$$

allowing the configuration integral to be written as

$$Q_\tau = \sum_{(n_l)} g(n_l) \prod_l (B_lV)^{n_l} \tag{13.70}$$

where the sum is over all values of n_l consistent with the restrictive condition $\sum_l ln_l = N$. The number g remains to be determined.

Inspection of Table 13.1 shows that many of the coefficients are binomial coefficients, such as 5, 10, or 15 for the 5 molecule case. Others, however, are not, such as $3B_2V^2$ or $6B_2V^3$. More generally, the number $g(n_l)$ is the number of times the product $\prod_l (B_lV)^{n_l}$ occurs in the configuration integral. This combinatorial problem is

Table 13.1. Configuration integral, clusters and coefficients for N up to 5.

N	Q_τ	$\sum_l l n_l = N$					Term in Q_τ
1	V	$n_1 = 1$					V
		n_1					
		1					
2	$V^2 + B_2 V$	$n_1 + 2n_2 = 2$					V^2
		n_1	n_2				$B_2 V$
		2	0				
		0	1				
3	$V^3 + 3B_2 V^2 + B_3 V$	$n_1 + 2n_2 + 3n_3 = 3$					V^3
		n_1	n_2	n_3			$3B_2 V^2$
		3	0	0			$B_3 V$
		1	1	0			
		0	0	1			
4	$V^4 + 6B_2 V^3$ $+(4B_3 + 3B_2^2)V^2$ $+B_4 V$	$n_1 + 2n_2 + 3n_3 + 4n_4 = 4$					V^4
		n_1	n_2	n_3	n_4		$4B_2 V^3$
		4	0	0	0		$3B_2^2 V^2$
		2	1	0	0		$4B_3 V^2$
		0	2	0	0		$B_4 V$
		1	0	1	0		
		0	0	0	1		
5	$V^5 + 10B_2 V^4 + 15B_2^2 V^3$ $+15B_3 V^3 + 10B_2 B_3 V^2$ $+5B_4 V^2 + B_5 V$	$n_1 + 2n_2 + 3n_3 + 4n_4 + 5n_5 = 5$					V^5
		n_1	n_2	n_3	n_4	n_5	$10B_2 V^4$
		5	0	0	0	0	$15B_2^2 V^3$
		3	1	0	0	0	$15B_3 V^3$
		1	2	0	0	0	$10B_2 B_3 V^2$
		2	0	1	0	0	$5B_4 V^2$
		0	1	1	0	0	$B_5 V$
		1	0	0	1	0	
		0	0	0	0	1	

rigorously solved in Mayer and Mayer Appendix X, and is briefly summarized below.

The N molecules can be arranged in order in any of $N!$ different ways. The order in which the clusters occur does not matter. Consider a given distribution, that is, a specification of the n_l. For example, it might be $n_1 = 17$, $n_2 = 6$, $n_3 = 78$ and so on, meaning that there are

17 lone molecules, 6 binary clusters, 78 ternary clusters, and so on. The n_l measure the number of clusters containing l molecules each. The order of clusters in the N molecules does not matter, and $N!$ is too great by the factor $n_l!$ for each value of l. The number of different ways of arranging the molecules that is different is then $N!/n_l!$ for each value of l. In addition, the number of ways of arranging the numbered molecules in a cluster is immaterial, and there are $l!$ ways of doing this. Suppose $n_l = 2$ for some l. Then for each arrangement of the l molecules in one of the two clusters there are $l!$ ways of arranging the molecules in the other, a total of $l!l!$ or $l!^2$ ways that are too many. In general, for any value of l, the number of ways of arranging molecules in these clusters is $(l!)^{n_l}$. Then in general the number of ways of arranging N molecules into $n_1, n_2, n_3 \ldots n_l \ldots$ clusters is

$$g(n_l) = \frac{N!}{\prod_l n_l! \prod_l (l!)^{n_l}} \tag{13.71}$$

For example, consider the case $N = 4$, with $n_2 = 2$ and all other $n_l = 0$. The entry in the configuration integral is $3B_2^2 V^2$. The coefficient g is

$$g(2) = \frac{4!}{2!2!2} = \frac{4 \times 3 \times 2 \times 1}{2 \times 1 \times (2 \times 1)^2} = 3$$

The configuration integral becomes

$$Q_\tau = \sum_{(n_l)} \frac{N!}{\prod_l n_l!(l!)^{n_l}} \prod_l (B_l V)^{n_l} \tag{13.72}$$

subject to $\sum_l l n_l = N$. The exact solution generated a great deal of work calculating virial coefficients from potential functions, and in the process the lth cluster integral was defined. It is slightly different from the B_l used in this development, The lth cluster integral is defined as

$$b_l = \frac{1}{l!V} \int \sum \prod_{p<q} f_{pq} \prod_{i=1}^l d\tau_i \tag{13.73}$$

and the configuration integral is

$$\frac{Q_\tau}{N!} = \sum_{(n_l)} \frac{1}{\prod_l n_l!} \prod_l (b_l V)^{n_l} \qquad (13.74)$$

subject to the restrictive condition

$$\sum_l l n_l = N \qquad (13.75)$$

It is important to note that the cluster integral does not contain a volume dependence, and is a function of temperature alone, the gas remaining sufficiently dilute. Unfortunately, the u_{ij} are functions of volume when the pressure is made high enough, and this prohibits use of this theory to treat liquids. In principle, the pressure is now available using equation (13.12), in the form

$$\frac{p}{k_B T} = \left(\frac{\partial ln Q_\tau}{\partial V} \right)_{T,N}$$

but the nearly infinite nature of the sum makes the differentiation intractable. However, the sum contains a most probable term, and once this term is isolated the differentiation becomes tractable. Both the Boltzmann and the Darwin-Fowler approach have been used to determine the most probable term and from this, the thermodynamics. These give identical results, and as the Boltzmann method is algebraically less burdensome, it will be used here.

In the canonical ensemble the partition function is

$$Q = \left(\frac{2\pi m k_B T}{h^2} \right)^{3N/2} \frac{Q_\tau}{N!} \qquad (13.76)$$

with, of course, $N = \sum_l l n_l$. Then, more explicitly

$$Q = \left[\left(\frac{2\pi m k_B T}{h^2} \right)^{3/2} \right]^{\sum_l l n_l} \sum_{(n_l)} \prod_l \frac{(b_l V)^{n_l}}{n_l!} \qquad (13.77)$$

This rearranges to

$$Q = \sum_{(n_l)} \prod_l \left[\left(\frac{2\pi m k_B T}{h^2} \right)^{3l/2} b_l V \right]^{n_l} \Big/ n_l! \qquad (13.78)$$

and on introducing the definition of a factor $g_l = (\frac{2\pi m k_B T}{h^2})^{3l/2} b_l$ this becomes

$$Q = \sum_{(n_l)} \prod_l (g_l V)^{n_l}/n_l! \qquad (13.79)$$

The typical term in the sum is

$$\tau(n_l) = \prod_l \frac{(g_l V)^{n_l}}{n_l!} \qquad (13.80)$$

where $N = \sum_l l n_l$ is constant. Then

$$\ln \tau(n_l) = \sum_l (n_l \ln g_l V - n_l \ln n_l + n_l) \qquad (13.81)$$

The differential is

$$d \ln \tau(n_l) = 0 = \sum_l (\ln g_l V - \ln n_l) dn_l \qquad (13.82)$$

The condition of restraint introduces the undetermined multiplier α ($\sum_l \alpha l) dn_l = 0$ as N is constant), and adding this in

$$\sum_l (\ln g_l V - \ln n_l + \alpha l) dn_l = 0 \qquad (13.83)$$

It is worth remarking that α is a function of V, since $dA = -SdT - pdV + \mu dN$, and N and T are constant, the only way the populations n_l can change is if the volume changes. The most probable n_l is

$$n_l^* = g_l V e^{\alpha l} = g_l V \lambda^l \qquad (13.84)$$

where $\lambda = e^\alpha$. The most probable value of the logarithm of the canonical ensemble partition function is

$$\ln Q^* = \sum_l (g_l V \lambda^l \ln g_l V - g_l V \lambda^l \ln g_l V \lambda^l + g_l V \lambda^l) \qquad (13.85)$$

This may be more compactly expressed as

$$\ln Q^* = \sum_l (-g_l V \lambda^l \ln \lambda^l + n_l^*) \qquad (13.86)$$

Now $\ln \lambda^l = l \ln e^\alpha = \alpha l$, and $\ln Q$ becomes

$$\ln Q^* = \sum_l (n_l^* - \alpha l n_l^*) = \sum_l n_l^* - \alpha N \tag{13.87}$$

The Helmholz free energy is $A = -k_B T \ln Q$. Then

$$A = \alpha N k_B T - k_B T \sum_l n_l^* \tag{13.88}$$

and the pressure is

$$p = -\left(\frac{\partial A}{\partial V}\right)_{T,N} = -N k_B T \left(\frac{\partial \alpha}{\partial V}\right)_{T,N} + k_B T \sum_l \left(\frac{\partial n_l^*}{\partial V}\right)_{T,N} \tag{13.89}$$

Since $n_l^* = g_l V e^{\alpha l}$, and the g_l have no volume dependence, the volume derivative has two terms

$$\left(\frac{\partial n_l^*}{\partial V}\right)_{T,N} = g_l e^{\alpha l} + l g_l V e^{\alpha l} \left(\frac{\partial \alpha}{\partial V}\right)_{T,N} \tag{13.90}$$

Summing over l

$$\sum_l \left(\frac{\partial n_l^*}{\partial V}\right)_{T,N} = \sum_l g_l e^{\alpha l} + \sum_l l g_l V e^{\alpha l} \left(\frac{\partial \alpha}{\partial V}\right)_{T,N} \tag{13.91}$$

or

$$\sum_l \left(\frac{\partial n_l^*}{\partial V}\right)_{T,N} = \sum_l g_l e^{\alpha l} + \left(\sum_l l n_l^*\right)\left(\frac{\partial \alpha}{\partial V}\right)_{T,N} \tag{13.92}$$

and therefore

$$\sum_l \left(\frac{\partial n_l^*}{\partial V}\right)_{T,N} = \sum_l g_l e^{\alpha l} + N \left(\frac{\partial \alpha}{\partial V}\right)_{T,N} \tag{13.93}$$

Substituting this sum into the expression for p in equation (13.89) yields

$$p = -N k_B T \left(\frac{\partial \alpha}{\partial V}\right)_{T,N} + k_B T \sum_l \left(\frac{\partial n_l^*}{\partial V}\right)_{T,N} \tag{13.94}$$

Then

$$p = -Nk_BT \left(\frac{\partial \alpha}{\partial V}\right)_{T,N} + k_BT \left(\sum_l g_l e^{\alpha l} + N \left(\frac{\partial \alpha}{\partial V}\right)_{T,N}\right) \quad (13.95)$$

Cancelling terms gives

$$p = k_BT \sum_l g_l e^{\alpha l} \quad (13.96)$$

Since $n_l^* = g_l V e^{\alpha l}$, the pressure may be written as

$$p = k_BT \sum_l \frac{n_l^*}{V} \quad (13.97)$$

and the pV product is simply

$$pV = k_BT \sum_l n_l^* \quad (13.98)$$

This shows that the pV product is proportional, at constant temperature, to an essentially infinite sum of cluster integrals of increasing order, since n_l^* is proportional to g_l, which is itself proportional to the cluster integral b_l.

The nature of the undetermined multiplier α is apparent when the Gibbs free energy is calculated. By definition, $G = A + pV$, hence

$$G = \left(\alpha Nk_BT - k_BT \sum_l n_l^*\right) + k_BT \sum_l n_l^* = \alpha Nk_BT \quad (13.99)$$

or, since $\lambda = e^\alpha$, $\alpha = \ln \lambda$ and $G = Nk_BT \ln \lambda$. Since $G = \mu N$, $\mu = k_BT \ln \lambda$, and λ emerges as the activity of a single molecule in the gas, while α is its logarithm.

The main results of the theory are summarized in Table 13.2. The theory essentially presents the equation of state as a sum of mildly decorated cluster integrals, themselves determined entirely by the potential energy of interaction between colliding molecules.

The Darwin-Fowler method produces the same results with greater rigor and mathematical burden. Briefly, the approach to the

Table 13.2. Table of results.

$$N = \sum_l l n_l^*$$

$$pV = k_B T \sum_l n_l^*$$

$$\lambda^l n_l^* = \lambda^l V g_l = \lambda^l V \left(\frac{2\pi m k_B T}{h^2}\right)^{3l/2} b_l$$

$$b_l = \frac{1}{l!V} \int \sum \prod_{p<q} f_{pq} \prod_{i=1}^{l} d\tau_i$$

problem requires two essential steps. The first is identifying the configuration integral as the coefficient of z^N in the expansion of the exponential $e^{V \sum_l g_l z^l}$. The second is picking this out by writing

$$Q_\tau = \frac{1}{2\pi i} \oint \frac{e^{V \sum_l g_l z^l}}{z^{N+1}} dz \qquad (13.100)$$

leading to results identical to those presented above.

13.4. Liquids and Dense Gases

Liquids and dense gases differ fundamentally from the relatively dilute gases treated briefly above. The molecules all have inertia, and are tightly packed. Collisions are far more frequent, occurring at the almost unbelievably high rates called Brownian. In these circumstances, the molecules cannot translate the long distances that characterize the dilute gas, and as the mean free path is very short, molecules tend to stay locked into a small region of space, essentially by the fact that at these densities the molecules cannot travel far without colliding. The effect is somewhat colorfully known as the Cage Effect, in which a central molecule remains near its original position even after many collisions, as though the other molecules packed about it formed a cage. Of course, all the molecules in a liquid are mobile, and diffusion rates are finite, not infinitesimal. The problem that statistical mechanics faces is how to describe the liquid usefully, relating molecular fundamentals to thermodynamic functions. Such considerations lead naturally to an average distribution of molecules about a central molecule called the radial distribution function, to correlation functions and to a correlation length.

The density of molecules about a central molecule must obey certain limiting conditions. Since molecules cannot interpenetrate without undergoing nuclear reaction, the density of molecules near the center of mass of the central molecule must be zero. The mean density at large distances from the central molecule must be the bulk density of the fluid in any direction from the center. That is, the distribution of densities about the central molecule must not only converge to the bulk density at large distances from the center, it must also be spherically symmetric. Fluids do not preferentially gather at one end or another of the container. Near the central molecule, however, there may be on average relatively stable layers of other molecules, those that form the cage, and in that region the local mean density can be greater than it is elsewhere. Several layers will arise, rather like a smeared out onion, as each cage is held in place by a larger cage. The radial distribution function is defined as the ratio of the mean local density to the bulk density of the fluid. That is, the radial distribution function $g(\boldsymbol{r})$ is defined as

$$g(\boldsymbol{r}) = \frac{\langle \rho(\boldsymbol{r}) \rangle}{\langle \rho \rangle} \tag{13.101}$$

where $\boldsymbol{r}$ is the vector distance from the center of mass of the central molecule to the point at which the local density is sensed, and $\langle \rho \rangle$ is the bulk density. So defined, the radial distribution function is a dimensionless function that describes how the molecules in a fluid are distributed around a central molecule. If probabilities are used, Born and Green[3] define the radial distribution function a little differently. Let $n_1(\boldsymbol{r}_1)$ be the probability that a molecule numbered for convenience 1 be in a volume element centered at the tip of the location vector $\boldsymbol{r}_1$. Let $n_2(\boldsymbol{r}_1, \boldsymbol{r}_2)$ be the probability that simultaneously molecule 1 is in a volume element located by $\boldsymbol{r}_1$ while molecule 2 is in another located by vector $\boldsymbol{r}_2$. The number density ρ is identically n_1. The radial distribution function is then defined by n_2/n_1^2, which is a little different from the ratio of densities given above.

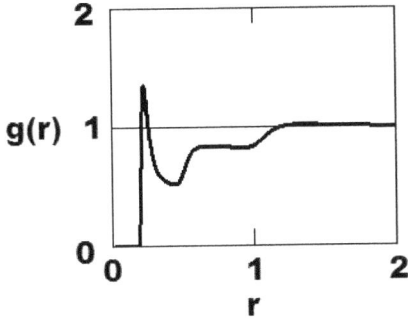

Fig. 13.4. Simulated radial distribution function for a dense fluid. Intermolecular distances are given in arbitrary units. The first maximum would be expected in the 2 to 10 Angstrom range for small molecules such as methane or ethane.

At low densities, the radial distribution function reverts to a single maximum that occurs at the most probable intermolecular collision distance. At higher densities, the cage effect becomes increasingly prominent and secondary maxima form at the most probable intermolecular distance of next nearest and next next nearest neighbors, as sketched informally in Figure 13.4.

The radial distribution function is important for several reasons. First, it may be experimentally observed in a scattering experiment. Depending on the size of the particles that compose the scattering entities, light, hot neutrons, electrons or X-rays, may be used. X-rays are typically used to study small molecules of the order of $1 - 2$ Angstroms in diameter, while colloidal particles would be best studied by the longer wavelengths afforded by visible light. Second, the radial distribution function may be used to determine the thermodynamic properties of the fluid. A more remote comparison with experiment may be achieved through comparison of thermodynamic functions calculated from the radial distribution function with those measured directly.

The main statistical jobs are first, to provide understanding of the radial distribution function, second, to provide means of computing radial distribution functions, and third, to provide its relation to thermodynamics. Let the local density r distant from the central molecule be $\rho(r)$ and the bulk density be $\langle \rho \rangle$. If the volume element at r is dr then the number of molecules at r is $dn = \rho(r)dr$. The density is a number density. By definition, $g(r) = \langle\rho(r)\rangle/\langle\rho\rangle$. The mean number of molecules in the volume element dr is

$$dn = \langle\rho\rangle g(r)dr \qquad (13.102)$$

One way of getting at a visualization of the radial distribution function is to use Dirac delta functions to count molecules, regarded as occupying points in space. For use in the following, the delta function has the properties

$$\delta(x - x_i) = 0 \; x \neq x_i$$
$$= 1 \; x = x_i$$
$$\int_{-\infty}^{\infty} \delta(x - x_i)dx = 1$$
$$\int_{-\infty}^{\infty} f(x)\delta(x - x_i)dx = f(x_i)$$

The property of greatest interest here is that an integral of a sum of delta functions counts them. Thus, for example

$$\int_{-\infty}^{\infty} \left[\delta(x - x_1) + \delta(x - x_2) + \delta(x - x_3) + \delta(x - x_4)\right] dx = 4$$

regardless of the values of x_i $(i = 1, 2, 3, 4)$.

The theory to be developed requires a three dimensional delta function. In Cartesian coordinates,

$$\delta(r - r_i) = \delta(x - x_i)\delta(y - y_i)\delta(z - z_i)$$

Given these properties, the number density at r may be written as

$$\rho(r) = \sum_{i=1}^{N} \delta(r - r_i) \qquad (13.103)$$

with the property

$$\int_{-\infty}^{\infty} \rho(r)dr = N \qquad (13.104)$$

where N is the total number of molecules in the specified volume V. Now let the averaging process refer to the ith molecule, that is, let

$$\langle f(\boldsymbol{r} - \boldsymbol{r}_i)\rangle = \int f(\boldsymbol{r} - \boldsymbol{r}_i)P(\boldsymbol{r}_i)d\boldsymbol{r}_i \qquad (13.105)$$

The probability $P(\boldsymbol{r}_i)$ must obey the relation

$$\int P(\boldsymbol{r}_i)d\boldsymbol{r}_i = 1 \qquad (13.106)$$

$P(\boldsymbol{r}_i)$ is the probability that the ith molecule is in the volume element $d\boldsymbol{r}_i$, normalized to unity.

Born and Green define similar probabilities that they call n_i. In their work, molecules numbered $1, 2, 3 \ldots h$ are instantaneously located in the liquid by position vectors $\boldsymbol{x}_1, \boldsymbol{x}_2, \boldsymbol{x}_3 \ldots \boldsymbol{x}_h$ and the probability that molecule 1 is in volume element $d\boldsymbol{x}_1$ at position $\boldsymbol{x}_1$, molecule 2 is in volume element $d\boldsymbol{x}_2$ at position $\boldsymbol{x}_2$, etc., is $n_h(\boldsymbol{x}_1, \boldsymbol{x}_2, \ldots \boldsymbol{x}_h)d\boldsymbol{x}_1 d\boldsymbol{x}_2 \ldots d\boldsymbol{x}_h$. These probabilities are not normalized to unity, but instead follow the rule

$$\int n_{h+1}(t, \boldsymbol{x})d\boldsymbol{x}_{h+1} = (N - h)n_h(t, \boldsymbol{x})$$

$$\int n_1(t, \boldsymbol{x}_1)d\boldsymbol{x}_1 = N$$

$$\int n_2(t, \boldsymbol{x}_1, \boldsymbol{x}_2)d\boldsymbol{x}_2 = (N - 1)n_1(t, \boldsymbol{x}_1)$$

The second of these relations establishes n_1 as the number density. The radial distribution function is n_2/n_1^2. The lack of normalization of these functions n_i requires care in direct comparison between the development here and that in Born and Green's paper.

Now consider the density correlation function

$$\langle \rho(\boldsymbol{r}_1)\rho(\boldsymbol{r}_2)\rangle = \left\langle \sum_{i=1}^{N}\sum_{j=1}^{N}\delta(\boldsymbol{r}_1 - \boldsymbol{r}_i)\delta(\boldsymbol{r}_2 - \boldsymbol{r}_j) \right\rangle$$

$$= \sum_{i=1}^{N}\sum_{j=1}^{N}\langle\delta(\boldsymbol{r}_1 - \boldsymbol{r}_i)\delta(\boldsymbol{r}_2 - \boldsymbol{r}_j)\rangle \qquad (13.107)$$

In evaluating the sum, two cases arise: $i = j$ and $i \neq j$. Consider $i = j$ first,

$$\langle \delta(r_1 - r_i)\delta(r_2 - r_i) \rangle = \int \delta(r_1 - r_i)\delta(r_2 - r_i)\frac{dr_i}{V}$$

$$= \delta(r_2 - r_1)/V \qquad (13.108)$$

V is the volume of the liquid, and the factor $1/V$ arises as the probability that the ith molecule occurs in the volume element dr_i.

The other case, i different from j, involves the probability product $P(r_i)P(r_j)$. This product implies that simultaneously the ith molecule is in dr_i and the jth is in dr_j. The overall probability is

$$P(r_i)P(r_j)dr_idr_j = \frac{g(r_{ij})}{V^2}dr_idr_j \qquad (13.109)$$

A factor $1/V$ arises for each $P(r)$, and $g(r_{ij})$ is the additional probability that the ith molecule is in dr_i while the jth is in dr_j and the distance between them is r_{ij}.

The radial distribution function is related to the part of the energy that depends only on the configuration that is determined by the intermolecular potential $u(r_{ij})$. On average,

$$\langle E_{configuration} \rangle = \frac{1}{2}\sum_{i \neq j}\langle u(r_{ij}) \rangle = \sum_{i<j}\langle u(r_{ij}) \rangle$$

$$\langle E_{configuration} \rangle = \frac{1}{2}\sum_{i \neq j}\iint u(r_{ij})g(r_{ij})\frac{dr_i}{V}\frac{dr_j}{V} \qquad (13.110)$$

$$\langle E_{configuration} \rangle = \frac{V}{2}\sum_{i \neq j}\int \frac{u(r)g(r)dr}{V^2}$$

$$\langle E_{configuration} \rangle = \frac{N(N-1)}{2V}\int u(r)g(r)dr$$

$$\langle E_{configuration} \rangle = N\frac{\rho}{2}\int u(r)g(r)dr \qquad (13.111)$$

Since N is of the order of Avogadro's number, $N-1 = N$, in excellent approximation. This completes demonstration of the relation of the radial distribution function to the part of the energy that depends on configuration.

13.5. Pressure-Compressibility Fluctuation Theory

In the Grand Canonical Ensemble, fluctuations in N are given by

$$\langle N^2 \rangle = k_B T \left(\frac{\partial \langle N \rangle}{\partial \mu} \right)_{T,V} \tag{13.112}$$

The systems of the ensemble are open, N fluctuates, and $\langle N \rangle$ is the mean number of molecules in a system of the ensemble. The immediate objective is to express the derivative $(\frac{\partial \langle N \rangle}{\partial \mu})_{T,V}$ in terms involving the compressibility. From the thermodynamic Gibbs-Duhem equation at constant temperature

$$N d\mu = V dp$$

Then

$$N \left(\frac{\partial \mu}{\partial N} \right)_{T,V} = V \left(\frac{\partial p}{\partial N} \right)_{T,V} = \left(\frac{\partial p}{\partial \rho} \right)_{T,V}$$

Identify the thermodynamic N with the mean number of molecules $\langle N \rangle$ and substitute in the fluctuation relation

$$\langle N^2 \rangle = \langle N \rangle k_B T \left(\frac{\partial \rho}{\partial p} \right)_{T,V}$$

The compressibility is essentially $(\frac{\partial \rho}{\partial p})_{T,V}$. Now $\rho = \langle N \rangle / V$ and $V = \int d\mathbf{r} = \int d\mathbf{r}_1 = \int d\mathbf{r}_2$. Thus

$$\langle N \rangle = \rho V = \rho \int d\mathbf{r}_1 = \rho \int d\mathbf{r}_2 = \rho \int d\mathbf{r}_1 \int \frac{d\mathbf{r}_2}{V} = \rho \iint \frac{d\mathbf{r}_1 d\mathbf{r}_2}{V} \tag{13.113}$$

The square is therefore

$$\langle N \rangle^2 = \rho^2 \int d\mathbf{r}_1 \int d\mathbf{r}_2 = \rho^2 \iint d\mathbf{r}_1 d\mathbf{r}_2 = \rho^2 \iint d\mathbf{r}_1 d\mathbf{r}_2 \iint \frac{d\mathbf{r}_1}{V} \frac{d\mathbf{r}_2}{V} \tag{13.114}$$

However, since $\langle N \rangle = \int \rho(\boldsymbol{r})d\boldsymbol{r}$ and $\langle N^2 \rangle = \int \rho(r_1)dr_1 \int \rho(r_2)dr_2 = \int \int \rho(r_1)\rho(r_2)dr_1 dr_2$ according to equation (13.101)

$$\langle N^2 \rangle = \int\!\!\int \rho(\boldsymbol{r}_1)\rho(\boldsymbol{r}_2)d\boldsymbol{r}_1 d\boldsymbol{r}_2 = \int\!\!\int \left[\rho^2 g(\boldsymbol{r}_{12}) + \rho(\boldsymbol{r}_{12})\right]d\boldsymbol{r}_1 d\boldsymbol{r}_2$$
$$(13.115)$$

The fluctuation in N is

$$\langle \Delta N^2 \rangle = \langle N^2 \rangle - \langle N \rangle^2 \qquad (13.116)$$

Using equations (13.114) and (13.115), the fluctuation is

$$\langle \Delta N^2 \rangle = \int\!\!\int \left[\rho^2 g(\boldsymbol{r}_{12}) + \rho(\boldsymbol{r}_{12}) - \rho^2\right]d\boldsymbol{r}_1 d\boldsymbol{r}_2 \qquad (13.117)$$

which becomes

$$\langle \Delta N^2 \rangle = V\rho^2 \int \left[g(\boldsymbol{r}_{12}) - 1\right]d\boldsymbol{r}_{12} + \rho V \qquad (13.118)$$

and since $\rho = \langle N \rangle / V$ and $\langle \Delta N^2 \rangle = \langle N \rangle k_B T (\frac{\partial \rho}{\partial p})_{T,V}$, this may be rearranged to

$$k_B T(\frac{\partial \rho}{\partial p})_{T,V} = \rho \int \left[g(\boldsymbol{r}) - 1\right]d\boldsymbol{r} + 1 \qquad (13.119)$$

Rearranging and taking advantage of the spherical symmetry, the Ornstein-Zernike fluctuation formula results in the form

$$\left(\frac{\partial p}{\partial \rho}\right)_{T,V} = \frac{k_B T}{1 + \rho \int [g(r) - 1]4\pi r^2 dr} \qquad (13.120)$$

As the fluid becomes less and less dense, the flickering cage structure that effectively localizes the central molecule relaxes, and the central molecule becomes increasingly mobile. The effect on $g(r)$ is to smooth out the oscillations at large values of r increasingly, until a value of density is reached at which only the nearest neighbor maximum remains, something like the grin of the Cheshire cat. The central molecule then becomes very mobile, mean free paths having lengthened, and $g(r)$ closely resembles the exponential $e^{-u(r)/k_B T}$. In this low density region, the ratio in the previous equation may be expanded, giving a power series in ascending powers of the density. The leading terms are

$$\left(\frac{\partial p}{\partial \rho}\right)_{T,V} = k_B T \left[1 - \rho \int [g(r) - 1] 4\pi r^2 dr + \cdots \right] \qquad (13.121)$$

Integrating to get the pressure as a function of density,

$$p = \rho k_B T \left[1 - \frac{\rho}{2} \int [g(r) - 1] 4\pi r^2 dr + \cdots \right] \qquad (13.122)$$

Even better,

$$\frac{p}{\rho k_B T} = 1 - \frac{\rho}{2} \int [g(r) - 1] 4\pi r^2 dr + \cdots \qquad (13.123)$$

This expansion is of the form of the virial expansion

$$\frac{p}{\rho k_B T} = 1 + \rho \frac{B_2(T)}{N_0} + \cdots \qquad (13.124)$$

Since the second virial coefficient is $B_2(T) = 2\pi N_0 \int_0^\infty (1 - e^{-u(r)/k_B T}) r^2 dr$ this suggests that if $g(r)$ is identified with $e^{-u(r)/k_B T}$, the low density expansion of the Ornstein-Zernike formula is the virial expansion, giving a correct expression for the second virial coefficient.

In the usual plot, the pressure in a real gas is plotted against volume, and the critical point is defined in part by the relation $(\frac{\partial p}{\partial V})_{T_c} = 0$. When density is used instead of volume, the plot is of p against ρ^{-1}, and the first derivative that partially defines the critical point is $(\frac{\partial p}{\partial \rho^{-1}})_{T_c} = 0$. However, $(\frac{\partial p}{\partial \rho^{-1}})_T = -\rho^2 (\frac{\partial p}{\partial \rho})_T$, and the critical point is also partially defined by the first derivative $(\frac{\partial p}{\partial \rho})_{T_c} = 0$. The definition is completed by requiring the second derivative to vanish. The derivative is still given by the Ornstein-Zernike relation, equation (13.120), and evidently the integral in this relation diverges at the critical point. Some insight into what this means physically is provided by consideration of the Ornstein-Zernike expression for the two-body time independent correlation function

$$g(r) - 1 = \frac{Ae^{-\kappa r}}{r} \tag{13.125}$$

where κ is the inverse two body correlation length ξ. A is a constant. In the critical region, $(\frac{\partial p}{\partial \rho})_T$ is extremely sensitive to the long range behavior of g(r), and as the critical point is approached, ξ grows and κ tends towards zero, until in the critical limit the correlation length becomes indefinitely long and κ essentially vanishes. Away from the critical region, κ is determined by the square root of the derivative$(\frac{\partial p}{\partial \rho})_T$.

A general expression for the equation of state of the liquid was given in 1946 by Born and Green[3] as

$$p = \rho k_B T \left[1 - \frac{\rho}{6k_B T} \int r u'(r) g(r) dr \right] \tag{13.126}$$

albeit in a different notation. A simple derivation of this relation begins with the canonical ensemble relations $A = -k_B T ln Q_L$, where Q_L is the partition function for the liquid. The pressure is $p = -k_B T(\frac{\partial A}{\partial V})_{T,N}$ and the partition function is

$$Q_L = \frac{1}{N! h^{3N}} \int \cdots \int e^{-\left[\sum_i \frac{p_i^2}{2m} + \sum_{i<j} u(r_{ij}) \right]/k_B T}$$
$$\times d\mathbf{r}_1 \cdots d\mathbf{r}_N d\mathbf{p}_1 \cdots d\mathbf{p}_N \tag{13.127}$$

At this point the derivation diverges from that for the non-ideal gas. The configuration integral is formally the same

$$Q_c = \int e^{-\sum_{i<j} u(r_{ij})/k_B T} d\mathbf{r}_1 \cdots d\mathbf{r}_N = Z \tag{13.128}$$

but now the variables are changed. Define $\mathbf{R}_i$ to lie between 0 and 1 by defining an L such that $\mathbf{r}_i = L^3 \mathbf{R}_i$ where L is such that $0 < \mathbf{R}_i < 1$. Here $L^3 = V$, the volume. The pressure is then

$$p = \frac{k_B T}{Z} \frac{\partial}{\partial V} \left[V^N \int e^{-\sum_{i<j} u(LR_{ij})/k_B T} d\mathbf{R}_1 \cdots d\mathbf{R}_N \right] \tag{13.129}$$

The derivative is rather lengthy,

$$p = \frac{k_B T}{Z} \left[NV^{N-1} \int e^{-\sum_{i<j} u(LR_{ij})/k_B T} d\mathbf{R}_1 \cdots d\mathbf{R}_N \right.$$

$$\left. + V^N \int \frac{-1}{k_B T} \sum_{i<j} \frac{\partial u(r_{ij})}{\partial r_{ij}} R_{ij} \frac{\partial L}{\partial V} e^{-\sum_{i<j} u(LR_{ij})/k_B T} d\mathbf{R}_1 \cdots d\mathbf{R}_N \right]$$

$$(13.130)$$

Since $V = L^3, L = V^{1/3}$ and $\frac{\partial L}{\partial V} = \frac{1}{3} V^{-2/3}$. At this point, revert to the original variables. Note that $V^N d\mathbf{R}_1 \cdots d\mathbf{R}_N = d\mathbf{r}_1 \cdots d\mathbf{r}_N$.

$$p = k_B T \left[\rho + \frac{-1}{3 k_B T V} \frac{\sum_{i<j} \int r_{ij} u'(r_{ij}) e^{-\sum_{i<j} u(r_{ij})/k_B T} d\mathbf{r}_1 \cdots d\mathbf{r}_N}{Z} \right]$$

$$(13.131)$$

Introducing the average

$$p = k_B T \left[\rho - \frac{1}{3 k_B T V} \sum_{i<j} \langle r_{ij} u'(r_{ij}) \rangle \right] \qquad (13.132)$$

The average evaluates as

$$\langle r_{ij} u'(r_{ij}) \rangle = \iint r_{ij} u'(r_{ij}) \frac{d\mathbf{r}_i}{V} \frac{d\mathbf{r}_j}{V} g(r_{ij}) \qquad (13.133)$$

$$\langle r_{ij} u'(r_{ij}) \rangle = V^{-2} V \int r u'(r) g(r) dr \qquad (13.134)$$

Then

$$p = k_B T \left[\rho - \frac{1}{3 k_B T V} \frac{N(N-1)}{2V} \int r u'(r) g(r) dr \right] \qquad (13.135)$$

Since N is of the order of Avogadro's number $N - 1$ is very nearly N, and as $\rho = N/V$, the result is the desired relation,

$$p = k_B T \left[\rho - \frac{\rho^2}{6 k_B T} \int r u'(r) g(r) dr \right] \qquad (13.136)$$

showing how the radial distribution function contributes to the pressure and thus the compressibility, through the derivative $(\frac{\partial \rho}{\partial p})_{T,V}$.

Problems

1. Use the potential energy of Figure 13.1 to calculate the second virial coefficient and from this the equation of state assuming there are no higher virial coefficients.
2. Use the potential energy of Figure 13.2 to calculate the second virial coefficient. How is this related to the van der Waals equation?
3. Expand $\prod_{i<j=1}^{4}(1+f_{ij}) = (1+f_{12})(1+f_{13})(1+f_{14})\,(1+f_{23})(1+f_{24})(1+f_{34})$.

References

1. J. E. Mayer and M. G. Mayer, *Statistical Mechanics*, John Wiley and Sons, New York, 1940.
2. H. D. Ursell, Proc. Cambridge Phil. Soc. **23**, 685 (1927).
3. M. Born and H. S. Green, Proc. Roy. Soc. London **188**, 10–18 (1946).

Chapter 14

Bose-Einstein, Fermi-Dirac and Boltzmann Statistics, Quantum Gas

14.1. Quantum Statistics

In quantum mechanics, identical particles are indistinguishable. Let the combined space and spin coordinates of two such particles be represented by ξ_i, where i is either 1 or 2. The probability amplitude, or wavefunction, of the two particles is $\Psi(\xi_1, \xi_2)$. If the particles are truly indistinguishable, then the observable consequences of exchanging coordinates must be undetectable. Thus

$$\psi * (\xi_{1,2})\psi(\xi_{1,2}) = \psi * (\xi_{2,1})\psi(\xi_{2,1}) \tag{14.1}$$

This condition is generally satisfied by writing

$$\psi(\xi_{1,2}) = e^{i\alpha}\psi(\xi_{2,1}) \tag{14.2}$$

where α is real, but otherwise unrestricted. The exponential $e^{i\alpha}$ is called a phase factor. A second exchange returns the probability $\psi^*\psi$ to its original condition, giving rise to the equality

$$\psi * (\xi_{1,2})\psi(\xi_{1,2}) = e^{2i\alpha}\psi * (\xi_{1,2})\psi(\xi_{1,2}) \tag{14.3}$$

Since a thing is always identically itself, one is forced to conclude that

$$e^{2i\alpha} = 1 \tag{14.4}$$

and as $\sqrt{1} = \pm 1$, it appears that there are only two kinds of wave-function. Either

$$\psi(\xi_1, \xi_2) = +\psi(\xi_2, \xi_1) \tag{14.5}$$

or

$$\psi(\xi_1, \xi_2) = -\psi(\xi_2, \xi_1) \tag{14.6}$$

As long as the particles remain truly indistinguishable, only one of these two behaviors is possible. Those that follow the plus sign are said to be bosons and are governed by Bose-Einstein statistics. The other kind are said to be fermions and follow Fermi-Dirac statistics.

14.2. Comparison of Statistics

Boltzmann, Bose-Einstein and Fermi-Dirac statistics are compared in the following discussion, using the grand canonical ensemble. Initially, consider a crystal. The grand canonical ensemble has the partition function

$$\Xi(T, V, \mu) = \sum_N \sum_E \Omega(E, V, N) e^{-E/k_B T} e^{\mu N/k_B T} \tag{14.7}$$

Now, the total energy $E = \sum_i \varepsilon_i n_i$ and the absolute activity $\lambda = e^{\mu/k_B T}$. Using the notation (n_i) to mean all possible distribution numbers consistent with the number of particles $N = \sum_i n_i$ allows rewriting the partition function as

$$\Xi = \sum_N \sum_{(n_i)} \frac{N!}{\prod_i n_i!} e^{-\sum_i \varepsilon_i n_i/k_B T} \lambda^{\sum_i n_i/k_B T} \tag{14.8}$$

The coefficient $\frac{N!}{\prod_i n_i!}$ counts the number of states accessible for one specification of the distribution numbers n_i. There are many such distributions consistent with N. This may be rewritten as

$$\Xi = \sum_N \sum_{(n_i)} \frac{N!}{\prod_i n_i!} \prod_i (\lambda e^{-\varepsilon_i/k_B T})^{n_i} \tag{14.9}$$

But the multinomial theorem states

$$\left(\sum_i \lambda e^{-\varepsilon_i/k_B T}\right)^N = \sum_{(n_i)} \frac{N!}{\prod_i n_i!} \prod_i (\lambda e^{-\varepsilon_i/k_B T})^{n_i} \qquad (14.10)$$

and therefore

$$\Xi = \sum_N \left(\sum_i \lambda e^{-\varepsilon_i/k_B T}\right)^N \qquad (14.11)$$

Now $q = \sum_i e^{-\varepsilon_i/k_B T}$ and λ factors, leaving

$$\Xi = \sum_N q^N \lambda^N \qquad (14.12)$$

To the extent that this is an effectively infinite sum, this is very nearly

$$\Xi = \frac{1}{1 - q\lambda} \qquad (14.13)$$

provided $q\lambda < 1$. This can be shown in this way. First, calculate the mean number of particles assuming that equation (14.13) is valid

$$\langle N \rangle = \frac{\partial \ln \Xi}{\partial \ln \lambda} \qquad (14.14)$$

The result is

$$\langle N \rangle = \frac{q\lambda}{1 - q\lambda} \qquad (14.15)$$

Solving this for $q\lambda$, one finds that

$$q\lambda = \frac{\langle N \rangle}{\langle N \rangle + 1} < 1 \qquad (14.16)$$

for every possible finite value of the mean number of particles.

The mean occupation numbers are also available from this analysis. Note that $q = \sum_i e^{-\varepsilon_i/k_B T}$ and the mean number of particles is

$$\langle N \rangle = \sum_i \langle n_i \rangle = \frac{q\lambda}{1 - q\lambda} = \frac{\sum_i \lambda e^{-\varepsilon_i/k_BT}}{1 - q\lambda} \tag{14.17}$$

This implies that

$$\langle n_i \rangle = \frac{\lambda e^{-\varepsilon_i/k_BT}}{1 - q\lambda} \tag{14.18}$$

in Boltzmann statistics. Now, $1 - q\lambda = q\lambda/\langle N \rangle$ and thus

$$\langle n_i \rangle = \langle N \rangle \frac{e^{-\varepsilon_i/k_BT}}{q} \tag{14.19}$$

which compares interestingly with the more familiar expression from the microcanonical ensemble, specifically

$$n_i^* = N \frac{e^{-\varepsilon_i/k_BT}}{\sum_i e^{-\varepsilon_i/k_BT}} \tag{14.20}$$

Now, in forming the grand partition function, both E and N were summed over all possible values. Exactly the same result can be achieved by summing over the occupation numbers n_i, and expressing N and E in terms of the occupation numbers. The result is

$$\Xi = \sum_{(n_i)} \frac{(\sum_i n_i)!}{\prod_i n_i!} \prod_i (\lambda e^{-\varepsilon_i/k_BT})^{n_i} \tag{14.21}$$

Equivalently

$$\Xi = \sum_{n_0=0}^{\infty} \sum_{n_1=0}^{\infty} \sum_{n_2=0}^{\infty} \cdots \frac{(\sum_i n_i)!}{\prod_i n_i!} \prod_i (\lambda e^{-\varepsilon_i/k_BT})^{n_i} \tag{14.22}$$

The number of accessible states is different for different symmetries. The Boltzmann method of counting states differs from the method that must be used to count states in either the Fermi-Dirac or Bose-Einstein statistics. To illustrate this point count states for a gas of three independent and indistinguishable oscillators sharing a total energy of $7\frac{1}{2} h\nu$. That is, there are 6 quanta to distribute over the

Table 14.1.   States of three distinguishable oscillators sharing
6 quanta of energy. The number of states contributed by each
variation is shown below the variation set.

600	510	420	411	321	330	222
060	051	042	141	132	033	
006	105	204	114	213	303	
	150	240		231		
	015	024		123		
	501	402		231		
3	6	6	3	6	3	1

states of this super-microcanonical ensemble. In Boltzmann statistics, states are counted for the three oscillators regarded as distinguishable, and conversion to indistinguishability made by division of the count by $N!$ at the end. For Boltzmann statistics, the assignment of energy to oscillators is unrestricted, and all 6 quanta may be given to any one of the three oscillators, resulting in a total of three states. Alternatively, one oscillator may have 5 quanta and one of the others one quantum of energy. There are six ways to do this, for six more states. In an obvious notation, the various possibilities are shown in Table 14.1. There are a total of 28 states when the oscillators are distinguishable, and $28/3! = 4\ 2/3$ states when the oscillators are indistinguishable.

Obviously, two thirds of a state cannot exist for states are counted, not measured. This may be taken as an indication that there is something wrong with Boltzmann statistics. Let us see what. For particles (oscillators in this example) obeying Fermi-Dirac statistics, an exchange of one pair of particles follows the rule $R_{ex}\psi = -\psi$. The symmetrized wavefunction is of the form

$$\psi = \frac{1}{\sqrt{3!}} \begin{vmatrix} \alpha(\xi_1) & \alpha(\xi_2) & \alpha(\xi_3) \\ \beta(\xi_1) & \beta(\xi_2) & \beta(\xi_3) \\ \gamma(\xi_1) & \gamma(\xi_2) & \gamma(\xi_3) \end{vmatrix} \tag{14.23}$$

where the ξ_i are in general the combined spin and space coordinates for the problem and α, β, γ are elementary functions for different states, cast as the elements of a determinant. It is a property of determinants that the numerical value vanishes if any two rows or any two columns are equal. This means both that no two particles can have the same coordinates and no two can have the same wavefunction. If $\alpha = \beta$, the wavefunction for the system vanishes. Inspection of Table 14.1 shows that only three states survive, namely 510, 420 and 321. The only possible values of the occupation numbers n_i are 0 and 1 (for 510, $n_5 = 1$, $n_1 = 1$ and all other $n_i = 0$).

For Bose-Einstein statistics, the exchange rule is $R_{ex}\psi = +\psi$. The wavefunction itself is the same as that for Fermi-Dirac statistics, but with omission of the minus signs that occur when the determinant is expanded. That is, for bosons

$$\psi = \frac{1}{\sqrt{3!}}(\alpha(\xi_1)\beta(\xi_2)\gamma(\xi_3) + \alpha(\xi_2)\beta(\xi_3)\gamma(\xi_1) + \alpha(\xi_3)\beta(\xi_1)\gamma(\xi_2)$$

$$+ \alpha(\xi_3)\beta(\xi_2)\gamma(\xi_1) + \alpha(\xi_2)\beta(\xi_1)\gamma(\xi_3) + \alpha(\xi_1)\beta(\xi_3)\gamma(\xi_2))$$

$$(14.24)$$

Now if two particles have the same set of quantum numbers and the same elementary wavefunction $\alpha = \beta$ and nothing untoward occurs: the state of the ensemble exists. There is no difference between 600, 060 and 006, and the total number of Bose-Einstein states is the number of sets, which reference to Table 14.1 will show is 7. The number of states is Ω, and in general as well as in this example

$$\Omega_{FD} \leq \Omega_B \leq \Omega_{EB} \qquad (14.25)$$

As the energy increases, the number of Fermi-Dirac states converges towards the number of states estimated by Boltzmann statistics, as do the number of Bose-Einstein states.

In Boltzmann statistics for localized identifiable particles (in the above, disembodied oscillators), the number of states Ω_B is given by

$$\Omega_B = \sum_{(n_i)} \frac{N!}{\prod_i n_i!} \prod_i \omega_i^{n_i} \qquad (14.26)$$

where ω_i is the degeneracy of the energy levels ε_i and the sum is constrained by the conditions $N = \sum_i n_i$ and $E = \sum_i n_i \varepsilon_i$, the sums running over the states. For comparison with indistinguishable

Fermi-Dirac or Bose-Einstein particles, the result is divided by $N!$ to correct approximately for the massive over count that otherwise results. The multinomial theorem allows evaluation in a sense, for the theorem states that in general

$$\left(\sum_i a_i \right)^N = \sum_{(n_i)} \frac{N!}{\prod_i n_i!} \prod_i a_i^{n_i} \tag{14.27}$$

subject to the single restriction $N = \sum_i n_i$. If the a_i are set equal to $\omega_i e^{-\varepsilon_i/k_B T}$ the result is

$$\left(\sum_i \omega_i e^{-\varepsilon_i/k_B T} \right)^N = \sum_{(n_i)} \frac{N!}{\prod_i n_i!} \prod_i (\omega_i e^{-\varepsilon_i/k_B T})^{n_i} \tag{14.28}$$

This may be rewritten

$$\left(\sum_i \omega_i e^{-\varepsilon_i/k_B T} \right)^N = \sum_{(n_i)} \frac{N!}{\prod_i n_i!} e^{-\sum_i \varepsilon_i n_i/k_B T} \prod_i \omega_i^{n_i} \tag{14.29}$$

The sum is unrestricted in E: every possible energy $E = \sum_i n_i \varepsilon_i$ occurs somewhere in the sum, which is thus of the form

$$\left(\sum_i \omega_i e^{-\varepsilon_i/k_B T} \right)^N = 1 + \Omega_1 e^{-E_1/k_B T} + \Omega_2 e^{-E_2/k_B T} + \Omega_3 e^{-E_3/k_B T} + \cdots \tag{14.30}$$

An example may help. Suppose a_i is set equal to z^i. Then

$$\sum_i a_i = \sum_{i=0}^{\infty} z^i = \frac{1}{1-z} \tag{14.31}$$

if $|z| < 1$. Then the multinomial theorem gives the result

$$\left(\frac{1}{1-z} \right)^N = 1 + \Omega_1 z + \Omega_2 z^2 + \Omega_3 z^3 + \cdots \tag{14.32}$$

Consider an "ensemble" consisting of 3 indistinguishable harmonic oscillators with total energy 7 1/2 hν. Since 1 1/2 hν is zero point energy, 6 $h\nu$ are shared, and the coefficient of z^6 is the total number of states for distinguishable oscillators. The expansion is

$$\left(\frac{1}{1-z}\right)^3 = 1 + 3z + 6z^2 + 10z^3 + 15z^4 + 21z^5 + 28z^6 + 36z^7 + 45z^8 + \cdots$$

(14.33)

Table 14.1 confirms that indeed 28 is the correct number of states for $N = 3$ and 6 quanta shared amongst them.

14.3. Calculation of Mean Occupation Numbers

The mean occupation numbers can be calculated directly from the grand partition function written in the form

$$\Xi = \sum_{n_0} \sum_{n_1} \cdots \sum_{n_j} \cdots \prod_i (\lambda e^{-\varepsilon_i/k_B T})^{n_i}$$

(14.34)

By definition of average

$$\langle n_j \rangle = \frac{\sum_{n_0} \sum_{n_1} \cdots \sum_{n_j} \cdots n_j (\lambda e^{-\varepsilon_j/k_B T})^{n_j} \prod_{i \neq j} (\lambda e^{-\varepsilon_i/k_B T})^{n_i}}{\sum_{n_0} \sum_{n_1} \cdots \sum_{n_j} \cdots (\lambda e^{-\varepsilon_j/k_B T})^{n_j} \prod_{i \neq j} (\lambda e^{-\varepsilon_i/k_B T})^{n_i}}$$

(14.35)

For Boltzmann statistics, the sums run over all possible distribution numbers, from zero to infinity. All of the sums except that over n_j are identical in both numerator and denominator, leaving just the sums over j.

$$\langle n_j \rangle = \frac{\sum_{n_j=0}^{\infty} n_j (\lambda e^{-\varepsilon_j/k_B T})^{n_j}}{\sum_{n_j=0}^{\infty} (\lambda e^{-\varepsilon_j/k_B T})^{n_j}}$$

(14.36)

This evaluates easily, using $\sum_{n=0}^{\infty} x^n = \frac{1}{1-x}$ and $\sum_{n=0}^{\infty} nx^n = \frac{x}{(1-x)^2}$ for $|x| < 1$.

$$\frac{\sum_{n=0}^{\infty} nx^n}{\sum_{n=0}^{\infty} x^n} = \frac{x}{1-x}$$

(14.37)

The mean occupation number evaluates exactly as

$$\langle n_j \rangle = \frac{\lambda e^{-\varepsilon_j/k_B T}}{1 - \lambda e^{-\varepsilon_j/k_B T}}$$

(14.38)

Note that nowhere did the nature of the energy levels enter. It is only necessary that these exist.'

In Fermi-Dirac statistics, the occupation numbers are restricted by symmetry to values of either 0 or 1. No other values may occur. In an ensemble of fermions, there is only one wavefunction for each possible selection of the occupation numbers. Therefore for each such selection, there is only one state and $\Omega = 1$. The grand partition function becomes

$$\Xi = \sum_{n_0=0}^{1} \sum_{n_1=0}^{1} \cdots 1 e^{-\sum_i \varepsilon_i n_i / k_B T} \lambda^{\sum_i n_i} \tag{14.39}$$

This may be written as

$$\Xi = \sum_{n_0=0}^{1} \sum_{n_1=0}^{1} \cdots \prod_i (\lambda e^{-\varepsilon_i / k_B T})^{n_i} \tag{14.40}$$

Interchanging the sum and product is rigorous since N is very large but finite, and results in

$$\Xi = \prod_i \sum_{n_i=0}^{1} (\lambda e^{-\varepsilon_i / k_B T})^{n_i} \tag{14.41}$$

The sums have only two terms, and expand to

$$\Xi = \prod_i (1 + \lambda e^{-\varepsilon_i / k_B T}) \tag{14.42}$$

The mean number N of fermions in a system of the ensemble is $\langle N \rangle = \lambda \frac{\partial \ln \Xi}{\partial \lambda}$, as shown in Chapter 2. Then

$$\langle N \rangle = \lambda \frac{\partial}{\partial \lambda} \sum \ln(1 + \lambda e^{-\varepsilon_i / k_B T}) \tag{14.43}$$

The derivative is

$$\langle N \rangle = \sum_i \frac{\lambda e^{-\varepsilon_i / k_B T}}{1 + \lambda e^{-\varepsilon_i / k_B T}} \tag{14.44}$$

However, since $\langle N \rangle = \sum_i \langle n_i \rangle$ must also hold, it immediately follows that

$$\langle n_i \rangle = \frac{\lambda e^{-\varepsilon_i/k_B T}}{1 + \lambda e^{-\varepsilon_i/k_B T}} \tag{14.45}$$

Alternatively, equation (14.45) may be calculated directly. By definition of average,

$$\langle n_i \rangle = \frac{\sum_{n_0=0}^{1} \sum_{n_1=0}^{1} \cdots \sum_{n_i=0}^{1} \cdots n_i (\lambda e^{-\varepsilon_i/k_B T})^{n_i} \prod_{j \neq i} (\lambda e^{-\varepsilon_j/k_B T})^{n_j}}{\sum_{n_0=0}^{1} \sum_{n_1=0}^{1} \cdots \sum_{n_i=0}^{1} \cdots (\lambda e^{-\varepsilon_i/k_B T})^{n_i} \prod_{j \neq i} (\lambda e^{-\varepsilon_j/k_B T})^{n_j}} \tag{14.46}$$

Cancelling identical sums gives

$$\langle n_i \rangle = \frac{\sum_{n_i=0}^{1} n_i (\lambda e^{-\varepsilon_i/k_B T})^{n_i}}{\sum_{n_i=0}^{1} (\lambda e^{-\varepsilon_i/k_B T})^{n_i}} \tag{14.47}$$

Carrying out the sums gives equation (14.45) again. The mean occupation numbers and grand partition functions for all three statistics are compared in Table 14.2.

For all three statistics, in the high temperature limit, the mean occupation numbers tend toward $\langle n_i \rangle \cong \lambda e^{-\varepsilon_i/k_B T}$. However, as the temperature tends towards the absolute zero, the behaviors become very different. Both Bose and Boltzmann statistics require all N particles to condense into the lowest accessible energy state, but Fermi statistics does not do this. Instead, the N lowest lying energy states are occupied. When the systems are warmed from absolute zero, the heat capacity increases initially linearly for Bose and Boltzmann statistics, but rises only slowly from zero slope for Fermi statistics.

14.4. Behavior of Occupation Numbers

In the grand ensemble, T, V and $\lambda = e^{\mu/k_B T}$ are specified. The mean occupation numbers for fermions may be alternatively expressed as

$$\langle n_i \rangle = \frac{1}{e^{-(\mu-\varepsilon_i)/k_B T} + 1} \tag{14.48}$$

If μ is specified at some value above the ground energy state, the value of the mean occupation number is determined by the specification

Table 14.2. Comparison of mean occupation numbers and grand partition functions for Bose-Einstein, Boltzmann and Fermi-Dirac statistics.

	Bose Einstein	Boltzmann	Fermi-Dirac
mean occupation number	$\langle n_i \rangle = \dfrac{\lambda e^{-\varepsilon_i/k_B T}}{1 - \lambda e^{-\varepsilon_i/k_B T}}$	$\langle n_i \rangle = \dfrac{\lambda e^{-\varepsilon_i/k_B T}}{1 - \lambda q}$	$\langle n_i \rangle = \dfrac{\lambda e^{-\varepsilon_i/k_B T}}{1 + \lambda e^{-\varepsilon_i/k_B T}}$
grand partition function	$\Xi = \displaystyle\prod_i \dfrac{1}{1 - \lambda e^{-\varepsilon_i/k_B T}}$	$\Xi = \dfrac{1}{1 - \lambda q} \quad q = \displaystyle\sum_i e^{-\varepsilon_i/k_B T}$	$\Xi = \displaystyle\prod_i 1 + \lambda e^{-\varepsilon_i/k_B T}$

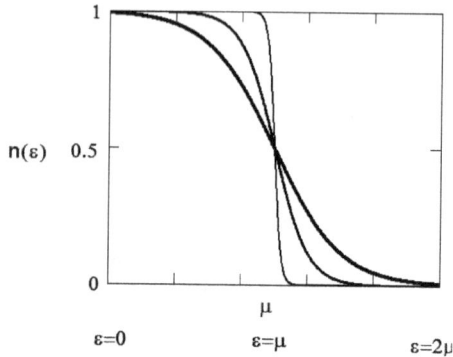

Fig. 14.1.   Mean occupation numbers for Fermi-Dirac statistics as a function of energy ε at three different temperatures. Heavy line: high temperature. Thin line: low temperature. The other line is for an intermediate temperature.

of temperature. If the temperature tends toward absolute zero, the occupation numbers for $\varepsilon_i < \mu$ all tend toward one and those above μ toward zero. This behavior is visualized in Figure 14.1.

Since only one fermion may occupy a state as the temperature tends toward zero, specification of the chemical potential in an ensemble of fermions may be regarded as specifying the density. For fixed T and ε_i allowing μ to decrease indefinitely to minus infinity implies that all the occupation numbers become zero, a circumstance that can only mean there are no fermions, that is, that the density is zero. On the other hand, allowing the chemical potential to increase boundlessly forces all the occupation numbers to become one, which can only occur if the density is infinite. Thus, for fermions, $-\infty < \mu < \infty$ specifies the density in the range zero to infinity.

In Bose-Einstein statistics, specification of the chemical potential again specifies the density, but the range of chemical potential is sharply defined by the energy in the ground state and severely constrained by the condition of series summability. When the average is taken to calculate the mean occupation number, equation (14.36) results. Evaluating the sums as shown above is valid provided $|\lambda e^{-\varepsilon_i/k_B T}| = |e^{(\mu - \varepsilon_i)/k_B T}| < 1$. This means that $\mu < \varepsilon_i$ for all i, including the ground state. Then select $\mu < \varepsilon_i$ for all i and let the temperature approach the absolute zero. The occupation numbers follow the relation $\langle n_i \rangle = \frac{1}{e^{(\varepsilon_i - \mu)/k_B T} - 1}$. The argument of the exponential must be positive for every state i. Then as T approaches zero the exponential grows boundlessly, sending every n_i to zero. The density is zero for every selection of chemical potential, the series remaining summable. Interestingly, if the chemical potential is selected equal to the ground state energy, the n_i all vanish but the population of the ground state is indefinitely large. The possible specifications of the chemical potential are summarized in the short table

$\mu > \varepsilon_0$	series cannot be summed
$\mu = \varepsilon_0$	infinite density
$\mu < \varepsilon_0$	zero density

Evidently the range of finite densities must lie in the narrow range of chemical potentials between the ground state and just below it. If the chemical potential is selected nearly infinitesimally less than the ground state energy, such as $(\varepsilon_0 - \mu)/k_B T \cong \frac{1}{\langle N \rangle}$ then the population of the ground state is given by

$$\langle n_0 \rangle = \frac{1}{e^{1/\langle N \rangle} - 1} \tag{14.49}$$

On expanding the exponential, one finds

$$\langle n_0 \rangle = \frac{1}{1 + \frac{1}{\langle N \rangle} + \frac{1}{2\langle N \rangle^2} + \cdots - 1} \cong \langle N \rangle \tag{14.50}$$

As temperature goes to absolute zero, the population of the ground state goes to the mean number of bosons and all states of higher energy have vanishing population. The behavior at higher temperatures is shown in Figure 14.2.

A comparison of the three statistics is shown in Figure 14.3. Evidently, Boltzmann statistics essentially splits the difference between the two quantum statistics, especially at high energy, and all three behaviors approach a common limit at very high energies.

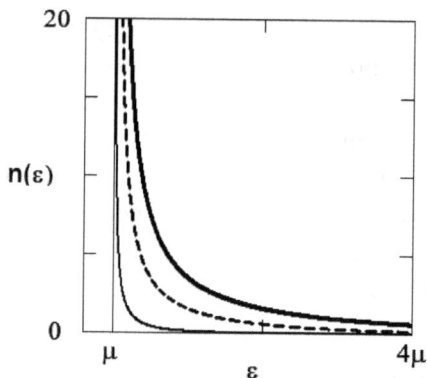

Fig. 14.2.   Bose-Einstein populations as a function of energy ε for the same three different temperatures used in Figure 14.1. Heavy line: highest temperature used. Light line, lowest temperature. Dashed line: intermediate temperature. All are asymptotic to the thin line $\varepsilon = \mu$.

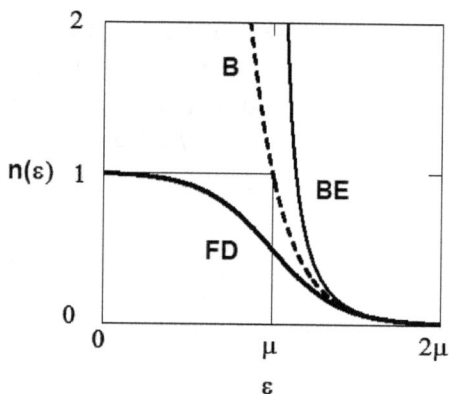

Fig. 14.3.   Comparison of Bose-Einstein (thin solid line marked BE), Fermi-Dirac (thick solid line marked FD) and Boltzmann (dashed line marked B) statistics. The Bose-Einstein curve approaches the vertical line $\varepsilon = \mu$ asymptotically. Boltzmann curve crosses this line at $n = 1$.

14.5. Gas of Fermions

The mean number of fermions per quantum state in a gas of these fermions is

$$\langle n_i \rangle = \frac{1}{e^{(\varepsilon_i - \mu)/k_B T} + 1} \tag{14.51}$$

Assume the gas exists. The mean number of fermions per unit momentum interval N_p is first required. In the x direction, the momentum is

$$p_x = n_x \frac{h}{L} \quad n_x = 1, 2, 3, \cdots \infty \tag{14.52}$$

when the gas is retained by a cubical box with sides of length L. The total number of fermions in the assumed gas is either

$$\langle N \rangle = \sum_i \langle n_i \rangle \tag{14.53}$$

or alternatively

$$\langle N \rangle = \int N_p dp_x dp_y dp_z \tag{14.54}$$

The problem is to find a suitable expression for the integrand. Formally,

$$dn_x = \frac{L}{h} dp_x \tag{14.55}$$

then including degeneracy ω due to spin

$$\omega dn_x dn_y dn_z = \omega \frac{V}{h^3} dp_x dp_y dp_z \tag{14.56}$$

where $V = L^3$ is the volume of the container. This is the number of states in the interval $dn_x dn_y dn_z$. This number times the average number of fermions per state is the total number of fermions in the momentum interval $dp_x dp_y dp_z$. Then

$$N_p dp_x dp_y dp_z = \langle n_i \rangle \frac{V}{h^3} dp_x dp_y dp_z \tag{14.57}$$

Each energy ε_i is of the form $\frac{p_i^2}{2m}$ and the energy levels in any box of laboratory size essentially form a continuum, as do the momenta.

This allows writing

$$N_p = \frac{V}{h^3} \frac{1}{e^{(\frac{p^2}{2m} - \mu)/k_B T} + 1} \tag{14.58}$$

The total number of fermions is

$$N = \frac{V}{h^3} \int \frac{dp_x dp_y dp_z}{e^{(\frac{p^2}{2m} - \mu)/k_B T} + 1} \tag{14.59}$$

The integral is intractable. Another approach is needed, and will be supplied later. The only hope of evaluation in this case is to allow the temperature to become so high that there is no appreciable distinction between Fermi-Dirac and Boltzmann statistics. In this limit, every $\langle n_i \rangle \ll 1$. Let the ground energy state be $\varepsilon_1 = 0$. Then the mean occupation number for the ground state is

$$\langle n_0 \rangle = \frac{1}{e^{-\mu/k_B T} + 1} \ll 1 \tag{14.60}$$

This implies that $\mu/k_B T \ll 0$ and this allows the integration to proceed. The integral becomes approximately

$$N \cong \frac{V}{h^3} e^{\mu/k_B T} \int_{-\infty}^{\infty} e^{-p^2/2mk_B T} dp_x dp_y dp_z \tag{14.61}$$

The integral is Gaussian and therefore

$$N \cong \frac{V}{h^3} e^{\mu/k_B T} (2\pi m k_B T)^{\frac{3}{2}} \tag{14.62}$$

Solving for the chemical potential yields

$$\frac{\mu}{k_B T} = \ln \left[\frac{N h^3}{V \omega} \frac{1}{(2\pi m k_B T)^{\frac{3}{2}}} \right] \tag{14.63}$$

The number concentration $c = \frac{N}{V}$ and the thermal wavelength $\Lambda = \frac{h}{(2\pi m k_B T)^{\frac{1}{2}}}$. In these terms

$$\frac{\mu}{k_B T} = \ln \left(\frac{c\Lambda^3}{\omega} \right) \tag{14.64}$$

where the criterion that Boltzmann statistics apply, $0 < c\Lambda^3 \ll 1$ is implied. The temperature has to be thousands of degrees Kelvin for this criterion to be met.

In thermodynamics, a variation in the Helmholz free energy is given by $dA = -SdT - pdV + \mu dN$ and the pressure is therefore $p = -(\frac{\partial A}{\partial V})_{N,T}$. At constant temperature and volume, $dA = \mu dN$. Integrating produces

$$A(N) = A(0) + k_B T \int_0^N \left(\ln \left(\frac{\Lambda^3}{V\omega} \right) + \ln N \right) dN \quad (14.65)$$

$$A(N) = A(0) + N k_B T \left(\ln \left(\frac{N\Lambda^3}{V\omega} \right) - 1 \right) \quad (14.66)$$

The pressure is

$$p = - \left(\frac{\partial A}{\partial V} \right)_{N,T} = \frac{N k_B T}{V} \quad (14.67)$$

which is the ideal; gas law. This confirms the essential correctness of the approach, but does not supply an evaluation of N or E. These require a slightly different approach.

14.6. Electron Gas

Instead of integrating over n_x, n_y and n_z, the integration is conducted over all energies. The first step is to write a function expressing the number of fermions in unit energy interval. This is[1]

$$n(\varepsilon) = \frac{g(\varepsilon)}{e^{(\varepsilon - \mu)/k_B T} + 1} \quad (14.68)$$

where $g(\varepsilon)$ is the number of states accessible to a fermion in unit energy interval. This is calculable for a gas of fermions by noting that the number of states available to a representative point is proportional to the volume of phase space divided by h for each infinitesimal space and conjugate momentum product in the volume element.

This means that

$$g(\varepsilon) = \frac{1}{h^3} \iiint\!\!\iiint dp_x dp_y dp_z dx dy dz \qquad (14.69)$$

The space integral is over the volume of the container and the momentum integral is over momenta consistent with energies lying between ε and $\varepsilon + d\varepsilon$. The kinetic energy is $\varepsilon = \frac{p^2}{2m}$ for a fermion of mass m. The Cartesian momenta are inconvenient for integration, and a change in integration variables from Cartesian to spherical is more useful. The volume element is $p^2 \sin \Theta dp d\Theta d\varphi$, and $g(\varepsilon)$ becomes

$$g(\varepsilon) = \frac{4\pi(2m)^{3/2}V}{2h^3} \int_{\varepsilon}^{\varepsilon+d\varepsilon} \varepsilon^{1/2} d\varepsilon \qquad (14.70)$$

By the trapezoidal rule, the value of the remaining integral is simply the integrand, and the final result is

$$g(\varepsilon) = \frac{2^{5/2}\pi m^{3/2}V}{h^3} \sqrt{\varepsilon} \qquad (14.71)$$

For electrons, a spin degeneracy of 2 should be included, resulting in

$$g(\varepsilon) = \frac{2^{7/2}\pi m^{3/2}V}{h^3} \sqrt{\varepsilon} = 4\pi \left(\frac{2m}{h^2}\right)^{3/2} V\varepsilon^{1/2} \qquad (14.72)$$

as the number of states between ε and $\varepsilon + d\varepsilon$ for electrons in a gas. It is useful to note that p^2, m and 2 are all positive. Hence, the kinetic energy can only be positive, and therefore $g(\varepsilon) = 0$ if $\varepsilon < 0$. The number of fermions N in a system of the ensemble is then

$$N = \int_{-\infty}^{\infty} n(\varepsilon)d\varepsilon = \int_{0}^{\infty} \frac{g(\varepsilon)d\varepsilon}{e^{(\varepsilon-\mu)/k_B T} + 1} \qquad (14.73)$$

and for the energy E

$$E = \int_{-\infty}^{\infty} \varepsilon n(\varepsilon)d\varepsilon = \int_{0}^{\infty} \frac{\varepsilon g(\varepsilon)d\varepsilon}{e^{(\varepsilon-\mu)/k_B T} + 1} \qquad (14.74)$$

When the fermions are electrons in a conducting metal, there is an additional complication. Conducting metals conduct because somewhere in their electronic structure there is a band of energy levels that is only partially filled. When the metal is cooled to absolute zero, all the electrons in the conducting band will occupy the lowest energy levels, two per level, until the supply of electrons is depleted, and the metal ceases to conduct as there are no electrons in the upper levels of the unfilled band. The highest occupied energy level is called the Fermi level, and as the metal is gradually warmed from zero, the level drops slowly as the electrons in energy levels close to the Fermi level begin to evaporate into upper unfilled energy levels. The Fermi level is a slowly varying function of temperature. Its specification at zero temperature determines the number of electrons in the upper levels as zero, and it is therefore the chemical potential. At higher temperatures, some electrons are in the upper levels, and these are mobile to the extent of the metal, constituting effectively a gas of relatively free electrons. For fermions, recall that specification of the chemical potential in the range $-\infty \leq \mu \leq \infty$ specifies the density. The number of electrons in the gas N, that is, not condensed in the low energy levels of the band, is determined by the Fermi level. Setting the chemical potential μ equal to the Fermi level η, N and E for mobile electrons in a conducting metal may be written as

$$N = 4\pi \left(\frac{2m}{h^2}\right)^{\frac{3}{2}} V \int_0^\infty \frac{\varepsilon^{1/2} d\varepsilon}{e^{(\varepsilon-\eta)/k_B T} + 1} \tag{14.75}$$

$$E = 4\pi \left(\frac{2m}{h^2}\right)^{\frac{3}{2}} V \int_0^\infty \frac{\varepsilon^{3/2} d\varepsilon}{e^{(\varepsilon-\eta)/k_B T} + 1} \tag{14.76}$$

Let the coefficients multiplying the integrals be represented by C, with

$$C = 4\pi \left(\frac{2m}{h^2}\right)^{\frac{3}{2}} V \tag{14.77}$$

The number of states accessible to a fermion in unit energy becomes simply

$$g(\varepsilon) = C\varepsilon^{1/2} \quad 0 \leq \varepsilon \leq \infty \tag{14.78}$$

Now further define

$$x = \frac{\varepsilon - \eta}{k_B T} \rightarrow \varepsilon = k_B T x + \eta \rightarrow d\varepsilon = k_B T dx \tag{14.79}$$

and

$$F(x) = \frac{1}{e^x + 1} \tag{14.80}$$

The number of gaseous electrons in a band is then

$$N = \int_{-\infty}^{\infty} g(\varepsilon) F\left(\frac{\varepsilon - \eta}{k_B T}\right) d\varepsilon \tag{14.81}$$

and in these terms the energy is

$$E = \int_{-\infty}^{\infty} \varepsilon g(\varepsilon) F\left(\frac{\varepsilon - \eta}{k_B T}\right) d\varepsilon \tag{14.82}$$

These relations give both N and E as functions of T and η. Generally, the interest lies in experimentally measurable quantities, specifically, the heat capacity at constant pressure. The heat capacity at constant volume can be calculated from these relations, and if it differs appreciably from the heat capacity at constant pressure, a conversion can be made using a well known thermodynamic relation between the two. The heat capacity at constant volume is defined as $C_V = (\frac{\partial E}{\partial T})_{V,N}$ As $E = E(T, \eta)$, a general variation in E is

$$dE = \left(\frac{\partial E}{\partial T}\right)_\eta dT + \left(\frac{\partial E}{\partial \eta}\right)_T d\eta \tag{14.83}$$

Holding V constant throughout this analysis, C_v is

$$\left(\frac{\partial E}{\partial T}\right)_N = \left(\frac{\partial E}{\partial T}\right)_\eta + \left(\frac{\partial E}{\partial \eta}\right)_T \left(\frac{\partial \eta}{\partial T}\right)_N \tag{14.84}$$

Since $N = N(T, \eta)$ it must be that $\eta = \eta(T, N)$. The triad relation allows evaluation as

$$\left(\frac{\partial \eta}{\partial T}\right)_N = -\frac{\left(\frac{\partial N}{\partial T}\right)_\eta}{\left(\frac{\partial N}{\partial \eta}\right)_T} \tag{14.85}$$

This allows the specific heat to be written in general as

$$\left(\frac{\partial E}{\partial T}\right)_N = \left(\frac{\partial E}{\partial T}\right)_\eta - \frac{\left(\frac{\partial E}{\partial \eta}\right)_T \left(\frac{\partial N}{\partial T}\right)_\eta}{\left(\frac{\partial N}{\partial \eta}\right)_T} \qquad (14.86)$$

All these derivatives may be evaluated from the expressions for E and N given above. The four required derivatives are

$$\left(\frac{\partial N}{\partial T}\right)_\eta = -\frac{1}{k_B T^2}\int_{-\infty}^{\infty} g(\varepsilon) F'\left(\frac{\varepsilon - \eta}{k_B T}\right)(\varepsilon - \eta) d\varepsilon \qquad (14.87)$$

$$\left(\frac{\partial N}{\partial \eta}\right)_T = -\frac{1}{k_B T}\int_{-\infty}^{\infty} g(\varepsilon) F'\left(\frac{\varepsilon - \eta}{k_B T}\right) d\varepsilon \qquad (14.88)$$

$$\left(\frac{\partial E}{\partial T}\right)_\eta = -\frac{1}{k_B T^2}\int_{-\infty}^{\infty} \varepsilon(\varepsilon - \eta) g(\varepsilon) F'\left(\frac{\varepsilon - \eta}{k_B T}\right) d\varepsilon \qquad (14.89)$$

$$\left(\frac{\partial E}{\partial \eta}\right)_T = -\frac{1}{k_B T}\int_{-\infty}^{\infty} \varepsilon g(\varepsilon) F'\left(\frac{\varepsilon - \eta}{k_B T}\right) d\varepsilon \qquad (14.90)$$

where $F'(x) = \frac{dF}{dx}$. The heat capacity is

$$C_V = -\frac{1}{k_B T^2}\int_{-\infty}^{\infty} \varepsilon(\varepsilon - \eta) g(\varepsilon) F'\left(\frac{\varepsilon - \eta}{k_B T}\right) d\varepsilon$$

$$+ \frac{1}{k_B T^2} \frac{\int_{-\infty}^{\infty} \varepsilon g(\varepsilon) F'\left(\frac{\varepsilon - \eta}{k_B T}\right) d\varepsilon \int_{-\infty}^{\infty} (\varepsilon - \eta) g(\varepsilon) F'\left(\frac{\varepsilon - \eta}{k_B T}\right) d\varepsilon}{\int_{-\infty}^{\infty} g(\varepsilon) F'\left(\frac{\varepsilon - \eta}{k_B T}\right) d\varepsilon}$$

$$(14.91)$$

Now add and subtract $\frac{\eta}{k_B T^2}$ to the first integrand. The result is

$$-\frac{1}{k_B T^2}\int_{-\infty}^{\infty} (\varepsilon - \eta)^2 g(\varepsilon) F'\left(\frac{\varepsilon - \eta}{k_B T}\right) d\varepsilon$$

$$-\frac{\eta}{k_B T^2}\int_{-\infty}^{\infty} (\varepsilon - \eta) g(\varepsilon) F'\left(\frac{\varepsilon - \eta}{k_B T}\right) d\varepsilon$$

Now multiply the second term and divide the product by $\int_{-\infty}^{\infty} g(\varepsilon) F'(\frac{\varepsilon - \eta}{k_B T}) d\varepsilon$. This results in

$$-\frac{1}{k_B T^2} \int_{-\infty}^{\infty} (\varepsilon - \eta)^2 g(\varepsilon) F' \left(\frac{\varepsilon - \eta}{k_B T} \right) d\varepsilon$$

$$-\frac{1}{k_B T^2} \frac{\eta \int_{-\infty}^{\infty} (\varepsilon - \eta) g(\varepsilon) F' \left(\frac{\varepsilon - \eta}{k_B T} \right) d\varepsilon \int_{-\infty}^{\infty} g(\varepsilon) F' \left(\frac{\varepsilon - \eta}{k_B T} \right) d\varepsilon}{\int_{-\infty}^{\infty} g(\varepsilon) F' \left(\frac{\varepsilon - \eta}{k_B T} \right) d\varepsilon}$$

Now add in the second term in C_v. The heat capacity may then be expressed in three terms. The first two are

$$C_v = -\frac{1}{k_B T^2} \int_{-\infty}^{\infty} (\varepsilon - \eta)^2 g(\varepsilon) F' \left(\frac{\varepsilon - \eta}{k_B T} \right) d\varepsilon$$

$$-\frac{1}{k_B T^2} \frac{\eta \int_{-\infty}^{\infty} (\varepsilon - \eta) g(\varepsilon) F' \left(\frac{\varepsilon - \eta}{k_B T} \right) d\varepsilon \int_{-\infty}^{\infty} g(\varepsilon) F' \left(\frac{\varepsilon - \eta}{k_B T} \right) d\varepsilon}{\int_{-\infty}^{\infty} g(\varepsilon) F' \left(\frac{\varepsilon - \eta}{k_B T} \right) d\varepsilon}$$

and the third is

$$+\frac{1}{k_B T^2} \frac{\int_{-\infty}^{\infty} \varepsilon g(\varepsilon) F' \left(\frac{\varepsilon - \eta}{k_B T} \right) d\varepsilon \int_{-\infty}^{\infty} (\varepsilon - \eta) g(\varepsilon) F' \left(\frac{\varepsilon - \eta}{k_B T} \right) d\varepsilon}{\int_{-\infty}^{\infty} g(\varepsilon) F' \left(\frac{\varepsilon - \eta}{k_B T} \right) d\varepsilon}$$

This formidable expression may be rewritten in shorter form

$$C_v = -\frac{1}{k_B T^2} \int_{-\infty}^{\infty} (\varepsilon - \eta)^2 g(\varepsilon) F' \left(\frac{\varepsilon - \eta}{k_B T} \right) d\varepsilon$$

$$+\frac{1}{k_B T^2} \frac{\left[\int_{-\infty}^{\infty} (\varepsilon - \eta) g(\varepsilon) F' \left(\frac{\varepsilon - \eta}{k_B T} \right) d\varepsilon \right]^2}{\int_{-\infty}^{\infty} g(\varepsilon) F' \left(\frac{\varepsilon - \eta}{k_B T} \right) d\varepsilon} \qquad (14.92)$$

Now substitute $x = \frac{\varepsilon - \eta}{k_B T}$ and obtain the more compact expression

$$C_v = -k_B^2 T \int_{-\infty}^{\infty} x^2 g(\eta + k_B T x) F'(x) dx$$

$$+k_B^2 T \frac{\left[\int_{-\infty}^{\infty} x g(\eta + k_B T x) F'(x) dx \right]^2}{\int_{-\infty}^{\infty} g(\eta + k_B T x) F'(x) dx} \qquad (14.93)$$

The occurrence of the derivative $F'(x)$ in all integrals guarantees that the dominant majority of electrons contributing to the specific heat are those near the Fermi level, as is expected on physical grounds. These integrals are too difficult to evaluate in general, but a limiting value may be obtained that is valid at low temperatures, where it is experimentally possible to disentangle the electronic contribution from the vibrational. The number of states in unit energy interval $g(\eta + k_B T x)$ is expanded in Taylor's series as a power series in x about $x = 0$. This gives a series of powers of $k_B T x$ and at low temperatures only the leading constant term need be retained to achieve an approximate expression for the heat capacity. Then the second term is an odd function integrated between symmetric limits and thus vanishes, while the first term gives

$$C_v \cong -g(\eta) k_B^2 T \int_{-\infty}^{\infty} x^2 F'(x) dx \qquad (14.94)$$

This integral may be integrated by parts. First observe that the integrand is symmetric, so

$$\int_{-\infty}^{\infty} x^2 F'(x) dx = 2 \int_0^{\infty} x^2 F'(x) dx$$

and then integrating by parts

$$2 \int_0^{\infty} x^2 F(x) dx = -4 \int_0^{\infty} x F(x) dx$$

Next note that

$$F(x) = \frac{1}{e^x + 1} = \frac{e^{-x}}{1 + e^{-x}}$$

and for small x this may be expanded as

$$F(x) = e^{-x}(1 - e^{-x} + e^{-2x} - e^{-3x} + e^{-4x} - \cdots)$$

which is the infinite alternating sign series

$$F(x) = e^{-x} - e^{-2x} + e^{-3x} - e^{-4x} + \cdots$$

This allows us to write

$$\int_0^\infty xF(x)dx = \int_0^\infty (xe^{-x} - xe^{-2x} + xe^{-3x} - xe^{-4x} + \cdots)dx$$

Since

$$\int_0^\infty xe^{-nx}dx = \frac{1}{n}\int_0^\infty e^{-nx}dx = \frac{1}{n^2} \tag{14.95}$$

the integral becomes

$$-4\int_0^\infty xF(x)dx = -4\sum_{n=1}^\infty \frac{(-1)^n}{n^2} \tag{14.96}$$

and the heat capacity at low temperatures is approximately

$$C_v \cong -g(\eta)k_B^2 T\left[-4\sum_{n=1}^\infty \frac{(-1)^n}{n^2}\right] \tag{14.97}$$

The sum is tabulated and is $\frac{\pi^2}{12}$. The heat capacity is approximately

$$C_v \cong \frac{g(\eta)\pi^2 k_B^2 T}{3} \tag{14.98}$$

The linear dependence of the electronic contribution to the heat capacity is very different from the low temperature behavior of lattice vibrational contributions. The Debye theory led to a T^3 contribution, and the sum of a linear and a cubic contribution suggests that plotting C_v/T against T^2 at low temperatures should be linear, with the Debye Θ_D available from the slope and the characteristic $g(\eta)$ of the electronic contribution available from the intercept.

Can N be evaluated at low temperatures? It is

$$N = \int_{-\infty}^\infty g(\varepsilon)F\left(\frac{\varepsilon - \mu}{k_B T}\right)d\varepsilon \tag{14.99}$$

Since $g(\varepsilon)$ is zero in the range $-\infty \le \varepsilon \le 0$, this is equally well written

$$N = \int_0^\infty g(\varepsilon)F\left(\frac{\varepsilon - \mu}{k_B T}\right)d\varepsilon \tag{14.100}$$

for any gas of fermions. Once again, define $x = \frac{\varepsilon - \mu}{k_B T}$, and as $\varepsilon = \mu + k_B T x$, $d\varepsilon = k_B T dx$ and N becomes

$$N = k_B T \int_0^\infty g(\mu + k_B T x) \frac{e^{-x} dx}{1 + e^{-x}} \qquad (14.101)$$

Note that

$$\frac{e^{-x}}{1 + e^{-x}} = \sum_{n=1}^\infty (-1)^{n+1} e^{-nx}$$

The expansion is valid if x is large, as will be the case for sufficiently low temperature. The next step is to expand $g(\mu + k_B T x)$ in a power series in $k_B T x$, which will be small when $\varepsilon - \mu$ is small. Retaining only the leading term, and setting $g(\varepsilon) = C\varepsilon^{1/2} = C(\mu + k_B T x)^{1/2}$ leads to the leading term in the series as $g(\mu) = C\mu^{1/2}$, which is constant in the x integration. Then

$$N = C\mu^{1/2} \int_0^\infty \sum_{n=1}^\infty (-1)^{n+1} e^{-nx} dx \qquad (14.102)$$

Now, the integral $\int_0^\infty e^{-nx} dx = |-\frac{1}{n} e^{-nx}|_0^\infty = \frac{1}{n}$ and the expression becomes

$$N = C\mu^{1/2} \sum_{n=1}^\infty \frac{(-1)^{n+1}}{n} \qquad (14.103)$$

Expanding the series

$$N = C\mu^{1/2} \left(1 - \frac{1}{2} + \frac{1}{3} - \frac{1}{4} + \cdots \right)$$

Clearly, the even numbers in the denominator have a minus sign while the odd numbers have a plus sign, suggesting the series be rewritten

$$N = C\mu^{1/2} \sum_{n=0}^\infty \left(\frac{1}{2n + 1} - \frac{1}{2n + 2}\right)$$

which can be rewritten as

$$N = C\mu^{1/2} \sum_{n=0}^\infty \frac{1}{4n^2 + 6n + 2}$$

The series evaluates as $\ln(2)$. Substituting the value of C gives the result

$$N = \frac{4\pi}{\mu}\left(\frac{2m\mu}{h^2}\right)^{\frac{3}{2}} V \ln(2) \tag{14.104}$$

On the other hand, the number of electrons at absolute zero is somewhat different. At this temperature, the distribution $n(\varepsilon)$ is 1 for energies $0 \le \varepsilon \le \mu$ and 0 for $\varepsilon > \mu$, where μ is the Fermi level. Then N at absolute zero becomes

$$N(T = 0) = 4\pi\left(\frac{2m}{h^2}\right)^{\frac{3}{2}} V \int_0^\mu \varepsilon^{1/2} d\varepsilon \tag{14.105}$$

Completing the integration

$$N = \frac{8\pi}{3}\left(\frac{2m\mu}{h^2}\right)^{\frac{3}{2}} V \tag{14.106}$$

By the same argument, the energy at the absolute zero is

$$E(T = 0) = 4\pi\left(\frac{2m}{h^2}\right)^{\frac{3}{2}} V \int_0^\mu \varepsilon^{3/2} d\varepsilon$$

Evaluating the integral

$$E = \frac{8\pi}{5}\left(\frac{2m}{h^2}\right)^{\frac{3}{2}} V \mu^{5/2} \tag{14.107}$$

The ratio of E to N at absolute zero shows that

$$E(T = 0) = \frac{3}{5}\mu N(T = 0) \tag{14.108}$$

As temperature rises from zero, initially the electronic energy is given by

$$E(T) = \int_0^T C_v dT$$

Taking C_v from equation (14.98)

$$E = \int_0^T \frac{g(\mu)\pi^2 k_B^2 T}{3} dT \tag{14.109}$$

Finally

$$E = \frac{3}{5}\mu N(T=0) + \frac{\pi^2 k_B^2}{4\mu}\left[\frac{8\pi}{3}\left(\frac{2m\mu}{h^2}\right)^{\frac{3}{2}} V\right] T^2 + \cdots$$

This can be written alternatively as

$$E(T) = \frac{3}{5}\mu N(T=0)\left[1 + \frac{5\pi^2}{12}\left(\frac{k_B T}{\mu}\right)^2 + \cdots\right] \qquad (14.110)$$

In all of the foregoing, the electrons were regarded as occupying doubly degenerate energy levels. The question arises, how high a temperature is needed to break the degeneracy. As the temperature rises, more and more electrons leave the lower energy levels and enter the higher energy levels. As this process advances, the Fermi level drops until at a high enough temperature the Fermi level lies at the bottom of the energy levels in the conduction band. At higher temperatures, the electrons behave as if the degeneracy has been lifted. The critical temperature T_c produces a critical number of electrons N_c, which may be calculated by setting the Fermi level to zero. The critical number is

$$N_c = \int_0^\infty \frac{g(\varepsilon)d\varepsilon}{e^{\varepsilon/k_B T_c} + 1} \qquad (14.111)$$

This may also be written as

$$N_c = \int_0^\infty g(\varepsilon)d\varepsilon(e^{-\varepsilon/k_B T_c} - e^{-2\varepsilon/k_B T_c} + e^{-3\varepsilon/k_B T_c} + e^{-4\varepsilon/k_B T_c} - \cdots) \qquad (14.112)$$

Defining n as $\frac{i}{k_B T_c}$ where i is a positive integer, the typical integral is

$$\int_0^\infty \varepsilon^{1/2} e^{-n\varepsilon} d\varepsilon = 2\int_0^\infty x^2 e^{-nx^2} dx = \int_{-\infty}^\infty x^2 e^{-nx^2} dx \qquad (14.113)$$

after making the substitution $\varepsilon = x^2$. The integral can be evaluated from the basic integral

$$\int_{-\infty}^\infty e^{-a^2 x^2} dx = \frac{\sqrt{\pi}}{a} \qquad (14.114)$$

The first derivative of the integral with respect to a gives, after rearrangement and substitution of n for a^2

$$\int_{-\infty}^{\infty} x^2 e^{-nx^2} dx = \frac{\sqrt{\pi}}{2n^{3/2}} = \frac{\sqrt{\pi}}{2i^{3/2}} (k_B T_c)^{3/2} \qquad (14.115)$$

The critical number N_c is

$$N_c = \frac{k_B T_c g(k_B T_c) \pi^{1/2}}{2} \sum_{n=1}^{\infty} \frac{(-1)^{n+1}}{n^{3/2}} \qquad (14.116)$$

The sum converges to 0.765147, which multiplied by $\sqrt{\pi}/2$ gives

$$N_c = 0.678094 k_B T_c g(k_B T_c) \qquad (14.117)$$

The critical concentration is

$$\frac{N_c}{V} = (0.678094) 4\pi \left(\frac{2m}{h^2} \right)^{\frac{3}{2}} (k_B T_c)^{3/2} \qquad (14.118)$$

Substituting numerical values for h, k_B and the rest mass of the electron gives the result

$$\frac{N_c}{V} = 3.695 \times 10^{15} T_e^{3/2} \text{ electrons per cubic cm} \qquad (14.119)$$

For any reasonable density, this gives a very large temperature. For example, the density of sodium metal at 20 degree Centigrade is 0.971, and one mole weighs 22.997 g, so the volume available to the valence electrons in the metal is about 23.7 cubic centimeters. The electron density in the valence band is N_0/V or about 2.54×10^{22} electrons per cubic centimeter. The critical temperature is then about 3.6×10^4 degrees Kelvin. Sodium boils at $880°C$, and will have evaporated entirely into a gas thousands of degrees cooler than the critical temperature.

However, in heavy metals, electron speeds near the nuclei become relativistic and the average mass is much greater than the rest mass of the electron, The relativistic momentum p is

$$p = \frac{mv}{\sqrt{1 - \frac{v^2}{c^2}}} \tag{14.120}$$

where m is the rest mass of the particle moving with velocity v and c is the velocity of light in vacuo. This may be interpreted to mean

$$m^* = \frac{m}{\sqrt{1 - \frac{v^2}{c^2}}} \tag{14.121}$$

In such cases the critical temperature may be lowered below room temperature. In these terms, the equation determining the critical temperature may be written

$$\frac{N_c}{V} = (0.678094)4\pi m^{\frac{3}{2}} \left(\frac{2(m^*/m)}{h^2} \right)^{\frac{3}{2}} (k_B T_c)^{3/2} \tag{14.122}$$

The critical density is

$$\frac{N_c}{V} = 3.695 \times 10^{15} (m^*/m)^{3/2} T_e^{3/2} \text{ electrons per cubic cm} \tag{14.123}$$

For Bismuth, the density is an order of magnitude greater than sodium (9.747 g/cm^3) and m^*/m may be somewhere near 0.1. However, bismuth is a poor conductor of electricity and the electron density is much less than the 10^{22} per cc of sodium. Taking the electron density to be about 5×10^{16}, the result is a temperature at which the degeneracy is lifted that is near 60 K.

Reference

1. G. H. Wannier, Elements of Solid State Theory, Cambridge at the University Press, GB, 1959.

Chapter 15

Adsorption

Adsorption problems occur frequently in both the physical and biological sciences, and for that reason require fairly extensive presentation. Generally, adsorption refers to a situation in which there is a greater concentration of adsorbed molecules on the surface of an adsorbing solid than there is in a fluid phase with which the surface molecules are in equilibrium. This chapter begins with the most primitive of analyses, the Langmuir adsorption, proceeds through more complex adsorption problems and ends with an extension of basic statistical mechanical theory applied to a typical biological problem called the McGhee-von Hippel binding problem. Along the way, the Ising problem is introduced and treated in linear approximation, leading to a clear exposition of the nature and origin of hysteresis in magnetization-demagnetization in particular and by extension to hysteresis in adsorption problems in general.

15.1. Langmuir Adsorption

The Langmuir adsorption isotherm occupies the same theoretical position in the theory of adsorption that the ideal gas equation of state does in the theory of real gases. It is a limiting law to which stable theories must condense in the limit of low concentrations. The effect may be routinely demonstrated experimentally in the adsorption of acetic acid from aqueous solution on activated charcoal, a high surface area solid.[1]

Consider a solid with M independent identical distinguishable adsorption sites, and let there be N molecules adsorbed onto some of these sites. Note that $M > N$. If all the sites were occupied, then $N = M$ and the partition function for the N adsorbed molecules in the canonical ensemble would be

$$Q = q^N \quad \text{for } N = M \tag{15.1}$$

When fewer molecules adsorb so that $N < M$, a configurational degeneracy is introduced. There are M ways the first molecule may adsorb onto the solid, and if there were only one adsorbed molecule the canonical ensemble partition function would be

$$Q(1) = Mq \quad N = 1 \tag{15.2}$$

Now let one more molecule adsorb. This time, there are only $N - 1$ ways the second molecule can adsorb, and the number of ways the first two molecules can adsorb is $M(M-1)/2$, for if the first molecule adsorbs onto site A and the second on site B, the result is the same as it would have been if the first molecule adsorbed on site B and the second on site A. The partition function is now

$$Q(2) = \frac{M(M-1)}{2}q^2 \quad N = 2 \tag{15.3}$$

Let another molecule adsorb onto one of the $N - 2$ remaining sites. The total number of ways of doing this is $M(M-1)(M-2)/2 \times 3$. Continue this until the supply of N molecules is exhausted. The result is the partition function

$$Q(N) = \frac{M(M-1)(M-2)\cdots(M-N+1)}{N!}q^N \tag{15.4}$$

This may be rewritten as

$$Q(N) = \frac{M!}{N!(M-N)!}q^N \tag{15.5}$$

The Helmholz free energy is $A = -k_B T \ln Q$, which, on applying Stirling's approximation, in this case is

$$A = -k_B T \left[M \ln M - N \ln N - (M-N)\ln(M-N) + N \ln q \right] \tag{15.6}$$

A variation in the energy of the adsorbed phase is $dE = TdS - \Phi dM + \mu dN$ where the Greek letter Φ represents the surface pressure. The Helmholtz free energy $A = E - TS$, and as a result a variation in this free energy of the adsorbed phase is $dA = -SdT - \Phi dM + \mu dN$ The surface pressure is therefore

$$\Phi = -\left(\frac{\partial A}{\partial M}\right)_{T,N} = k_B T \left(\frac{\partial \ln Q}{\partial M}\right)_{T,N} \tag{15.7}$$

After rearrangement, this is

$$\Phi = -k_B T \ln\left(\frac{M - N}{M}\right) \tag{15.8}$$

Defining the fraction of surface covered as $\vartheta \equiv \frac{N}{M}$, the surface pressure is

$$\Phi = -k_B T \ln(1 - \vartheta) \tag{15.9}$$

This is the equation of state of the Langmuir adsorption, corresponding to $p = k_B T \frac{N}{V}$ for the ideal gas. At low coverages, the logarithm may be expanded in power series, giving

$$\frac{\Phi}{k_B T} \cong \vartheta + \frac{1}{2}\vartheta^2 + \frac{1}{3}\vartheta^3 + \cdots \tag{15.10}$$

The chemical potential provides a connection between the particle density in the fluid phase and the fraction covered, since at equilibrium the chemical potentials in the fluid phase and in the adsorbed phase must be equal.[2] In the adsorbed phase, the chemical potential is

$$\frac{\mu}{k_B T} = \left(\frac{\partial \ln Q}{\partial N}\right)_{T,M} \tag{15.11}$$

Applying this to the Langmuir partition function gives

$$\frac{\mu}{k_B T} = \ln N - \ln(M - N) - \ln q \tag{15.12}$$

which is, in terms of the fraction covered,

$$\frac{\mu}{k_B T} = \ln\left(\frac{\vartheta}{q(1 - \vartheta)}\right) \tag{15.13}$$

In the fluid phase, the chemical potential is $\mu = \mu_0(T) + k_B T \ln p$, where p measures the particle density as pressure if the fluid phase is a gas or as elementary concentration (e.g., moles per liter) if a solution is involved. Equating chemical potentials

$$\frac{\mu_0(T)}{k_B T} + \ln p = \ln\left(\frac{\vartheta}{q(1-\vartheta)}\right) \tag{15.14}$$

Now q is a function of T alone, certainly not of N or M. Collecting all the temperature dependences in one symbol, the function is defined by taking the antilogarithm

$$e^{\frac{\mu_0(T)}{k_B T}} q(T) = \frac{\vartheta}{p(1-\vartheta)} \tag{15.15}$$

Define the function of temperature

$$\chi(T) = e^{\frac{\mu_0(T)}{k_B T}} q(T) \tag{15.16}$$

There are two forms of the isotherm. Solving for the fraction covered, one finds

$$\vartheta = \frac{p\chi(T)}{1 + p\chi(T)} \tag{15.17}$$

$$p\chi(T) = \frac{\vartheta}{1 - \vartheta} \tag{15.18}$$

These relations hold quite accurately as long as the fraction covered remains small, much less than 1. At higher coverages, departures from these is to be expected. A popular approximation that accommodates some divergence at somewhat higher coverages was originally proposed in a paper by Brunauer, Emmet and Teller, and is therefore called the BET isotherm. This is covered in the next section.

So far, the above quantifies only two of the three coefficients that measure a variation in the Helmholz free energy. The temperature dependence remains to be considered. It is the entropy of the adsorbed phase, defined by $S = -(\frac{\partial A}{\partial T})_{M,N}$. There follows immediately

$$S = k_B(M \ln M - N \ln N - (M - N) \ln(M - N))$$
$$+ Nk_B \ln q + Nk_B T \frac{d \ln q}{dT} \tag{15.19}$$

A moment's reflection will show that this is broken into two quite different parts. The terms not involving q are a configurational entropy, which remains as the temperature is lowered to absolute zero. The other is composed of temperature dependent terms involving the vibrational contributions to the partition function q, as all of the rotational and translational contributions are not present for an immobile adsorbed phase. The vibrational contributions to the entropy vanish at absolute zero. This is not a violation of the third law, as the immobile adsorbed phase is very far from a perfect crystal, as in fact all real crystals are owing to the unavoidable inclusion of voids in any crystallization.

The grand partition function is easily formed and evaluated, but leads to no new results. It is

$$\Xi = \sum_{N=0}^{M} \frac{M!}{N!(M-N)!}(q\lambda)^N \tag{15.20}$$

This evaluates immediately and exactly. According to the binomial theorem

$$\Xi = (1 + q\lambda)^M \tag{15.21}$$

The fraction covered is given by $\langle N \rangle / M$, and the mean number adsorbed is

$$\langle N \rangle = \left(\frac{\partial \ln \Xi}{\partial \ln \lambda}\right)_{M,T} = M\frac{q\lambda}{1 + q\lambda} \tag{15.22}$$

Hence

$$\vartheta = \frac{q\lambda}{1 + q\lambda} \qquad q\lambda = \frac{\vartheta}{1 - \vartheta} \tag{15.23}$$

The equation of state is immediate from the formula

$$\Phi M = k_B T \ln \Xi = M k_B T \ln(1 + q\lambda) \qquad (15.24)$$

Substituting $q\lambda$, the equation of state, equation (15.9) is recovered

$$\frac{\Phi}{k_B T} = -\ln(1 - \vartheta) \qquad (15.25)$$

Finally, substitution of $e^{\mu/k_B T}$ for λ leads, by the same route as in the canonical ensemble, to the Langmuir adsorption isotherm.

   All of the foregoing applies to immobile adsorption, in which an adsorbed molecule may desorb to maintain equilibrium with the fluid phase, but, once adsorbed, it continues indefinitely to occupy the same adsorption site. It may happen that the adsorption forces are large enough for adsorption onto a surface to occur, but these forces are not so great that the adsorbed molecule is forced to occupy just one site. The adsorbed phase may become mobile in the surface in these circumstances. Then q will consist of internal vibrations of the adsorbed molecule plus a vibration in the direction normal to the plane of adsorption, but essentially free along the surface to the extremities of the crystal face it occupies. The vibration normal to the surface involves a large mass (that of the whole molecule) and a weak force constant. As a result, at ambient temperatures, the partition function for this vibration has achieved its classical value $k_B T/h\nu$. Assuming the translations in the crystal face are free, or essentially so, the partition function for translation is $\frac{2\pi m k_B T}{h^2} A$, where m is the mass of an adsorbed molecule and A is the surface area. The partition function is then

$$Q_{mobile} = \frac{M!}{N!(M-N)!} \left(\frac{k_B T}{h\nu_z}\right)^N \left(\frac{2\pi m k_B T}{h^2} A\right)^N q_{vib}^N \qquad (15.26)$$

whereas that for the immobile phase is

$$Q_{immobile} = \frac{M!}{N!(M-N)!} \left(\frac{k_B T}{h\nu_x}\right)^N \left(\frac{k_B T}{h\nu_y}\right)^N \left(\frac{k_B T}{h\nu_z}\right)^N q_{vib}^N$$
$$(15.27)$$

In either case, the grand partition function leads to $q\lambda = \frac{\vartheta}{1-\vartheta}$.

At low pressures in the fluid phase, coverages will be small, $\vartheta \ll 1$ and $q\lambda \cong \vartheta$, while also $\lambda = e^{\mu_0/k_B T} p$. At these low pressures and high temperatures, approximately

$$\frac{\vartheta_{mobile}}{\vartheta_{immobile}} \cong \frac{q_{mobile}}{q_{immobile}} \tag{15.28}$$

Assuming the frequency of vibration normal to the surface is the same in both mobile and immobile phases, the ratio is very approximately

$$\frac{\vartheta_{mobile}}{\vartheta_{immobile}} \cong \frac{\sqrt{k_x k_y}}{2\pi k_B T} A \tag{15.29}$$

where the force constants for the vibrations polarized parallel to the plane of the adsorption area are k_x and k_y. While the fraction is composed of small quantities, the area A is typically huge, of the order of hundreds of square meters per gram. This suggests that the fraction covered in a mobile phase greatly exceeds the coverage for an immobile phase, other things being equal.

15.2. Brunauer Emmett Teller (BET) Adsorption

The BET adsorption isotherm is named for its first theoretical proposers, Brunauer, Emmett and Teller.[3] The theory may be developed in several ways, and is useful in explaining departures from the Langmuir adsorption isotherm at somewhat higher pressures than the low pressure regime in which the Langmuir isotherm always serves adequately. An apparatus suitable for adsorption studies of gases is described by Garland, Nibler and Shoemaker.[4] The theoretical development presented here is based on Hill.[5] The theory captures the initial phases of multilayer adsorption, but fails at high adsorption levels.

The theory begins by making some definitions. Let there be

B independent localized sites per unit area of adsorbent surface

X molecules adsorbed on the B sites in the first layer

A molecules adsorbed on B sites in all layers

Then there must be $A - X$ molecules adsorbed on top of the X molecules in the first layer. The potential energy of a molecule in the first layer is $-\varepsilon_1$ and that for molecules in the second and higher layers is $-\varepsilon_L$. The zero of energy to which all energies are referred is the infinitely separated molecules. The frequencies of vibration of an adsorbed molecule in the first layer against the adsorption site is taken as the same as the frequency of the vibration of the whole molecule in directions parallel to the adsorption surface. There are three of these very small molecule-surface frequencies, and at any temperature much above absolute zero this contribution to the partition function will have achieved classical values. Similar statements may be made about the molecules in the second and higher layers, but with a slightly different frequency. The partition functions in the first and higher layers may be approximated as

$$q_1 = \left(\frac{k_B T}{h\nu_1}\right)^3 q_{1,int} e^{\varepsilon_1/k_B T} \tag{15.30}$$

$$q_L = \left(\frac{k_B T}{h\nu_L}\right)^3 q_{L,int} e^{\varepsilon_L/k_B T} \tag{15.31}$$

In these equations, $q_{1,int}$ and $q_{L,int}$ are partition functions containing contributions from the internal vibrational structure of molecules respectively in the first layer and in all higher layers. These will generally not be much different, as frequency shifts on condensation from gas to liquid are usually not much more than 10 cm^{-1} for typical frequencies in the mid infrared, except for structural changes such as changes in hydrogen bonding. The surroundings of a molecule do in fact affect the vibrational frequencies by minor modifications of the electronic structure and thus of the force constants of the molecule, but the surroundings of a molecule in the first layer are not very different from the surroundings of molecules in higher layers, and frequency shifts from one layer to the next should be relatively small. In case this is not so, a different partition function is postulated for the first and for the higher layers.

The canonical ensemble partition function for the first layer is the same as for the Langmuir case. However, the canonical ensemble partition function for the higher layers requires some combinatorial work. The principle is that a molecule can only deposit on one that is already there. This corresponds to the placement of N indistinguishable objects in C numbered (and therefore distinguishable) boxes, a combinatorial problem solved in the book by Mayer and Mayer[6] and presented here in modified form. There are no restrictions on the number of molecules that may be placed in a box: placing no object in a box is as good as placing all or any other number between 0 and N. Since the number of objects (molecules) that may be placed in boxes (on occupied adsorption sites) is not limited, the number of objects (molecules) may exceed the number of boxes (occupied sites) and even the total number of sites. This means that A may be greater than B, or B may be greater than A, which is different from the Langmuir case, where the B sites always exceed the A adsorbed molecules.

An arrangement is characterized by the number of indistinguishable objects placed in the numbered boxes. Let the boxes be represented by the numbered letters z_1, z_2, z_3 and so on. Also let the N indistinguishable objects be represented by the letters a_1, a_2, a_3 and so on. Let these symbols be arranged in a straight line randomly, but insist that the first letter in the line be z_1. One such arrangement might be

$$z_1 a_3 z_6 a_8 a_{35} a_{53} a_2 z_{34} a_{27} z_{89} z_{76} a_{54} \ldots$$

In this arrangement, the a's written to the right of a z belong to that z up to the next z. If no a is written to the right of a z, then that box is empty in this arrangement. Since the first symbol must be z_1, and there are C numbered boxes and N molecules in all, there are $C + N - 1$ symbols remaining, and there are $(C+N-1)!$ ways of arranging these symbols. There are $(C-1)!$ ways of rearranging the remaining boxes with their contents, These arrangements are all equivalent, in that the order of boxes is immaterial, only the contents mattering. For example, an arrangement $\ldots z_6 a_7 a_8 z_9 a_5 a_4 a_3 \ldots$ is the same as $\ldots z_9 a_5 a_4 a_3 z_6 a_7 a_8 \ldots$. Furthermore, all $N!$ permutations of the N objects in the boxes are equivalent, so the total number of really different ways of placing N indistinguishable objects in C numbered boxes with no restriction on the arrangement of boxes is

$$\frac{(C+N-1)!}{N!(C-1)!} = \frac{(X+A-X-1)!}{(A-X)!(X-1)!} = \frac{(A-1)!}{(A-X)!(X-1)!} \quad (15.32)$$

since in the notation adopted for this problem $N = A - X$ and $C = X$.

The canonical ensemble partition function for the first layer Q_1 is the Langmuir partition function

$$Q_1 = \frac{B!}{X!(B-X)!}q_1^X \quad (15.33)$$

and that for the second and higher layers is

$$Q_L = \frac{(A-1)!}{(A-X)!(X-1)!}q_L^{A-X} \quad (15.34)$$

The subscript L is used to suggest that these higher layers are more liquid-like than the first, which is more distorted by adsorption forces than the higher layers. The higher layers are distorted only by van der Waals type forces. The complete partition function for the whole system is

$$Q = \sum_{X=1}^{A} Q_1 Q_L \quad (15.35)$$

if A is greater than B. If B is greater than A, then the upper limit on the sum is B. More explicitly

$$Q = \sum_{X=1}^{A>B} \frac{(A-1)!B!}{X!(B-X)!(A-X)!(X-1)!}q_1^X q_L^{A-X} \quad (15.36)$$

Since the numbers of molecules and sites are very large, 1 is negligible in comparison. With this modification, the partition function becomes almost exactly

$$Q = \sum_{X=1}^{A>B} \frac{A!B!}{(A-X)!(B-X)!(X!)^2}q_1^X q_L^{A-X} \quad (15.37)$$

The sum is too difficult to perform, but progress can be made by seeking the maximum term and assigning the dominant majority of the behavior to the maximum term. It is easier to deal with the logarithm of the typical term than it is to deal with the term itself. The logarithm expands as

$$\ln t(X) = \ln A! + \ln B! - \ln(A - X)! - \ln(B - X)! - 2\ln X!$$

$$+ X \ln q_1 + (A - X) \ln q_L \qquad (15.38)$$

where $t(X)$ is the typical term. The condition for an extremum is $\frac{d \ln t(X)}{dX} = 0$. Applying this

$$\frac{d \ln t(X)}{dX} = \ln(A - X) + \ln(B - X) - 2\ln X + \ln q_1 - \ln q_L \qquad (15.39)$$

Setting this equal to zero yields

$$\frac{(A - X)(B - X)}{X^2} = \frac{q_L}{q_1} \qquad (15.40)$$

A represents all the molecules adsorbed, and the chemical potential of A is

$$\frac{\mu_A}{k_B T} = -\frac{\partial \ln t}{\partial A} = \ln(A - X) - \ln A - \ln q_L \qquad (15.41)$$

In the ideal gas at pressure p and temperature T, $\mu_A = \mu_A^0(T) + k_B T \ln p$, and when the gas is at equilibrium with the liquid of vapor pressure p_0, the chemical potential is $\mu_A = \mu_A^0(T) + k_B T \ln p_0$. In the liquid, all the molecules are "adsorbed", and $X = 0$. Setting the chemical potentials equal in these circumstances yields

$$\frac{\mu_{A,liquid}}{k_B T} = \ln A - \ln A - \ln q_L = \frac{\mu_A^0(T)}{k_B T} + \ln p_0 \qquad (15.42)$$

Substituting this value of $\ln q_L$ in the expression for the chemical potential of the adsorbed phase yields

$$\frac{\mu_A}{k_B T} = \ln \frac{A - X}{A} + \frac{\mu_A^0(T)}{k_B T} + \ln p_0 = \frac{\mu_A^0(T)}{k_B T} + \ln p \qquad (15.43)$$

or

$$\frac{A - X}{A} = \frac{p}{p_0} \qquad (15.44)$$

Define the ratio $\frac{p}{p_0} \equiv x$ and the ratio $c = \frac{q_1}{q_L}$. Then

$$X = A(1 - x) \tag{15.45}$$

and

$$(A - X)(B - X) = \frac{1}{c}X^2 \tag{15.46}$$

Substituting X from equation (15.45)

$$(A - A(1 - x))(B - A(1 - x)) = \frac{1}{c}[A(1 - x)]^2 \tag{15.47}$$

After some straightforward algebra the BET isotherm, emerges

$$\frac{A}{B} = \frac{cx}{(1 - x)(1 - x + cx)} \tag{15.48}$$

where the ratio $\frac{A}{B}$ is the number of molecules adsorbed in all layers per adsorption site per unit area of adsorbent. It is essentially the fraction adsorbed, θ. The ratio c is a dimensionless function of temperature alone. As x approaches 1, the isotherm diverges, which real systems do not do. This is a serious defect of the theory, related to the relatively poor model of liquid-like behavior of linear stacks of molecules with no stack to stack interaction. More successfully, as x approaches zero, the fraction covered approaches $\frac{cx}{1+cx} = \frac{\chi(T)p}{1+\chi(T)p}$ with $\chi(T) = \frac{c}{p_0}$, the Langmuir result, as shown in Figure 15.1.

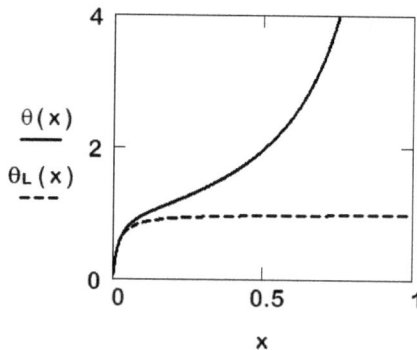

Fig. 15.1.   Solid line BET isotherm $\theta(x) = A/B$. Dashed line Langmuir isotherm $\theta_L(x)$. The value of c was taken as 60 and was used in both plots.

Garland *et al.*[4] point out that it is conventional to express adsorption results in terms of the volume measured in cubic centimeters that the adsorbate would occupy if considered as an ideal gas at standard conditions of temperature and pressure (273.15 K, 1 atm). If the volume required to form a complete monolayer coverage of the absorbent is v_m, the fraction covered at any pressure less than the vapor pressure is

$$\vartheta = \frac{v}{v_m} \qquad (15.49)$$

and the BET isotherm may be expressed in the form in which it is usually used

$$\frac{x}{v(1-x)} = \frac{1}{v_m c} + \frac{(c-1)x}{v_m c} \qquad (15.50)$$

This form has the advantage that a plot of $\frac{x}{v(1-x)}$ against x should be linear, and from the slope and intercept c and v_m can be obtained. The theory is good but not perfect and typically plots at very high coverages or very low coverages depart from linearity, and thus interpretability.

It is possible to derive the BET adsorption isotherm by using a more powerful method that allows avoiding the combinatorial problem of the number of really different ways the second and higher layers can form. However, first the method has to be established in general and checked on the Langmuir problem. That is next.

15.3. Single Site Grand Partition Function (The ξ Method)

This method does not have a generally accepted name, but when reference is needed, it will be called the ξ method. The quantity ξ is called the single site grand partition function.

In the Langmuir adsorption problem, the partition function in the canonical ensemble was

$$Q(N, M, T) = \frac{M!}{(M-N)!N!} q^N$$

for M equivalent, independent and distinguishable sites on which N indistinguishable molecules may condense up to a maximum of one molecule per site. That is, the number of molecules on a site is either 1 or 0. The total number of molecules condensed on sites is $a_1 = N$, and the total number not condensed on sites is $a_0 = M - N$. Obviously, $a_0 + a_1 = M$. Less obviously, $0a_0 + 1a_1 = N$, that is, $a_1 = N$. If in addition, partiton functions $q(0)$ and $q(1)$ are defined as $q(0) = 1, q(1) = q$, then the partition function may be defined as

$$Q(N, M, T) = \sum_a \frac{M! q(0)^{a_0} q(1)^{a_1}}{a_0! a_1!} \qquad (15.51)$$

where the sum is subject to the restrictive conditions $\sum_s a_s = N$ and $\sum_s s a_s = M$. The boldface type used for the summation index a means to sum over all values of the a_s consistent with the restrictive conditions. These are so restrictive in the Langmuir case that the sum has only one term.

More generally, when the number of molecules per adsorption site varies between 0 and a maximum number m, the partition function becomes

$$Q(N, M, T) = \sum_a \frac{M! q(0)^{a_0} q(1)^{a_1} q(2)^{a_2} \cdots q(m)^{a_m}}{a_0! a_1! a_2! \cdots a_m!}$$

$$M = \sum_{s=0}^{m} a_s, \quad N = \sum_{s=0}^{m} s a_s, \quad q(s) = \sum_i e^{-\varepsilon(s)_i / k_B T} \qquad (15.52)$$

The point of this generalization is to allow writing the grand partition function in the form

$$\Xi = \sum_{N=0}^{mM} Q(N, M, T) \lambda^N \qquad (15.53)$$

Substituting equation (15.3)

$$\Xi = \sum_a \frac{M! q(0)^{a_0} q(1)^{a_1} q(2)^{a_2} \cdots q(m)^{a_m}}{a_0! a_1! a_2! \cdots a_m!} \lambda^{\sum_{s=0}^{m} s a_s} \qquad (15.54)$$

Combining the activities with the q partition functions

$$\Xi = \sum_a \frac{M!(q(0)\lambda^0)^{a_0}(q(1)\lambda^1)^{a_1}(q(2)\lambda^2)^{a_2}\cdots(q(m)\lambda^m)^{a_m}}{a_0!a_1!a_2!\cdots a_m!}$$

which is more compactly

$$\Xi = \sum_a M! \prod_{s=0}^{m} \frac{(q(s)\lambda^s)^{a_s}}{a_s!} \tag{15.55}$$

Now consider the function ξ defined as

$$\xi \equiv q(0) + q(1)\lambda + q(2)\lambda^2 + \cdots q(m)\lambda^m \tag{15.56}$$

When this is raised to the power M, the result is

$$\xi^M = (q(0) + q(1)\lambda + q(2)\lambda^2 + \cdots q(m)\lambda^m)^M \tag{15.57}$$

By the multinomial theorem

$$\xi^M = \sum_{(a_s)} M! \prod_{s=0}^{m} \frac{(q(s)^s)^{a_s}}{a_s!} \tag{15.58}$$

This suggests that in general

$$\Xi(\lambda, M, T) = \xi(\lambda, T)^M \tag{15.59}$$

The quantity ξ has the form of a grand partition function in that it is a sum over a number of molecules up to a maximum, but with the difference that it applies to a single site rather than to an ensemble of them. In this case, all of the M sites are identical. Since the sum that defines ξ applies to a single site, if all the M sites are different, the grand partition function would be the product of sums $\Xi = \xi_1\xi_2\xi_3\cdots\xi_M$ and if there were only two kinds of sites, with N_1 sites of type 1 and N_2 of type 2, the grand partition function would be $\Xi = \xi_1^{N_1}\xi_2^{N_2}$ and so on.

When all the sites are the same, the average number of adsorbed molecules $\langle N \rangle$ is given by

$$\langle N \rangle = \lambda \left(\frac{\partial \ln \Xi}{\partial \lambda} \right)_{T,M}$$

Using equation (15.59)

$$\langle N \rangle = M\lambda \left(\frac{\partial \ln \xi}{\partial \lambda} \right)_T \tag{15.60}$$

This suggests defining the average number of molecules occupying a site as

$$\langle s \rangle = \lambda \left(\frac{\partial \ln \xi}{\partial \lambda} \right)_T \tag{15.61}$$

In cases in which $N < M$, such as the Langmuir problem, the fraction covered is exactly $\langle s \rangle$, but more generally it is simply the average number of molecules adsorbed on a site, all sites being equivalent. In adsorption problems, the mean number of molecules per site is the adsorption isotherm.

The equation of state for the adsorbed phase derives immediately from

$$e^{\Phi M/k_B T} = \Xi = \xi^M \tag{15.62}$$

Immediately

$$\Phi = k_B T \ln \xi \tag{15.63}$$

The approach has some generality, and thus power to solve some apparently intractable problems. Whenever the problem involves a set of independent, distinguishable and open sub systems, the single site grand partition function can be invoked. Further, note that the probability of occupation of a site by s molecules is given by

$$P(s) = \frac{q(s)\lambda^s}{\xi} \tag{15.64}$$

The application to the Langmuir problem has already been done above, with the exception that the appropriate single site grand partition function was not explicitly defined. It is

$$\xi = 1 + q\lambda \tag{15.65}$$

$$\langle s \rangle = \lambda \left(\frac{\partial \ln \xi}{\partial \lambda} \right)_T = \frac{q\lambda}{1 + q\lambda} = \vartheta \tag{15.66}$$

$$\Phi = k_B T \ln \xi = k_B T \ln(1 + q\lambda) \tag{15.67}$$

The algebra that leads to the adsorption isotherm and equation of state is given above in Section 15.1.

In the BET case, the single site grand partition function is almost indefinitely long. It depends on the observation that when two molecules are adsorbed, their energies are additive. This means that the energy levels of the molecule in the first level ε_{1i} and those in the second ε_{2i} are additive, and

$$q(2) = e^{-\varepsilon(1)_1/k_B T} \sum_i e^{-\varepsilon(2)_i/k_B T}$$

$$+ e^{-\varepsilon(1)_2/k_B T} \sum_i e^{-\varepsilon(2)_i/k_B T} + \cdots = q_1 q_2 \tag{15.68}$$

Similarly $q(2) = q_1 q_2 q_3$ and so on. The single site grand partition function is then

$$\xi = 1 + q_1\lambda + q_1 q_2\lambda^2 + q_1 q_2 q_3\lambda^3 + \cdots \tag{15.69}$$

The fraction adsorbed is

$$\langle s \rangle = \lambda \left(\frac{\partial \ln \xi}{\partial \lambda} \right)_T = \frac{q_1\lambda + 2q_1 q_2\lambda^2 + 3q_1 q_2 q_3\lambda^3 + \cdots}{1 + q_1\lambda + q_1 q_2\lambda^2 + q_1 q_2 q_3\lambda^3 + \cdots} \tag{15.70}$$

Now impose the characteristic BET restriction, $q_L = q_2 = q_3 = \cdots$, only q_1 being different. Then

$$\langle s \rangle = \frac{q_1\lambda + 2q_1 q_L\lambda^2 + 3q_1 q_L^2\lambda^3 + \cdots}{1 + q_1\lambda + q_1 q_L\lambda^2 + q_1 q_L^2\lambda^3 + \cdots} \tag{15.71}$$

It is convenient to factor this as

$$\langle s \rangle = \frac{q_1\lambda(1 + 2q_L\lambda + 3q_L^2\lambda^2 + \cdots)}{1 + q_1\lambda(1 + q_L\lambda + q_L^2\lambda^2 + \cdots)} \tag{15.72}$$

Inside the unit radius of convergence, the series $1 + z + z^2 + z^3 + \cdots$ defines the function $\frac{1}{1-z}$. The series is uniformly convergent inside the radius of convergence and therefore possesses derivatives of all orders. In particular, the first derivative gives the series representation

$$\frac{1}{(1-z)^2} = 1 + 2z + 3z^2 + \cdots$$

As long as $q\lambda < 1$, which will be the case for sufficiently low pressures relative to the vapor pressure, the mean occupation per site is given by

$$\langle s \rangle = \frac{q_1\lambda \left(\frac{1}{(1-q_L\lambda)^2} \right)}{1 + q_1\lambda \left(\frac{1}{1-q_L\lambda} \right)} = \frac{q_1\lambda}{(1 - q_L\lambda + q_1\lambda)(1 - q_L\lambda)} \tag{15.73}$$

Now set $x = q_L\lambda = q_L e^{\mu_0/k_B T} p$ and $c = q_1/q_L$. The mean occupation per site is

$$\langle s \rangle = \frac{cx}{(1 - x + cx)(1 - x)} = \vartheta \tag{15.74}$$

This is the form of the BET isotherm derived in Section 15.2. The expression is valid at low coverages and represents the experimental data well at higher pressures than the Langmuir isotherm. However, it is an approximate theory and cannot be pushed too far towards even higher pressures. Note that the combinatorial work that plagued the canonical ensemble approach is entirely avoided by use of the more powerful grand canonical ensemble.

The equation of state is almost immediate. From the basic theory

$$e^{\Phi M/k_B T} = \Xi = \xi^M \tag{15.75}$$

where

$$\xi = 1 + q_1\lambda(1 + q_L\lambda + q_L^2\lambda^2 + \cdots) = \frac{1 + q_1\lambda - q_L\lambda}{1 - q_L\lambda} \tag{15.76}$$

Immediately

$$\frac{\Phi}{k_B T} = \ln\left(\frac{1 + q_1\lambda - q_L\lambda}{1 - q_L\lambda} \right) = \ln\left(\frac{1 + cx - x}{1 - x} \right) \tag{15.77}$$

Evidently as x approaches 1, the surface pressure becomes indefinitely large. Thermodynamically, this is not the case: Φ remains finite. Of course, this difficulty arises because the theory is approximate and x approaching 1 pushes it beyond its limit of useful application, even though the condition of series summability has not been violated for x less than 1.

15.4. Independent Multiples of Sites

In this section, independent multiples of sites are considered. Here adsorption theory and chemical equilibrium become equivalent. For example, the stepwise addition of protons to a dibasic acid at decreasing pH, or the addition of protons to an aromatic base such as pyrazine or pyrimidine or ethylene diammine, all constitute examples of independent pairs of sites. The treatment here is limited to the pairwise cases. Extension to the case of three or four sites will be evident.

For purposes of discussion, let the two sites (say, the two carboxyl groups of a dibasic acid) be labeled 1 and 2. A molecule or ion adsorbed on site 1 has partition function q_1, and one adsorbed on the other site has partition function q_2. Additionally, suppose that when both sites are occupied, there is a further potential energy w acting between them. Now $q(s)$ represents a sum over all possible accessible states. The possible values of s are 0, 1 and 2. Then

$$q(0) = 1, \quad q(1) = q_1 + q_2 \quad q(2) = q_1 q_2 e^{-w/k_B T} \tag{15.78}$$

The single site partition function applicable to these independent pairs of sites is

$$\xi = q(0)\lambda^0 + q(1)\lambda^1 + q(2)\lambda^2 \tag{15.79}$$

Using the definitions in equations (15.78) gives

$$\xi = 1 + (q_1 + q_2)\lambda + q_1 q_2 e^{-w/k_B T}\lambda^2 \tag{15.80}$$

The adsorption isotherm is

$$\langle s \rangle = \frac{\partial \ln \xi}{\partial \ln \lambda} = \frac{(q_1 + q_2)\lambda + 2q_1 q_2 e^{-w/k_B T}\lambda^2}{1 + (q_1 + q_2)\lambda + q_1 q_2 e^{-w/k_B T}\lambda^2} \tag{15.81}$$

15.5. One Dimensional Lattice Gas

In this section, the interacting one dimensional lattice gas is treated. The problem is important because the solution is exact, and exact solutions are few and far between in statistical mechanics.

Let there be M sites and N molecules adsorbed on them as in the Langmuir case, but unlike the Langmuir problem let there be an interaction energy w between nearest neighbors. The maximum number of nearest neighbors in the one dimension lattice is 2. In 2 and 3 dimensional lattices, other maximum numbers of nearest neighbors apply, but in every case, there is a maximum number of nearest neighbors. The solution to the one dimensional lattice gas provides a basis for an approach to adsorption with interaction between nearest neighbors in the two dimensional case.

Start the theory by counting pairs. There are three kinds. Let there be

N_{00} pairs of voids

N_{11} pairs of occupants

N_{01} pairs of occupant-voids

Let o represent a void site and x represent an occupied site. To count pairs, draw a line from every x in both directions and no lines from each void. For a very small number M and a smaller number N, one configuration might be

$$o \; o \; o - x = x - o \; o \; o - x - o$$

To count pairs, one has only to count lines. In the figure, $N_{00} = 4$, $N_{01} = 4$ and $N_{11} = 1$. The total number of lines is 6, and the number of filled sites is the number of x symbols, which is 3. More generally, if there are N filled sites, there must be $2N$ lines, broken down as N_{01} plus $2N_{11}$. Hence

$$2N = N_{01} + 2N_{11} \tag{15.82}$$

Now let the absence of lines replace lines in the counting process. There are $2(M - N)$ absences, and by the same argument as the one for lines

$$2(M - N) = N_{01} + 2N_{00} \tag{15.83}$$

Since both M and N are specified, these two equations have three unknowns. Only one of them is independently variable. Once it is given a value, the other two are determined by these two equations. Any one of the three may be selected as independent variable, The usual choice is N_{01}. In a particular configuration, the interaction energy will be $N_{11}w$ Using the first of the two equations, one finds

$$N_{11}w = \left(N - \frac{1}{2}N_{01}\right)w \tag{15.84}$$

Next, suppose that there are $g(N, M, N_{01})$ configurations with N_{01} pairs of the type void-occupant. The contribution these make to the canonical partition function Q is

$$Q = \cdots + g(N, M, N_{01})e^{-(N-\frac{1}{2}N_{01})w/k_BT} + \cdots \tag{15.85}$$

if the adsorbed molecules have no internal energy structure. More usually, there is an internal energy structure for the adsorbed molecules that make a contribution to Q, and the partition function is

$$Q = q^N \sum_{N_{01}} g(N, M, N_{01})e^{-(N-\frac{1}{2}N_{01})w/k_BT} \tag{15.86}$$

When w vanishes, this becomes the Langmuir case, but now written as

$$Q_{Langmuir} = q^N \sum_{N_{01}} g(N, M, N_{01}) \tag{15.87}$$

The total number of configurations for given N and M is $\frac{M!}{N!(M-N!)}$, and it does not matter how this number is counted. Hence

$$\sum_{N_{01}} g(N, M, N_{01}) = \frac{M!}{N!(M - N!)} \tag{15.88}$$

Even though the individual values of the $g(N, M, N_{01})$ are not known yet, the sum is known. When w is not zero, the partition function may be written

$$Q = (qe^{-w/k_BT})^N \sum_{N_{01}} g(N, M, N_{01})(e^{w/2k_BT})^{N_{01}} \qquad (15.89)$$

The calculation of $g(N, M, N_{01})$ is not as hard as it might seem, for in fact the calculation has already been done in the treatment of the BET isotherm. However, the notation is different, and to avoid confusion the calculation will be repeated here in these terms. The calculation begins by observing that any particular configuration of the one dimensional lattice for given N, M and N_{01} has an appearance such as

$$x \mid oo \mid x \mid o \mid x\ x\ x\ x \mid o\ o \mid x \mid o\ o\ o \mid x \mid o\ o\ o\ o\ o \mid x \mid o \mid$$

The vertical slashes are boundaries between occupied and void regions, except for the one at the right end of the sequence. These are used to count the number of occupant void pairs, In this sequence, $N_{01} + 1 = 12$. The number of void groups is counted by noting that half the boundaries (including the rightmost) have at least one void site to the left. The reason for including the rightmost boundary is now clear: if it were omitted the count would not be correct. Consider the number of ways of arranging N adsorbed molecules into a given number of void groups or of filled groups. It is necessary to put at least one molecule into each filled group, for that molecule defines the group. The number R of remaining molecules is

$$R = N - \frac{N_{01} + 1}{2} \qquad (15.90)$$

These R molecules may be randomly arranged in the $\frac{N_{01}+1}{2}$ filled regions. Each of these filled regions is like a box into which the R molecules may be distributed. Let the boxes (filled regions) be labelled $B_1, B_2, B_3 \ldots$, and let the remaining molecules be labeled $R_1, R_2, R_3, \ldots$ One configuration might be

$$B_1 R_3 R_5 \mid B_3 R_1 R_2 R_4 \mid B_7 \mid B_2 R_{17} \cdots$$

The sequence must start with a box, since otherwise some of the remaining molecules would have no box to be placed in. B is the number of boxes, and the number of arrangements of boxes is $(B-1)!$ The number of arrangements of the remaining molecules is $R!$ and the number of arrangements of all symbols starting with a B is $(B-1+R)!$ The total number of ways of arranging the molecules for given N, M, and N_{01} is

$$\frac{(B-1+R)!}{(B-1)!R!} = \frac{\left(\frac{N_{01}+1}{2}-1+N-\frac{N_{01}+1}{2}\right)!}{\left(\frac{N_{01}+1}{2}-1\right)!\left(N-\frac{N_{01}+1}{2}\right)!} = \frac{N!}{\left(\frac{N_{01}}{2}\right)!\left(N-\frac{N_{01}}{2}\right)!}$$

$$(15.91)$$

when 1 is neglected in comparison with N_{01}. By the same argument, the number of ways of arranging $M-N$ voids into $\frac{N_{01}+1}{2}$ void regions is

$$\frac{(M-N)!}{\left(\frac{N_{01}}{2}\right)!\left(M-N-\frac{N_{01}}{2}\right)!}$$

The number of ways of placing N molecules on M sites with N_{01} pairs of the type ox is the product of these two if the leftmost site is filled. This site may equally well be void, and the final number is

$$g(N, M, N_{01}) = 2\frac{N!(M-N)!}{\left(\frac{N_{01}}{2}!\right)^2\left(N-\frac{N_{01}}{2}\right)!\left(M-N-\frac{N_{01}}{2}\right)!} \qquad (15.92)$$

The partition function is therefore

$$Q = 2N!(M-N)!(qe^{-w/k_BT})^N$$

$$\times \sum_{N_{01}} \frac{(e^{w/2k_BT})^{N_{01}}}{\left(\frac{N_{01}}{2}!\right)^2\left(N-\frac{N_{01}}{2}\right)!\left(M-N-\frac{N_{01}}{2}\right)!} \qquad (15.93)$$

The maximum term method is used to evaluate Q. The logarithm of the typical term t is

$$\ln t = \frac{N_{01}w}{2k_BT} - 2\ln\frac{N_{01}}{2}! - \ln\left(N-\frac{N_{01}}{2}\right)!$$

$$- \ln\left(M-N-\frac{N_{01}}{2}\right)! \qquad (15.94)$$

$$\frac{\partial \ln t}{\partial N_{01}} = \frac{w}{2k_B T} - \ln \frac{N_{01}}{2} + \frac{1}{2} \ln \left(N - \frac{N_{01}}{2} \right)$$

$$+ \frac{1}{2} \ln \left(M - N - \frac{N_{01}}{2} \right) = 0 \qquad (15.95)$$

$$e^{-w/2k_B T} = \frac{\left(N - \frac{N_{01}}{2} \right)^{1/2} \left(M - N - \frac{N_{01}}{2} \right)^{1/2}}{\frac{N_{01}}{2}} \qquad (15.96)$$

This already looks like an equilibrium constant expression. Squaring both sides and defining $\vartheta = \frac{N}{M}$ and $\alpha = \frac{N_{01}}{2M}$, the expression becomes

$$e^{-w/k_B T} = \frac{(\vartheta - \alpha)(1 - \vartheta - \alpha)}{\alpha^2} \qquad (15.97)$$

This equation is quadratic in α. It is easy to show that

$$\alpha^2 (1 - e^{-w/k_B T}) - \alpha + \vartheta(1 - \vartheta) = 0 \qquad (15.98)$$

Solving the quadratic for $\alpha = \frac{N_{01}}{2M}$ gives

$$\alpha = \frac{1 \pm \sqrt{1 - 4(1 - e^{-w/k_B T})(\vartheta)(1 - \vartheta)}}{2(1 - e^{-w/k_B T})} \qquad (15.99)$$

This suggests defining

$$\beta = \sqrt{1 - 4(1 - e^{-w/k_B T})(\vartheta)(1 - \vartheta)} \qquad (15.100)$$

Then

$$\alpha = \frac{1 \pm \beta}{2(1 - e^{-w/k_B T})} \qquad (15.101)$$

From equation (15.98)

$$1 - e^{-w/k_B T} = \frac{\alpha - \vartheta(1 - \vartheta)}{\alpha^2} \qquad (15.102)$$

Making this substitution

$$\alpha = \frac{1 \pm \beta}{2} \frac{\alpha^2}{\alpha - \vartheta(1 - \vartheta)}$$

After cancelling an alpha and rearranging the terms, there results

$$\alpha = 2\frac{\vartheta(1-\vartheta)}{1\pm\beta} \tag{15.103}$$

The sign of the function β is determined by considering the limiting case, $w = 0$. When this is so, $\beta = 1$, and if the minus sign is chosen, the function α diverges, which is unphysical. If the plus sign is selected, $\alpha = \vartheta(1-\vartheta)$, which is finite. The selection of the sign of β is not important, but is necessary to keep α finite for all values of the interaction potential w.

The value of the number of void-occupant pairs in the most probable configurations is defined as N_{01}^*, and similarly for the other two, N_{11}^*, N_{00}^*. Immediately, $N = N_{11}^* + \frac{N_{01}^*}{2}$, $M - N = N_{00}^* + \frac{N_{01}^*}{2}$, $\alpha = \frac{N_{01}^*}{2M}$, and

$$\vartheta - \alpha = \frac{N}{M} - \frac{N_{01}^*}{2M} = \frac{1}{M}\left(N_{11}^* + \frac{N_{01}^*}{2} - \frac{N_{01}^*}{2}\right) = \frac{N_{11}^*}{M} \tag{15.104}$$

$$1 - \vartheta - \alpha = 1 - \frac{N}{M} - \frac{N_{01}^*}{2M} = \frac{1}{M}\left(M - N - \frac{N_{01}^*}{2}\right)$$

$$= \frac{1}{M}\left(N_{00}^* + \frac{N_{01}^*}{2} - \frac{N_{01}^*}{2}\right) = \frac{N_{00}^*}{M} \tag{15.105}$$

Given these equalities, the equation $e^{-w/k_BT} = \frac{(\vartheta-\alpha)(1-\vartheta-\alpha)}{\alpha^2}$ becomes

$$e^{-w/k_BT} = \frac{(N_{11}^*/M)(N_{00}^*/M)}{(N_{01}^*/2M)^2} \tag{15.106}$$

Cancellation of M leads to

$$\frac{e^{-w/k_BT}}{4} = \frac{(N_{11}^*)(N_{00}^*)}{(N_{01}^*)^2} \tag{15.107}$$

This has the form of a law of mass action expression for the chemical reaction

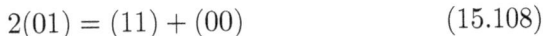

$$2(01) = (11) + (00) \tag{15.108}$$

with an equilibrium constant consistent with partition functions $q_{00} = 1$, $q_{11} = e^{-w/k_BT}$, and $q_{01} = 2$.

The factor of 2 for q_{01} occurs since there are two ways to make a 01 pair: either 01 or 10. The equilibrium constant is given by

$$K = \frac{q_{00}q_{11}}{q_{01}^2} = \frac{(N_{11}^*)(N_{00}^*)}{(N_{01}^*)^2} \tag{15.109}$$

Returning to the partition function evaluated by the maximum term method, the thermodynamic functions are available. The chemical potential is

$$\mu = -k_BT \left(\frac{\partial \ln Q}{\partial N} \right)_{T,M} \tag{15.110}$$

where

$$\ln Q \cong N \ln q e^{-w/k_BT} + \ln t(N_{01}^*, N, M, T) \tag{15.111}$$

When M and T are held constant, a general variation in the typical term t is

$$d\ln t = \frac{\partial \ln t}{\partial N_{01}^*} dN_{01}^* + \frac{\partial \ln t}{\partial N} dN \tag{15.112}$$

Then when M and T are held constant,

$$\left(\frac{\partial \ln t}{\partial N} \right)_{M,T} = \left(\frac{\partial \ln t}{\partial N_{01}^*} \right)_{M,N,T} \left(\frac{\partial N_{01}^*}{\partial N} \right)_{M,T} + \left(\frac{\partial \ln t}{\partial N} \right)_{N_{01}^*,M,T} \tag{15.113}$$

The condition that determines the maximum term and N_{01}^* is $(\frac{\partial \ln t}{\partial N_{01}^*})_{M,N,T} = 0$. Hence

$$\left(\frac{\partial \ln t}{\partial N} \right)_{M,T} = \left(\frac{\partial \ln t}{\partial N} \right)_{N_{01}^*,M,T} \tag{15.114}$$

This allows the chemical potential to be written as

$$-\frac{\mu}{k_BT} = \ln q e^{-w/k_BT} + \left(\frac{\partial \ln t}{\partial N} \right)_{N_{01}^*,M,T} \tag{15.115}$$

In terms of the number of different configurations for given N_{01} the typical term is $t = g(e^{w/2k_BT})^{N_{01}^*}$ and the derivative with respect to

N is

$$\left(\frac{\partial \ln t}{\partial N}\right)_{N_{01}^*,M,T} = \left(\frac{\partial \ln g}{\partial N}\right)_{N_{01}^*,M,T} \tag{15.116}$$

Since

$$g(N, M, N_{01}) = 2\frac{N!(M-N)!}{\left(\frac{N_{01}}{2}!\right)^2 \left(N - \frac{N_{01}}{2}\right)! \left(M - N - \frac{N_{01}}{2}\right)!} \tag{15.117}$$

the logarithm is rather lengthy, and several of the factors do not contribute terms to the chemical potential. Saving only those that do contribute,

$$\ln g = \ln N! + \ln(M-N)! - \ln\left(N - \frac{N_{01}}{2}\right)!$$

$$- \ln\left(M - N - \frac{N_{01}}{2}\right)! + \cdots \tag{15.118}$$

The derivative is

$$\frac{\partial \ln g}{\partial N} = \ln N - \ln(M-N) - \ln\left(N - \frac{N_{01}}{2}\right) + \ln\left(M - N - \frac{N_{01}}{2}\right)$$

$$= \ln\left[\frac{N\left(M - N - \frac{N_{01}}{2}\right)}{(M-N)\left(N - \frac{N_{01}}{2}\right)}\right] = \ln\left[\frac{\vartheta(1 - \vartheta - \alpha)}{(1-\vartheta)(\vartheta - \alpha)}\right] \tag{15.119}$$

The chemical potential is then

$$\frac{\mu}{k_B T} = -\ln(qe^{-w/k_B T}) - \left(\frac{\partial \ln g}{\partial N}\right)_{N_{01}^*,M,T}$$

$$= -\ln(qe^{-w/k_B T}) - \ln\left[\frac{\vartheta(1 - \vartheta - \alpha)}{(1-\vartheta)(\vartheta - \alpha)}\right] \tag{15.120}$$

The activity is $\lambda = e^{\mu/k_B T}$. In this case, it is

$$\lambda = \frac{1}{qe^{-w/k_B T}} \frac{(1-\vartheta)(\vartheta - \alpha)}{\vartheta(1 - \vartheta - \alpha)} \tag{15.121}$$

where the activity is proportional to the pressure if the gas in the fluid phase is ideal. Note that q is for the adsorbed molecules, in which the rotational manifold is fully suppressed, as are translational modes.

There remain only vibrational modes, and q is a function of T, as is the exponential for fixed w. Then, collecting all the functions of T in one place,

$$\lambda q e^{-w/k_B T} = \frac{(1 - \vartheta)(\vartheta - \alpha)}{\vartheta(1 - \vartheta - \alpha)} \propto p \qquad (15.122)$$

After some algebra using $\alpha = 2\frac{\vartheta(1-\vartheta)}{1+\beta}$, to eliminate α, the alternate expression is

$$\lambda q e^{-w/k_B T} = \frac{\beta - 1 + 2\vartheta}{\beta + 1 - 2\vartheta} \qquad (15.123)$$

When $w = 0$, $\beta = 1$ and $\lambda q e^{-w/k_B T} = \lambda q$. The expression becomes

$$\lambda q = \frac{\vartheta}{1 - \vartheta} \qquad (15.124)$$

which is the Langmuir result. Clearly, $\lambda q e^{-w/k_B T}$ is the adsorption isotherm. As this is proportional to the pressure, the logarithm of $\lambda q e^{-w/k_B T}$ is the logarithm of the pressure within a constant that is a function of temperature alone. It is conventional to plot ϑ against the logarithm of pressure to display the behavior of the isotherm. This is done in Figs. 15.2 and 15.3 for several values of $w/k_B T$.

The equation of state requires extensive algebra, straightforward but too lengthy for inclusion here, based on the observation that the thermodynamic variation in the energy of the adsorbed phase is

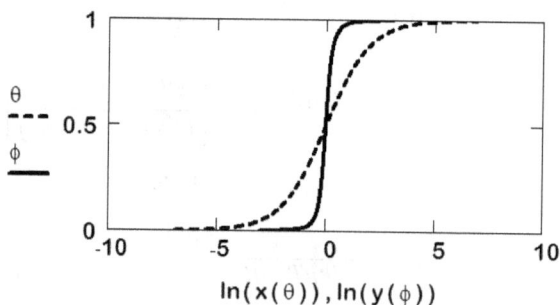

Fig. 15.2. Adsorption isotherms. Dashed line, Langmuir isotherm. Solid line, exact one dimensional adsorption isotherm with interaction potential between molecules adsorbed on adjacent sites, with $w/k_B T = -4$. Attractive potential.

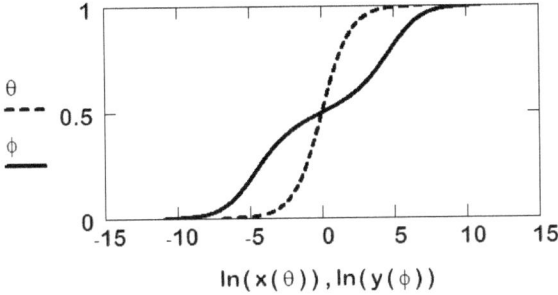

$$\ln(x(\theta)), \ln(y(\phi))$$

Fig. 15.3. Solid line: adsorption isotherm for $w/k_BT = +4$. Repulsive potential. When the repulsive potential becomes very large $(w/k_BT = +30)$ there is a separation of the isotherm into two isotherms, one at very negative $\ln y$, in the range $0 < \Theta < 1/2$, and one in the range $1/2 < \Theta < 1$. Dashed line: Langmuir adsorption isotherm.

controlled by the relation $dE = TdS - \Phi dM + \mu dN$. For a single species, the Gibbs free energy is by Euler's theorem on homogeneous functions $G = \mu N = E - TS + \Phi M = A + \Phi M = -k_BT \ln Q + \Phi M$ and therefore

$$\Phi M = \mu N + k_BT \ln Q \qquad (15.125)$$

Using $\alpha = \frac{N_{01}}{2M} = 2\frac{\vartheta(1-\vartheta)}{1\pm\beta}$ and from equation (15.123) to get $\mu = k_BT \ln \lambda$, and expressions for $\ln Q$ from the above work, one finally arrives at

$$\frac{\Phi}{k_BT} = \ln\left(\frac{\beta+1}{\beta+1-2\vartheta}\right) \qquad (15.126)$$

When $\beta \to 1$, this expression for $\frac{\Phi}{k_BT} \to -\ln(1-\vartheta)$, the Langmuir equation of state.

15.6. Bragg-Williams Approximation

The Bragg-Williams approximation is a very simple modification of the statistical mechanics of adsorption. It cannot possibly be close to physical reality, yet the approximation captures a surprising amount of real adsorption behavior. In this approximation, adsorption onto sites is handled as though adsorption were completely random The

Langmuir partition function is very simply modified by multiplication by $e^{-\langle N_{11}\rangle w/k_B T}$, where $\langle N_{11}\rangle w$ is the average interaction energy of nearest neighbor pairs. The calculation of $\langle N_{11}\rangle$ is done as follows. Let Θ be the fraction of sites occupied, and let there be c nearest neighbors, not just 2. Then the average number of nearest neighbors of one adsorbed molecule is $c\Theta$, and the total average number of nearest neighbors in the adsorbed phase is $c\Theta(\frac{N}{2})$. The factor of 2 arises because there are two ways to make any pair. As $\Theta = \frac{N}{M}$ the average number of nearest neighbors is

$$\langle N_{11}\rangle = c\frac{N}{M}\frac{N}{2} = \frac{cN^2}{2M} \tag{15.127}$$

The Bragg-Williams partition function is then

$$Q = \frac{M!}{N!(M-N)!}q^N e^{-cN^2 w/2Mk_B T} \tag{15.128}$$

The route to thermodynamics is straightforward. As in the Langmuir case

$$\ln Q = -\frac{A}{k_B T} = M\ln M - N\ln N - (M-N)\ln(M-N)$$
$$+ N\ln q - cN^2 w/2Mk_B T \tag{15.129}$$

The entropy is the same in the Bragg-Williams approximation as it is in the Langmuir adsorption. In effect

$$\frac{S}{k_B} = \ln Q + T\left(\frac{\partial \ln Q}{\partial T}\right)_{N,M} = \ln\left(\frac{M!}{N!(M-N)!}\right)$$
$$+ N\left(\ln q + T\frac{d\ln q}{dT}\right) \tag{15.130}$$

The equation of state follows from

$$\frac{\Phi}{k_B T} = \left(\frac{\partial \ln Q}{\partial M}\right)_{N,T} = \ln M - \ln(M-N) + \frac{cN^2 w}{2M^2 k_B T} \tag{15.131}$$

Applying the definition of the fraction covered yields an equation of state that is somewhat different from the Langmuir, namely

$$\frac{\Phi}{k_B T} = -\ln(1 - \Theta) + \frac{cw}{2k_B T}\Theta^2 \qquad (15.132)$$

When the fraction covered is small, the logarithm may be expanded in power series

$$\ln(1 - \Theta) = -\Theta - \frac{1}{2}\Theta^2 - \frac{1}{3}\Theta^3 - \cdots \quad |\Theta| \leq 1$$

and the equation of state at low coverage may be expanded in a Virial kind of expansion

$$\frac{\Phi}{k_B T} = \Theta + \left(\frac{1}{2} + \frac{cw}{2k_B T}\right)\Theta^2 + \frac{1}{3}\Theta^3 + \cdots \qquad (15.133)$$

which is of the form

$$\frac{\Phi}{k_B T} = \Theta + B_2^{BW}(T)\Theta^2 + \cdots \qquad (15.134)$$

The second virial coefficient for the exact solution to the one dimensional adsorption problem worked out above is different at low temperatures, but when the temperature is high enough that $w/k_B T$ becomes small, the exponential may be expanded. In first approximation,

$$B_2^{exact}(T) = \frac{c + 1 - ce^{-w/k_B T}}{2} \cong \frac{c + 1 - c(1 - w/k_B T + \cdots)}{2}$$

$$\cong \frac{1}{2} + \frac{cw}{2k_B T} + \cdots \qquad (15.135)$$

Both the Bragg-Williams approximation and the exact adsorption theory predict a Boyle temperature ,$T_B \cong \frac{-cw}{k_B}$. The Boyle temperature is defined in thermodynamics by the relation

$$\frac{\partial pV}{\partial p} = 0 \qquad (15.136)$$

the effect of which is to require the second virial coefficient to vanish at the Boyle temperature. For gases, this means that in the vicinity of the Boyle temperature, the gas becomes nearly ideal. For adsorption, it means that near the Boyle temperature, the surface pressure is essentially

$$\frac{\Phi}{k_B T} \approx \Theta \tag{15.137}$$

or $\Phi M \approx N k_B T$, a familiar form.

The chemical potential leads to the adsorption isotherm, and is

$$\frac{\mu}{k_B T} = -\left(\frac{\partial \ln Q}{\partial N}\right)_{M,T} = \ln\left(\frac{\Theta e^{cw\Theta/k_B T}}{(1-\Theta)q}\right) \tag{15.138}$$

from which the relative volatility y follows

$$y = q\lambda e^{-cw/2k_B T} = \frac{\Theta e^{cw(2\Theta-1)/2k_B T}}{1-\Theta} \tag{15.139}$$

The behavior of equation (15.139) is shown in Figure 15.4.

One of the most interesting properties of the Bragg-Williams approximation is that it gives a plot of Φ versus $1/\Theta$ that has the appearance of a van der Waals plot of p against V. The ordinate of the Φ plot is the reciprocal of the fraction occupied because M takes the place of V as Φ does for p, and of course Θ is proportional

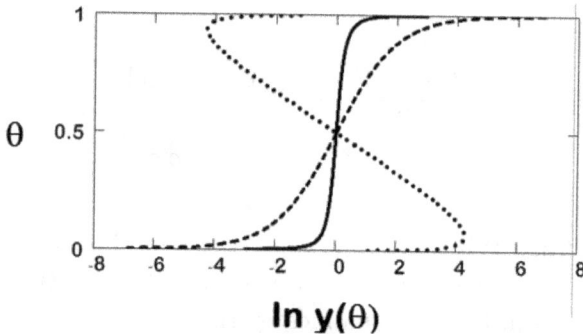

In y(θ)

Fig. 15.4.   Comparison of adsorption isotherms for Langmuir, exact one dimension adsorption and Bragg-Williams approximation. Dashed line: Langmuir. Solid line: exact one dimensioned adsorption. Dotted line: Bragg-Williams modification of Langmuir. In all plots, $w/k_B T = 4$ and in the Bragg-Williams plot, $c = 4$. Note the large region of instability in the Bragg-Williams plot (region of negative slope).

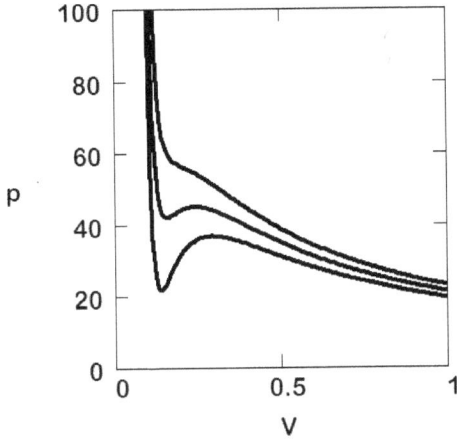

Fig. 15.5.   van der Waals plot for ethane. Temperatures used 320, 200 and 380 K. Constants $a = 5.563$ and $b = 0.0638$ pressures are in units of bar, volumes in liters per mole. $R = 0.08314$. The critical temperature is about 305 K.

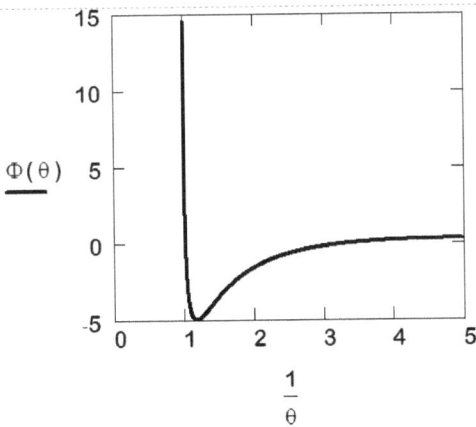

Fig. 15.6.   Bragg-Williams equation of state for the parameters of Figure 15.5.

to the reciprocal of M. For comparison, a van der Waals plot of a typical imperfect gas, ethane, is shown in Figure 15.5. Figure 15.6 is a plot of the Bragg-Williams approximation for the parameters of Figure 15.4.

Both the van der Waals and the Bragg-Williams plots suggest the separation into two phases below a certain critical temperature. The criterion in the Bragg-Williams case is the slope of the isotherm in Figure 15.5. When the slope $\frac{d\Theta}{dlny}$ is positive at $\Theta = 1/2$, there is no separation. When the slope at that point is negative, there are two regions, one of high coverage and one of low. If the pressure is set at a value that would produce a fraction covered in the region between the two turning points, some of the adsorbing molecules will occupy regions of high density and the others regions of low density, set by the turning points. The critical temperature is in principle available from the criterion

$$\left(\frac{d\Theta}{dlny} \right)_{\Theta=\frac{1}{2}} = \infty \qquad (15.140)$$

The more commonly quoted criterion determining the critical temperature is obtained from the van der Waals like plot of Figure 15.5. As $1/\Theta$ increases from small values (large Θ near 1) the slope $\frac{d\Phi}{d\frac{1}{\Theta}}$ increases from large negative values to zero, and thereafter becomes positive, at first increasing, then reversing to a decreasing trend at an inflection point where the second derivative vanishes. The positive slope diminishes until at length it becomes zero, after which it becomes negative again. At a higher temperature, the two points of zero slope move towards the inflection point, and at higher temperatures these three points move towards coalescence until at the critical point all three are combined in a single point at a value of $1/\Theta = 2$ At that point, both the first and second derivatives vanish, and at higher temperatures, the slope never becomes other than negative. The criteria that determines the critical temperature then is

$$\left(\frac{d\Phi}{d\frac{1}{\Theta}} \right)_{\frac{1}{\Theta}=2} = 0 \qquad (15.141)$$

and

$$\left(\frac{\partial^2 \Phi}{\partial \left(\frac{1}{\Theta} \right)^2} \right)_{\frac{1}{\Theta}=2} = 0 \qquad (15.142)$$

The criterion as stated in equation (15.141) is a little cumbersome to use, and the equivalent statement

$$\left(\frac{d\Phi}{d\Theta}\right)_{\Theta=\frac{1}{2}} = 0 \tag{15.143}$$

is usually used to calculate the critical temperature. The equation of state for the Bragg-Williams approximation is

$$\Phi == -k_B T \ln(1-\Theta) + \frac{cw}{2}\Theta^2 \tag{15.144}$$

and the derivative is

$$\frac{d\Phi}{d\Theta} = \frac{k_B T}{1-\Theta} + cw\Theta \tag{15.145}$$

Setting this to zero when $\Theta = 1/2$ gives the result $-\frac{cw}{4k_B T_c} = 1$. Since w is negative, this gives a positive critical temperature.

15.7. Ferromagnetism

Ferromagnetism is a widespread phenomenon, known to the ancients, though its use in a magnetic compass was not discovered until the first millennium in China. The Chinese thought imperial administration was more important than long range sea faring trade, and in fact did not develop it. The Europeans rediscovered it, apparently independently, and used it in world wide navigation. Ferromagnetism admits of a simple one dimensional interpretation, but two dimensional or three dimensional treatments remain to be presented, with the notable exception of Lars Onsager's treatment of the two dimensional problem for the special case $\Theta = 1/2$, a treatment that predicts a critical temperature and in the treatment requires the use of Lie algebras. The problem was originally proposed by Ising in his PhD thesis, and he could not solve the general case either, but proposed a correct treatment of the one dimensional case.

In the following, a magnetic material is placed in an externally applied magnetic field with a field direction that is vertical. The interaction is between the nuclear magnetic moment m formed in magnetic resonance by the spin of the nucleus, and in ferromagnetism by the net angular momentum of the electronic structure of the atoms that form the magnet. In either case, a magnetic field is generated by the charge circulation. Suppose there are only two orientational states of the effective magnets produced by the charge circulation. Either the effective magnet is oriented with the field or against it. Suppose there are

M effective magnets in all
N effective magnets aligned against the field with potential energy $+mH$
$M-N$ effective magnets aligned with the field with potential energy $-mH$

The magnetic dipole moment m is the energy per effective magnet per unit field strength H. The total potential energy V is

$$V = NmH - (M - N)mH = (2N - M)mH \qquad (15.146)$$

The average number of effective magnets aligned with the field is $\langle N \rangle$, which must lie between 0 and M. The work effect attending an increase in field strength dH is $dw = -IdH$, where I is the intensity of magnetization, that is, the energy of the excess of effective magnets aligned with the field. Mathematically,

$$I = (M - 2N)m \qquad (15.147)$$

or

$$I = \left(1 - 2\frac{\langle N \rangle}{M}\right)mM \qquad (15.148)$$

Defining $\Theta \equiv \frac{\langle N \rangle}{M}$, the intensity becomes

$$I = (1 - 2\Theta)mM \qquad (15.149)$$

Solving this for Θ yields

$$\Theta = \frac{1}{2}\left(1 - \frac{I}{mM}\right) \qquad (15.150)$$

The thermodynamic equation defining the energy function for this system is

$$dE = TdS - IdH + \mu dM \qquad (15.151)$$

For a given N, there are $\frac{M!}{N!(M-N)!}$ arrangements of aligned effective magnets on the M sites of the lattice. In the canonical ensemble the partition function is therefore

$$Q = q^M \sum_{N=0}^{M} \frac{M!}{N!(M-N)!} e^{-(2N-M)mH/k_BT} \qquad (15.152)$$

$$Q = q^M e^{MmH/k_BT} \sum_{N=0}^{M} \frac{M!}{N!(M-N)!} e^{-2NmH/k_BT}$$

$$Q = q^M e^{MmH/k_BT} (1 + e^{-2mH/k_BT})^M \qquad (15.153)$$

by the binomial theorem. Since $A = E - TS$, variations in the free energy are governed by

$$dA = -SdT - IdH + \mu dM \qquad (15.154)$$

Recalling that $A = -k_BT \ln Q$, immediately

$$A = -k_BT \ln \left[q^M e^{MmH/k_BT} (1 + e^{-2mH/k_BT})^M \right] \qquad (15.155)$$

Writing this out in full provides

$$A = -Mk_BT \ln q - MmH - Mk_BT \ln(1 + e^{-2mH/k_BT}) \qquad (15.156)$$

Inserting the definition of $I(\Theta)$ gives the intensity of magnetization as

$$mM(1 - 2\Theta) = I = -\left(\frac{\partial A}{\partial H}\right)_{T,M} = k_BT \left(\frac{\partial \ln Q}{\partial H}\right)_{T,M} \qquad (15.157)$$

Performing the indicated differentiation gives

$$I = Mm + Mk_BT \frac{e^{-2mH/k_BT}}{1 + e^{-2mH/k_BT}} \left(-\frac{2m}{k_BT}\right) \qquad (15.158)$$

or

$$I = Mm - 2Mm\frac{e^{-2mH/k_BT}}{1 + e^{-2mH/k_BT}}$$

Factoring mM produces

$$I = Mm\left(\frac{1 + e^{-2mH/k_BT} - 2e^{-2mH/k_BT}}{1 + e^{-2mH/k_BT}}\right) \qquad (15.159)$$

This condenses to

$$I = mM\left(\frac{1 - e^{-2mH/k_BT}}{1 + e^{-2mH/k_BT}}\right)$$

or

$$I = mM\left(\frac{e^{mH/k_BT} - e^{-mM/k_BT}}{e^{mH/k_BT} + e^{-mM/k_BT}}\right)$$

which is identically

$$I = mM\tanh(mM/k_BT) \qquad (15.160)$$

From equation (15.149), $I = (1 - 2\Theta)mM$, and setting these two expressions for the intensity of magnetization equal, mM cancels leaving

$$1 - 2\Theta = \tanh(mM/k_BT) \qquad (15.161)$$

All of the above assumes that there is no interaction between effective magnets, a not very realistic model. Each effective magnet represents circulating charge, which must generate a magnetic field. Two magnets brought into proximity will align themselves in one another's magnetic field, as is easy to demonstrate by lining up three or four identical magnetic compasses. All will point north independent of the direction the line of compasses is facing. Nuclear spins (or spins generated by the electronic field of a ferromagnetic element) are not compass needles, and some nearest neighbors in the lattice will align the same way (either both with or both against the externally imposed magnetic field), and some will be opposed. Although there is no limit to the extent of the magnetic field generated by charge circulation at one atom, and the interaction is really of unlimited extent, governed in magnitude by the inverse square law, the first approximate treatment assumes the interaction is limited to nearest neighbors. Let the interaction energy be $-J$ for nearest neighbor effective magnets aligned the same way (either both with or both against the applied field) and $+J$ otherwise (either the one with and the other against, or the one against and the other with the external field). While this will not capture all of the energetics, it may capture the larger part. The origin of the charge circulation and the resulting magnetic moment may be ignored for statistical purposes.

Let the subscript 1 refer to an effective magnet aligned against the external field and the subscript 0 refer to one aligned with the field. The total interaction energy for a given value of N, the number of effective magnets aligned against the external field, is

$$w = N_{11}w_{11} + N_{00}w_{00} + N_{01}w_{01} \qquad (15.162)$$

in an obvious notation. In equation (15.162), the energies

$$w_{00} = w_{11} = -J \qquad (15.163)$$

and

$$w_{01} = +J \qquad (15.164)$$

As in the Bragg-Williams approximation, on the average

$$\langle N_{11} \rangle = \frac{cN^2}{2M}, \quad \langle N_{01} \rangle = \frac{cN(M-N)}{M}, \quad \langle N_{00} \rangle = \frac{c(M-N)^2}{2M}$$

Substituting these factors into equation (15.162) produces

$$w = \frac{cN^2}{2M}(-J) + \frac{cN(M-N)}{M}(+J) + \frac{c(M-N)^2}{2M}(-J) \qquad (15.165)$$

After some algebra, there results

$$w = -\frac{cJ}{2M}(2N-M)^2 \qquad (15.166)$$

The partition function becomes

$$Q = q^M \sum_{N=0}^{M} \frac{M!}{N!(M-N)!} e^{-(2N-M)mH/k_BT} e^{\frac{cJ}{2Mk_BT}(2N-M)^2}$$

(15.167)

$$Q = (qe^{(2mH+cJ)/2k_BT})^M \sum_{N=0}^{M} \frac{M!}{N!(M-N)!} e^{\frac{2cJN^2}{Mk_BT}} \left[e^{-\frac{2(mH+cJ)}{k_BT}} \right]^M$$

(15.168)

Now, compare this result with the Bragg-Williams result

$$Q(N,M,T) = \frac{M!}{N!(M-N)!} q^N e^{-\frac{cN^2 w}{2Mk_BT}}$$

(15.169)

The grand partition function for the Bragg-Williams lattice gas is

$$\Xi = \sum_{N=0}^{M} Q(N,M,T)\lambda^N = \sum_{N=0}^{M} \frac{M!}{N!(M-N)!} (q\lambda)^N e^{-\frac{cN^2 w}{2Mk_BT}}$$

(15.170)

Comparing equations (15.168) and (15.170), the sums are seen to be the same if the identifications $-w \leftrightarrow 4J$ and $q \leftrightarrow e^{-2(mH+cJ)/k_BT}$ are made.

A principal result of the Bragg-Williams lattice gas was the relative volatility

$$y = q\lambda e^{-cw/2k_BT} = \frac{\Theta}{1-\Theta} e^{cw(2\Theta-1)/2k_BT}$$

Substituting the equivalences, the ferromagnet y becomes

$$y = e^{-2(mH+cJ)/k_BT} e^{-c(-4J)/2k_BT}$$

(15.171)

or

$$y = e^{-2(mH+cJ)/k_BT} e^{2cJ/k_BT} = e^{-2mH/k_BT}$$

If I/mM is plotted against $\ln(y)$, the plot is essentially intensity of magnetization against external magnetic field. However, since $\frac{I}{mM} = 1-2\Theta$, the limits of the abscissa are $+1$ and -1, not 0 and 1, as is the case when Θ is used as the dependent variable. Such a plot is shown

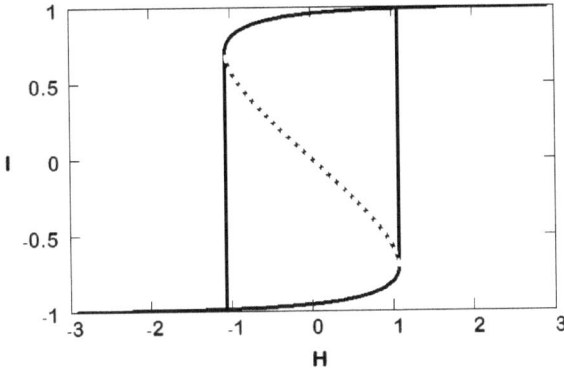

Fig. 15.7.   Plot of I/mM versus ln y showing intensity of magnetization I as a function of externally applied field strength H. The dotted region is the region of instability. In effect, as the field strength increases from -3 towards positive values, the intensity of magnetization encounters the turnaround point. Further increase in field strength enters the region of instability and the magnetization changes abruptly to the magnetization at the upper turnaround point, and thereafter follows the upper curve to saturation. When the saturated magnet is placed in a high positive field and the field is decreased, the lower turnaround point is reached and the magnet abruptly demagnetizes, thereafter following the lower curve.

in Figure 15.7, where the experimentally observed phenomenon of hysteresis is displayed as generated by the nearest neighbor only interaction.

Consider now the critical temperature. The thermodynamic equations that govern the lattice gas and the ferromagnet are

Lattice gas $d(\Phi M) = d(k_B T \ln \Xi) = SdT + Nd\mu + \Phi dM$

Ferromagnet $- dA = d(k_B T \ln Q) = SdT + IdH - \mu dM$

For the lattice gas, the critical temperature was found above to be $T_c = -\frac{cw}{4k_B}$. For the ferromagnet, the critical temperature (also called the Curie temperature) is easily obtained by using the equivalence $-w = 4J$. This gives immediately

$$T_c = \frac{cJ}{k_B} \tag{15.172}$$

Temperatures below T_c allow spontaneous magnetization I_s, at field strengths above about 1 in Figure 15.7. Experimentally, it is found that once spontaneous magnetization has taken place, the magnetization remains if the field strength is removed. Alternatively, the criterion

$$\left(\frac{d\mu}{d\Theta}\right)_{\Theta=\frac{1}{2}} = 0 \tag{15.173}$$

may be applied. This criterion is equivalent to $(\frac{d\Theta}{d\ln y})_{\Theta=\frac{1}{2}} = \infty$, since $\ln \lambda = \frac{\mu}{k_B T}$

15.8. Two Dimensional Lattice Gas Approximation

There have been no exact solutions to the problem of the lattice gas in more than one dimension, though many approximate expansions have appeared. As an example, the equation of state for a two dimensional lattice gas is treated below as a virial expansion of the form

$$\frac{\Phi}{k_B T} = \Theta + B_2(T)\Theta^2 + B_3(T)\Theta^3 + \cdots$$

For adsorption problems with M sites, the grand partition function may be written as

$$\Xi(T, M, \mu) = e^{\Phi M/k_B T} = 1 + Q(1, M, T)\lambda + Q(2, M, T)\lambda^2 + \cdots \tag{15.174}$$

where as usual $\lambda = e^{\mu/k_B T}$. The N body partition function $Q(N, M, T)$ is

$$Q(N, M, T) = (qe^{-cw/2k_B T})^N \sum_{N_{01}} g(N, M, N_{01}) x^{N_{01}} \tag{15.175}$$

where $x = e^{w/2k_B T}$ as shown above. The sum is restricted by relations that are an obvious modification of earlier work.

$$cN = N_{01} + 2N_{11} \tag{15.176}$$

$$c(M - N) = N_{01} + 2N_{00} \tag{15.177}$$

The number g is not so easily counted as it was in the one dimensional case, and the simplification provided by the expansion allows some progress to occur. When $N = 0$, there are no adsorbed molecules, and

$$Q(0, M, T) = (qe^{-cw/2k_BT})^0 g(0.M, 0)x^0 = 1$$

When $N = 1$, there are no nearest neighbors, and there are exactly M ways the molecule can adsorb. Further, N_{01} is exactly c. Then for one molecule adsorbed,

$$Q(1, M, T) = qe^{-cw/2k_BT} M x^c \tag{15.178}$$

When $N = 2$, there are two possibilities for adsorption. Either the two adsorb on distant sites, or the two adsorb on adjacent sites, as nearest neighbors. When adsorption is adjacent, N_{01} becomes two less (one less for each member of the pair) than is the case when adsorption is distant. For distant adsorption, the number of 01 pairs is $2c$, and when adsorption is adjacent, the number of 01 pairs is $2c - 2$. The number of ways of achieving distant adsorption is clearly M for the first, and $M - c - 1$ for the second. Since either may be adsorbed first, g for this case is

$$g(2, M, 2c) = \frac{M(M - c - 1)}{2}$$

while g for adjacent adsorption is

$$g(2, M, 2c - 2) = \frac{Mc}{2}$$

Thus

$$Q(2, M, T) = (qe^{-cw/2k_BT})^2 \left[\frac{Mc}{2} x^{2c-2} + \frac{M(M - c - 1)}{2} x^{2c} \right]$$

Combining terms

$$Q(2, M, T) = (qe^{-cw/2k_BT} x^c)^2 M \left(\frac{cx^{-2} + M - c - 1}{2} \right) \tag{15.179}$$

This establishes the grand partition function up to the quadratic term.

$$\Xi = 1 + (q\lambda e^{-cw/2k_BT}x^c)M + (q\lambda e^{-cw/2k_BT}x^c)^2 M$$
$$\times \left(\frac{cx^{-2} + M - c - 1}{2}\right) + \cdots \qquad (15.180)$$

In this form, the expression is somewhat cumbersome. To achieve a more compact expression, define an expansion parameter z as

$$z = qe^{-cw/2k_BT}x^c\lambda \qquad (15.181)$$

Then the grand partition function is

$$\Xi = 1 + Mz + Mz^2\frac{cx^{-2} + M - c - 1}{2} \qquad (15.182)$$

This can be written in the form

$$\Xi = 1 + M(z) \qquad (15.183)$$

The following depends on $M(z)$ being small. If it is,

$$\frac{\Phi M}{k_BT} = \ln\Xi = \ln(1 + M(z)) \cong M(z) \qquad (15.184)$$

neglecting quadratic and higher terms. The next term is $-\frac{1}{2}(M(z))^2 = -\frac{1}{2}M^2z^2 + \cdots$.

The surface pressure is the result, and is

$$\frac{\Phi M}{k_BT} = Mz + Mz^2\frac{cx^{-2} + M - c - 1}{2} - \frac{1}{2}M^2z^2 + \cdots \qquad (15.185)$$

Cancelling M reveals the compact expression

$$\frac{\Phi}{k_BT} = z + \frac{cx^{-2} - c - 1}{2}z^2 + \cdots \qquad (15.186)$$

Now, $x \equiv e^{w/2k_BT}$ by definition, so $x^c = e^{cw/2k_BT}$ and $e^{-cw/2k_BT}x^c = 1$, so $z = q\lambda e^{-cw/2k_BT}x^c = q\lambda$, where q is a function of T alone. Then

$$N = \left(\frac{\partial \ln\Xi}{\partial \ln\lambda}\right)_{T,M} = \left(\frac{\partial \ln\Xi}{\partial \ln z}\right)_{T,M} \qquad (15.187)$$

Since

$$\ln \Xi = \frac{\Phi M}{k_B T} = M z + M z^2 \frac{cx^{-2} - c - 1}{2} + \cdots \qquad (15.188)$$

the mean number of adsorbed molecules is

$$\langle N \rangle = M z + M z^2 (cx^{-2} - c - 1) + \cdots \qquad (15.189)$$

and as $\Theta = \frac{\langle N \rangle}{M}$, the fraction adsorbed is given as a power series in z

$$\Theta = z + z^2 (cx^{-2} - c - 1) + \cdots \qquad (15.190)$$

The next step is to invert the series and obtain z as a power series in Θ. This is done by first writing each as a general power series in the other

$$z = \Theta + A_2 \Theta^2 + A_3 \Theta^3 + \cdots$$
$$\Theta = z + B_2 z^2 + B_3 z^3 + \cdots$$

and substituting the first into the second of these equations. This yields

$$\Theta = (\Theta + A_2 \Theta^2 + A_3 \Theta^3 + \cdots) + B_2 (\Theta + A_2 \Theta^2 + A_3 \Theta^3 + \cdots)^2 + \cdots \qquad (15.191)$$

To quadratic terms, the result is

$$\Theta = \Theta + (A_2 + B_2) \Theta^2 + O(\Theta^3) \qquad (15.192)$$

where $O(\Theta^3)$ means "terms of cubic and higher order". Since a thing is always identically itself, the coefficient $A_2 + B_2$ must vanish. Similar restrictions apply to coefficients of higher order. Then B_2 must equal the negative of A_2, and up to cubic terms the desired series expansion is

$$z = \Theta + \Theta^2 (c + 1 - cx^{-2}) + \cdots \qquad (15.193)$$

The objective here is the second virial coefficient in the expansion of the surface pressure as a function of the fraction covered. Instead, the expansion of equation (15.186) is in z. Upon substituting the power series in the fraction covered for the expansion parameter z, the surface pressure is

$$\frac{\Phi}{k_B T} = \Theta + \frac{c + 1 - cx^{-2}}{2}\Theta^2 + \cdots \qquad (15.194)$$

The second virial coefficient is

$$B_2(T) = \frac{c + 1 - cx^{-2}}{2} = \frac{c + 1 - ce^{-w/k_B T}}{2} \qquad (15.195)$$

which appears to be exact, though this is of little value when coverages larger than nearly vanishingly small are encountered.

15.9. Quasichemical Approximation

The quasichemical approximation is occasionally useful and appears from time to time in the original literature. In the quasichemical approximation pairs of sites are treated as independent, rather than the sites themselves. Of course, this is not true, as pairs of sites overlap when a single site is common to two or more pairs. However, as long as the coverage is not great, the approximation offers a much better treatment than the Bragg-Williams at the cost of somewhat greater algebraic complexity. The mathematical burden is, however, manageable and results in surprisingly simple relationships. The notation used here is the same as in Hill's book, which set the standard in this and many other areas of statistical mechanics.

There are M sites on a lattice with c nearest neighbors for each site, neglecting edge effects, and N molecules adsorbed in a monolayer. Each site is either occupied, represented by a 1, or unoccupied, represented by a 0. There are then four different types of independent pairs of sites: 11, 10, 01, or 00. The total number of pairs of sites is $\frac{Mc}{2}$, and the number of pairs of sites of each type is calculated from the restrictive conditions. In the case of N_{11},

$$N_{11} = \frac{Nc}{2} - \frac{N_{01}}{2} \qquad (15.196)$$

which is just $Nc = 2N_{11} + N_{01}$ solved for N_{11}. Similarly, solving $(M - N)c = 2N_{00} + N_{01}$ for N_{00} yields

$$N_{00} = \frac{(M - N)c}{2} - \frac{N_{01}}{2} \tag{15.197}$$

Finally, there are as many pairs of the type 01 as there are of the type 10, and the total number must be N_{01}. Thus, the number of each must be $\frac{N_{01}}{2}$. The total number of ways of configuring the lattice for given N, M and N_{01} is then

$$\frac{\left(\frac{Mc}{2}\right)!}{\left(\frac{Nc}{2} - \frac{N_{01}}{2}\right)!\left(\frac{(M-N)c}{2} - \frac{N_{01}}{2}\right)!\left[\left(\frac{N_{01}}{2}\right)!\right]^2} = w \tag{15.198}$$

It would be nice if w were the same as $g(N, M, N_{01})$, but it clearly is not as it cannot satisfy the sum requirement

$$\sum_{N_{01}} g(N, M, N_{01}) = \frac{M!}{N!(M - N)!}$$

The reason is that w counts configurations that cannot logically occur. For example, consider the three contiguous sites abc, where ab is one nearest neighbor pair, bc is another, and both pairs have the common site b. One configuration that w counts has a occupied and b void while b is occupied and c is void. Site b cannot simultaneously be both occupied and void, yet w includes this impossible configuration. To rectify the over count, a correction factor is introduced in the form

$$g(N, M, N_{01}) = C(N, M)w(N, M, N_{01}) \tag{15.199}$$

for then

$$\sum_{N_{01}} g(N, M, N_{01}) = \frac{M!}{N!(M - N)!} = C(N, M) \sum_{N_{01}} w(N, M, N_{01}) \tag{15.200}$$

The sum of w over N_{01} is impossible to perform, but adsorption numbers are large and it is legitimate to replace the sum by its maximum term. To do this, maximize the logarithm instead of the typical term, making repeated use of the relation

$$\frac{\partial \ln x!}{\partial x} \cong \frac{\partial (x \ln x - x)}{\partial x} = \ln x$$

N and M are fixed in the differentiation, and the criterion

$$\frac{\partial \ln w}{\partial N_{01}} = 0$$

determines the maximum probability value of N_{01}^* as

$$\frac{N_{01}^*}{2} = \frac{cN(M - N)}{2M} \qquad (15.201)$$

If this is substituted into w and the logarithm taken, after lengthy but straightforward algebra, there results

$$\frac{1}{c} \ln w = M \ln M - N \ln N - (M - N) \ln(M - N) \qquad (15.202)$$

Taking the inverse of Stirling's approximation,

$$w(N, M, N_{01}^*) = \left(\frac{M!}{N!(M - N)!} \right)^c \qquad (15.203)$$

In order to satisfy the sum requirement, this requires $C(N, M)$ to be

$$C(N, M) = \left(\frac{M!}{N!(M - N)!} \right)^{1-c} \qquad (15.204)$$

From here, the path is the same as in the exact one dimensional case with interaction energy w, except that now $N_{01}^*/2M = \alpha$. The principal results are the relative volatility y and the equation of state. The volatility is

$$y = q e^{-cw/2k_B T} = \left(\frac{1 - \Theta}{\Theta} \right)^{c-1} \left(\frac{\Theta - \alpha}{1 - \Theta - \alpha} \right)^{c/2} \qquad (15.205)$$

Alternatively

$$y = \left[\frac{(\beta - 1 + 2\Theta)(1 - \Theta)}{(\beta + 1 - 2\Theta)\Theta} \right]^{c/2} \frac{\Theta}{1 - \Theta} \qquad (15.206)$$

The equation of state is

$$\frac{\Phi}{k_B T} = \ln\left\{ \left[\frac{(\beta+1)(1-\Theta)}{\beta+1-2\Theta} \right]^{c/2} \frac{1}{1-\Theta} \right\} \quad (15.207)$$

These functions behave qualitatively like the corresponding functions for the Bragg-Williams approximation. The equation of state gives a critical temperature at the critical coverage $\Theta_c = 1/2$,

$$e^{w/2k_B T_c} = \frac{c-2}{c} \quad (15.208)$$

This may be alternatively expressed as

$$\frac{cw}{k_B T_c} = 2c\ln\frac{c-2}{c} \quad (15.209)$$

By happenstance, if c is set equal to 2, these results are identical with those for the exact one dimensional adsorption problem. The critical temperature is zero as then $\frac{1}{k_B T_c} = \ln 0 = -\infty$. The square lattice ($c = 4$) was treated exactly for the case $\Theta = 1/2$ (the critical density) by Lars Onsager, and the critical temperature found by Kramers and Wannier[7] to result from the condition

$$e^{w/2k_B T_c} = \sqrt{2} - 1 \quad (15.210)$$

The Bragg-Williams treatment of the square lattice predicted a value of $\frac{4w}{k_B T_c} = -4$, the quasichemical approximation predicts $\frac{4w}{k_B T_c} = 2c\ln(\frac{c-2}{c}) = 8\ln(\frac{1}{2}) = -5.54$, while the exact square lattice has $\frac{4w}{k_B T_c} = 8\ln(\sqrt{2}-1) = -7.051$, fitting the quasichemical approximation between the Bragg-Williams result and the exact, demonstrating one aspect of the improved approximation afforded by the quasichemical approximation. The quasichemical approximation continues to appear occasionally in the literature, as various decorations or extensions are considered. For example, straight rod adsorption was considered by Pinto and co-workers.[8]

15.10. General Method for McGhee-von Hippel Adsorption

The last adsorption problem to be considered in this book is the McGhee-von Hippel adsorption problem.[9] McGhee and von Hippel were interested especially in the adsorption of proteins onto DNA, and derived their binding constant using a one dimensional substrate with M binding sites. Each bound r-mer occupies r consecutive sites on the substrate. When two r-mers occupy $2r$ consecutive sites, there is a potential energy w of interaction, but if the two r-mers do not occupy sequential sites and have one or more vacant sites between them, there is no potential energy of interaction. The McGhee-von Hippel equation was first derived by Zasedatelev, Gurskii and Volkenstein,[10] but is generally referred to as the McGhee-von Hippel equation since the thermodynamic and kinetic arguments they used to obtain it are regarded as more transparent than the combinatorial approach. The method used here is an extension of statistical methods due to Hill[11] that completely avoids combinatorial complexity.

The basic thermodynamics are familiar. The variations and relation between the chemical potential and the surface pressure are

$$dA = -SdT - \Phi dM + \mu d\langle N \rangle$$

$$d(\Phi M) = SdT + \Phi dM + \langle N \rangle d\mu$$

$$\mu \langle N \rangle = A + \Phi M$$

To form the grand partition function the sum over N is used

$$\Xi(M, T, \lambda) = e^{\Phi M / k_B T} = \sum_N Q(N, M, T) \lambda^N \qquad (15.211)$$

Nothing prevents summing again but this time over M producing a new partition function

$$\mathrm{T}(\Phi, T, \lambda) = \sum_M \Xi(M, T, \lambda) e^{-\Phi M / k_B T} \qquad (15.212)$$

For reasons of later mathematical convenience, define the functions γ and ϕ by

$$\gamma = \frac{1}{\varphi} = e^{\Phi / k_B T} \qquad (15.213)$$

The fraction of sites covered is $\Theta = \frac{\langle N \rangle}{M}$. The initial binding constant is

$$Kc = q\lambda = x \qquad (15.214)$$

where c is the concentration of ligand in solution, λ is its absolute activity and q is the effective partition function for the ligand bound to the substrate. The quantity x is defined by the relation. The strategy of the calculation that follows is to obtain $x(\gamma), \Theta(\gamma), \gamma(\Theta), x(\Theta) = Kc$. The latter is the McGhee-von Hippel equation.

First, it is necessary to establish an alternative method for certain very large systems. The method is due to Hill.[11] Consider a chain of M identical adsorption sites on which some ligands are adsorbed. Each site is either occupied, in state 2, or free, in state 1. One configuration might be

$$\dots 111221111121211111122222 \dots$$

Let N_{12} and N_{21} be the total number of boundaries of type 12 and 21 respectively and let N' be the total number of boundaries between 1 and 2 regions. Clearly, either $N_{12} = N_{21}$ or these are 1 different.

Now consider a substrate of M sites with N adsorbed ligands but with adsorption restricted to exactly N' boundaries. In a real system with N and M fixed, N' would fluctuate if the N ligands were mobile, or if desorption and adsorption occurred at the same rate to maintain N constant. The Helmholz free energy varies as

$$dA = -SdT - \Phi dM + \mu dN + \mu' dN' \tag{15.215}$$

where N' appears as a new extensive variable. Integrating produces

$$A = -\Phi M + \mu N + \mu' N' \tag{15.216}$$

The partition function T is related to the N' variable as

$$\mu' N' = -k_B T \ln T(\Phi, \mu, T, N') \tag{15.217}$$

The fact that the partition function T is summed over both N and M means that these quantities are not fixed but fluctuate, requiring the averages to be used in the thermodynamic equations. In a long chain with N' boundaries, there will be $N'/2$ regions of type 1 and $N'/2$ regions of type 2. Each type 1 sequence can fluctuate in size from $n_1 = 1$ to N as N approaches infinity, and similarly for regions of type 2. Now let ψ_1 and ψ_2 represent partition functions for single sequences of the two types. Note that n_1 and n_2 fluctuate. Let w_{12} represent a potential energy introduced at each boundary, and define $y_{12} = e^{-w_{12}/k_B T}$ and write

$$e^{-\mu' N'/k_B T} = \psi_1^{N'/2} \psi_2^{N'/2} y_{12}^{N'/2} \tag{15.218}$$

Eliminating N' and squaring results in the simplified relation

$$e^{-2\mu'/k_B T} = \psi_1 \psi_2 y_{12}^2 \tag{15.219}$$

This equation is a relation among the intensive variables of the system, specifically $T\Phi, \mu, \mu'$. At this point, the system is not at equilibrium due to the restraint on N', that the number of boundaries it counts are specified and remain constant. Remove the restraint. The system will then relax toward equilibrium and the Helmholz free energy will seek a minimum value, defined by

$$\left(\frac{\partial A}{\partial N'}\right)_{T,M,N} = 0 \tag{15.220}$$

As $dA = -SdT - \Phi dM + \mu dN + \mu' dN'$, the equilibrium condition requires $\mu' = 0$ and therefore

$$1 = \psi_1 \psi_2 y_{12}^2 \tag{15.221}$$

which is a relation between the intensive variables T, Φ, μ. The functions ψ_1 and ψ_2 are called the sequence partition functions.

The fundamental statistical mechanics has

$$e^{\Phi M/k_B T} = \Xi(T, M, \mu) = \sum_N Q(T, M, N) \lambda^N$$

Further summing over M gives the partition function $T = \sum_M \sum_N Q(T, M, N) \lambda^N e^{-\Phi M/k_B T}$. Summation over N' in $T(N')$ would also give the partition function T. This is the partition function for a completely open system with fluctuations in both N and M, and with independent intensive variables T, Φ, μ. The form of the equation suggests that each of the M sub-units contributes a factor $e^{-\Phi/k_B T} = \varphi$ to each term of T, and each bound ligand contributes a factor $q = K_0 c = x_0$. Each neighboring pair ij contributes y_{ij} as a factor. Both x_0 and y_{ij} appear in the grand partition function Ξ. The new feature in T is φ. The sequence partition functions are

$$\psi_1 = \sum_{n_1=1}^{\infty} \varphi^{n_1} y_{11}^{n_1-1} = \frac{\varphi}{1 - y_{11}\varphi} \tag{15.222}$$

$$\psi_2 = \sum_{n_2=1}^{\infty} \varphi^{n_2} x_0^{n_2} y_{22}^{n_2-1} = \frac{\varphi x_0}{1 - y_{22}\varphi x_0} \tag{15.223}$$

A sequence of n_1 subunits can only have $n_1 - 1$ pairs of subunits ...11..., and similarly for n_2. The mean number of subunits of each type is

$$\langle n_i \rangle = \frac{\partial \ln \psi_i}{\partial \ln \varphi} \tag{15.224}$$

Now, the equilibrium condition is $1 = \psi_1 \psi_2 y_{12}^2$, hence

$$1 = \frac{\varphi}{1 - y_{11}\varphi} \frac{\varphi x_0}{1 - y_{22}\varphi x_0} y_{12}^2 = \frac{\varphi^2 x_0 y_{12}^2}{(1 - y_{11}\varphi)(1 - y_{22}\varphi x_0)} \tag{15.225}$$

This equation is quadratic in φ, which can be solved. Since by definition $\gamma = \frac{1}{\varphi} = e^{\Phi/k_B T}$, the equation of state is

$$\Phi = -k_B T \ln \varphi = k_B T \ln \gamma \tag{15.226}$$

In order for the sums to converge, the criteria $y_{11}\varphi < 1, y_{22}x_0\varphi < 1$ must be met. Selection of the correct sign in the solution of the quadratic for φ will assure this. This completes the outline of the general method, which accommodates greater flexibility than is needed in treating the problem of the McGhee-von Hippel binding constant.

15.11. McGhee-von Hippel Adsorption

The method may be applied whenever the microscopic system consists of identical repeats, each repeat consists of one completely open sequence of each type (occupant, void; type A, type B; etc.) plus two boundaries. As a direct consequence of these conditions, the intensive properties of the macroscopic system can be deduced from analysis of a single unit.

In the McGhee–von Hippel–Zasedatelev problem, $y_{11} = y_{22} = 1$ and each adsorbed subsystem occupies r sequential sites. Then the sequence partition functions become

$$\psi_1 = \sum_{n_1=1}^{\infty} \varphi^{n_1} = \frac{\varphi}{1-\varphi} \tag{15.227}$$

$$\psi_2 = \sum_{n_2=1}^{\infty} \varphi^{rn_2} x^{n_2} y^{n_2-1} = \frac{\varphi^r x}{1 - xy\varphi^r} \tag{15.228}$$

The mean numbers of void and occupied sites are

$$\langle n_1 \rangle = \frac{\partial \ln \psi_1}{\partial \ln \varphi} = \frac{1}{1-\varphi} = \frac{1}{\gamma - 1} \tag{15.229}$$

$$\langle n_2 \rangle = \frac{\partial \ln \psi_2}{\partial \ln \varphi} = \frac{\gamma^r}{\gamma^r - xy} \tag{15.230}$$

$$1 = \psi_1 \psi_2 = \frac{x}{(\gamma - 1)(\gamma^r - xy)} \tag{15.231}$$

The first main result is found by solving equation (15.231) for x. When this is done, the result is

$$x = \frac{(\gamma - 1)^r}{y\gamma - y + 1} \tag{15.232}$$

which is the function $x(\gamma)$. Further, the mean values of the numbers of void and occupied sites are given above as functions of γ, which allows extraction of the value of the fraction covered Θ as

$$r\Theta(\gamma) = \frac{r\langle n_2 \rangle}{r\langle n_2 \rangle + \langle n_1 \rangle} \tag{15.233}$$

In terms of the ratio of the mean numbers of occupied and void sequences

$$\Theta = \frac{\langle n_2 \rangle / \langle n_1 \rangle}{1 + r\langle n_2 \rangle / \langle n_1 \rangle} \tag{15.234}$$

Solving this for the ratio $\langle n_2 \rangle / \langle n_1 \rangle$ yields the remarkably simple result

$$\langle n_2 \rangle / \langle n_1 \rangle = \frac{\Theta}{1 - r\Theta} \tag{15.235}$$

Considered as a function of γ, the ratio of numbers of void and occupied sites is

$$\langle n_2 \rangle / \langle n_1 \rangle = \frac{\gamma^r}{\gamma^r - xy} \frac{\gamma - 1}{\gamma} \tag{15.236}$$

$x = \frac{(\gamma - 1)\gamma^r}{y\gamma - y + 1}$ has to be substituted to get the ratio as a function of γ and y. The result is

$$\langle n_2 \rangle / \langle n_1 \rangle = \frac{(\gamma - 1)(y\gamma - y + 1)}{\gamma} \tag{15.237}$$

This gives Θ as a function of γ

$$\Theta = \frac{\langle n_2 \rangle / \langle n_1 \rangle}{1 + r\langle n_2 \rangle / \langle n_1 \rangle} = \frac{(\gamma - 1)(y\gamma - y + 1)}{r(\gamma - 1)(y\gamma - y + 1) + \gamma} \tag{15.238}$$

By inverting this, the function $\gamma(\Theta)$ can be obtained. The equation is quadratic in γ, and when solved for γ yields

$$\gamma = \frac{(2y - 1)(1 - r\Theta) + \Theta + R}{2y(1 - r\Theta)} \tag{15.239}$$

where

$$R = \sqrt{[1 - (r + 1)\Theta]^2 + 4y\Theta(1 - r\Theta)} \tag{15.240}$$

Last, solve for $x(\Theta)$. Since $\gamma - 1 = \frac{\Theta\gamma}{(1 - r\Theta)(y\gamma - y + 1)}$ (from the ratio $\langle n_2 \rangle / \langle n_2 \rangle$ above),

$$x = \frac{\gamma^r(\gamma - 1)}{y\gamma - y + 1} = \frac{\gamma^r \Theta\gamma}{(1 - r\Theta)(y\gamma - y + 1)^2} = \frac{\Theta\gamma^{r-1}}{1 - r\Theta}\left(\frac{\gamma}{y\gamma - y + 1}\right)^2 \tag{15.241}$$

Substituting γ produces the McGhee–von Hippel binding constant expression Kc

$$Kc = x(\Theta) = \frac{\Theta}{1 - r\Theta}\left[\frac{2(y - 1)(1 - r\Theta)}{(2y - 1)(1 - r\Theta) + \Theta - R}\right]^{r-1}$$

$$\times \left[\frac{2(1 - r\Theta)}{1 - (r + 1)\Theta + R}\right]^2 \tag{15.242}$$

Appendix

A good deal of the algebraic detail involved in the Zasedatelev-Gurskii-Volkenstein-McGhee-von Hippel binding constant was omitted to save space. Much of the omitted algebra is provided here, since some of it is not immediately transparent.

In the text, it was shown that the ratio of mean numbers of occupied to unoccupied sites is

$$\frac{\langle n_2 \rangle}{\langle n_1 \rangle} = \frac{\Theta}{1 - r\Theta} = \frac{\gamma^r(\gamma - 1)}{(\gamma^r - xy)}$$

and it was asserted that this ratio is also

$$\frac{\langle n_2 \rangle}{\langle n_1 \rangle} = \frac{(y\gamma - y + 1)(\gamma - 1)}{\gamma}$$

This result follows from $x = \frac{\gamma^r(\gamma-1)}{y\gamma-y+1}$. First, rewrite the ratio as a product of two ratios

$$\frac{\langle n_2 \rangle}{\langle n_1 \rangle} = \frac{1}{1 - \frac{y}{\gamma^r}x}\frac{\gamma - 1}{\gamma} = \frac{1}{1 - \frac{y}{\gamma^r}\frac{\gamma^r(\gamma-1)}{y\gamma-y+1}}\frac{\gamma - 1}{\gamma}$$

and note that γ^r cancels. The ratio becomes

$$\frac{\langle n_2 \rangle}{\langle n_1 \rangle} = \frac{y\gamma - y + 1}{y\gamma - y + 1 - y(\gamma - 1)}\frac{\gamma - 1}{\gamma} = \frac{(y\gamma - y + 1)(\gamma - 1)}{\gamma}$$

which is the result stated. From the two relations for the ratio, namely

$$\frac{\langle n_2 \rangle}{\langle n_1 \rangle} = \frac{\Theta}{1 - r\Theta} = \frac{(y\gamma - y + 1)(\gamma - 1)}{\gamma}$$

it is possible to extract γ as a function of Θ. Proceed as follows.

$$\gamma\Theta = (1 - r\Theta)(y\gamma - y + 1)(\gamma - 1)$$

Multiplying this out

$$\gamma^2\left[y(1 - r\Theta)\right] + \gamma\left[(1 - 2y)(1 - r\Theta) - \Theta\right] + \left[(y - 1)(1 - r\Theta)\right] = 0$$

Evidently γ is double valued, and is

$$\gamma = \frac{-\left[(1 - 2y)(1 - r\Theta) - \Theta\right] \pm R}{2y(1 - r\Theta)}$$

where
$$R = \sqrt{[(1 - 2y)(1 - r\Theta) - \Theta]^2 - 4[y(1 - r\Theta)][(y - 1)(1 - r\Theta)]}.$$ The
simplification of R to the final expression is left as an exercise for the
reader. The result is

$$R = \sqrt{[1 - (r + 1)\Theta]^2 + 4y\Theta(1 - r\Theta)}$$

As it stands, the radical R appears in the numerator. It may also be
placed in the denominator. Choose the $+$ sign for the radical. The
quantity γ is

$$\gamma = \frac{(2y - 1)(1 - r\Theta) + \Theta + R}{2y(1 - r\Theta)}$$

Then define a temporary quantity A as

$$A = (2y - 1)(1 - r\Theta) + \Theta$$

Then the denominator may be rewritten as

$$2y(1 - r\Theta) = (2y - 1)(1 - r\Theta) + 1 - r\Theta + \Theta - \Theta$$
$$= A + 1 - r\Theta - \Theta$$
$$= A + [1 - (r + 1)\Theta]$$

Now set $B = 1 - (r + 1)\Theta$. The function γ becomes

$$= \frac{A + R}{A + B} = \frac{z}{A - R}$$

where z is a function yet to be determined. Note that $A + B = 2y(1 - r\Theta)$ and that

$$R^2 = B^2 + 4y\Theta(1 - r\Theta)$$

Also using $z = \frac{A^2 - R^2}{A + B}$

$$z = \frac{A^2 - B^2 - 4y\Theta(1 - r\Theta)}{A + B}$$
$$= \frac{(A + B)(A - B)}{A + B} - \frac{4y\Theta(1 - r\Theta)}{2y(1 - r\Theta)}$$

since $A + B = 2y(1 - r\Theta)$. This reduces to

$$z = A - B - 2\Theta$$

Reintroducing the definitions of A and B, z is

$$z = (2y - 1)(1 - r\Theta) + \Theta - [1 - (r + 1)\Theta] - 2\Theta$$

After a little straightforward algebra, one finds

$$z = 2(y - 1)(1 - r\Theta)$$

and the alternate form is found by substituting A and z into $\frac{z}{A-R}$, giving

$$\gamma = \frac{z}{A - R} = \frac{2(y - 1)(1 - r\Theta)}{(2y - 1)(1 - r\Theta) + \Theta - R}$$

It was shown above that $x(\gamma)$ is known, these results may be used to get $x(\Theta)$, which is the final result. as $Kc = x(\Theta)$. Indeed, $x(\gamma)$ is

$$x(\gamma) = \frac{\gamma^r(\gamma - 1)}{y\gamma - y + 1}$$

From the above work

$$\frac{\langle n_2 \rangle}{\langle n_1 \rangle} = \frac{\Theta}{1 - r\Theta} = \frac{(\gamma - 1)(y\gamma - y + 1)}{\gamma}$$

and from this

$$\gamma - 1 = \frac{\Theta}{1 - r\Theta} \frac{\gamma}{y\gamma - y + 1}$$

and

$$x = \frac{\gamma^r}{y\gamma - y + 1} \frac{\Theta}{1 - r\Theta} \frac{\gamma}{y\gamma - y + 1}$$

$$x = \frac{\gamma^{r-1}\Theta}{1 - r\Theta} \left[\frac{\gamma}{y\gamma - y + 1} \right]^2$$

At this point, the job is half done as the above work showed that $\gamma = \frac{2(y-1)(1-r\Theta)}{(2y-1)(1-r\Theta)+\Theta-R}$. In principle, that could just be substituted into the expression for x. However, doing so leads to a less than compact expression and the following procedure leads to the conventional expression. Note that x occurs as two factors, the factor that is squared and the one that is not. This suggests that the quantity being squared should be investigated. Note that

$$\frac{\gamma}{y\gamma - y + 1} = \frac{\gamma}{(\gamma - 1)y + 1}$$

Rather than work with the full expression, work with A and B.

$$\gamma = \frac{A + R}{A + B}$$

$$\gamma - 1 = \frac{A + R}{A + B} - 1 = \frac{R - B}{A + B}$$

$$\frac{\gamma}{y\gamma - y + 1} = \left(\frac{A + R}{A + B}\right)\left(\frac{1}{y\frac{R-B}{A+B} + 1}\right)$$

$$= \frac{(A + R)(A + B)}{(A + B)(y(R - B) + A + B)}$$

$$= \frac{A + R}{A + B}\left(\frac{A + B}{yR - y[1 - (r + 1)\Theta] + 2y(1 - r\Theta)}\right)$$

$$= \frac{A + R}{A + B}\left(\frac{A + B}{y(R + 1 - r\Theta + \Theta)}\right)$$

$$= \frac{A + R}{A + B}\left(\frac{A + B}{y(R + B + 2\Theta)}\right)$$

$$= \frac{A + R + B - B}{A + B}\left(\frac{A + B}{y(R + B + 2\Theta)}\right)$$

$$= \frac{R - B + A + B}{A + B}\left(\frac{A + B}{y(R + B + 2\Theta)}\right)$$

$$= \frac{2\Theta\frac{A+B}{R+B} + A + B}{A + B}\left(\frac{A + B}{y(R + B + 2\Theta)}\right)$$

$$= \frac{R + B + 2\Theta}{R + B}\left(\frac{A + B}{y(R + B + 2\Theta)}\right)$$

$$= \frac{A+B}{y(R+B)}$$

$$= \frac{2(1-r\Theta)}{1-(r+1)\Theta+R}$$

This algebra shows that

$$\frac{\gamma}{y\gamma-y+1} = \frac{2(1-r\Theta)}{1-(r+1)\Theta+R}$$

Since

$$Kc = x = \frac{\gamma^{r-1}\Theta}{1-r\Theta}\left[\frac{\gamma}{y\gamma-y+1}\right]^2$$

the final result is the Zasedatelev-Gurskii-Volkenstein-McGhee-von Hippel binding constant expression.

$$Kc = \frac{\gamma^{r-1}\Theta}{1-r\Theta}\left[\frac{2(1-r\Theta)}{1-(r+1)\Theta+R}\right]^2$$

References

1. D. P. Shoemaker, C. W. Garland, and J. L. Steinfeld, "Experiments in Physical Chemistry," Ed. 3, McGraw-Hill, New York, NY, 1974, p. 364. This experiment does not appear in Edition 7 of this series.
2. J. W. Gibbs, "On the Equilibrium of Heterogeneous Substances," Collected Works, Vol. I, Yale Univ. Press, New Haven, Ct. 1950.
3. S. Brunauer, P. H. Emmett and E. Teller, J. Amer. Chem. Soc. 60, 309 (1938).
4. C. W. Garland, J. W. Nibler and D. P. Shoemaker, "Experiments in Physical Chemistry," Ed. 7, McGraw-Hill, New York, NY, 2003, p. 301.
5. T. L. Hill, J. Chem. Phys. **14**, 263 (1946).
6. Mayer and Mayer, p. 438.
7. H. A. Kramers and G. H. Wannier, Phys. Rev. **60**, 252, 263 (1941).
8. O. A. Pinto, F. Nieto and A. J. Ramirez-Pastor, Phys. Rev. E, **84**, 061142 (2011).
9. J. D. McGhee and P. H. von Hippel, J. Mol. Biol. **86**, 469 (1974).
10. A. S. Zasedatelev, G. V. Gurskii and M. V. Volkenstein, Molek. Biol. 5, 245 (1971).
11. T. L. Hill, Proc. Nat. Acad. Sci. USA **69**, 1165 (1972).

Chapter 16

Heat Capacities of Solids

16.1. Einstein's Heat Capacity

In the early 1900's, the allowed and disallowed regions of the energy were unknown. Schrodinger would not write his famous equation until twenty years later, and the vibrational energy

$$E_n = h\nu \left(n + \frac{1}{2} \right) \quad n = 0, 1, 2 \cdots \infty \tag{16.1}$$

was as yet unknown. However, Max Planck had published his analysis of black body radiation. One of the features of Planck's theory was the assumption that the walls of the black body cavity were covered with oscillators with energies

$$\varepsilon_n = nh\nu \tag{16.2}$$

where n is zero or a positive integer. In the end of the analysis, Planck intended initially to pass to the limit of vanishing h in order to recover the classical energy continuum. However, if he did not do so, his black body radiation curve matched experiment, and if he passed to the continuous limit it suffered the ultraviolet catastrophe. Apparently, then, the integer energy structure had to be retained. At this time X-ray crystallography had not been invented and there was no structural information available for solids. The one intriguing fact was that the law of Dulong and Petit seemed to be universal, that the heat capacity of metallic solids were uniformly $3R$ per mole at sufficiently high temperature.

Einstein had available to him the Planck oscillator energy and the Maxwell-Boltzmann energy distribution, that the probability a given molecule in a gas have an energy ε was proportional to the exponential $e^{-\varepsilon/k_B T}$. The average energy per molecule is given by ordinary statistics as

$$\langle \varepsilon \rangle = \sum_n P_n \varepsilon_n \tag{16.3}$$

where the probability P_n is given by

$$P_n = C e^{-\varepsilon_n/k_B T} \tag{16.4}$$

In common with all probabilities,

$$\sum_n P_n = 1 \tag{16.5}$$

The constant C is evaluated from this, and is

$$C = \frac{1}{\sum_n e^{-\varepsilon_n/k_B T}} \tag{16.6}$$

The average energy is now

$$\langle \varepsilon \rangle = \frac{\sum_n \varepsilon_n e^{-\varepsilon_n/k_B T}}{\sum_n e^{-\varepsilon_n/k_B T}} \tag{16.7}$$

Presumably, at this point Einstein must have assumed that vibrational energy among the oscillators in the solid was distributed the same way kinetic energy is distributed among the molecules of a gas. Then the average energy of one oscillator is given by

$$\langle \varepsilon \rangle = \frac{\sum_{n=0}^{\infty} nh\nu e^{-nh\nu/k_B T}}{\sum_{n=0}^{\infty} e^{-nh\nu/k_B T}} \tag{16.8}$$

and the energy of the solid is $3N$ times as much

$$\langle E \rangle = 3N \langle \varepsilon \rangle \tag{16.9}$$

Provided that $e^{-h\nu/k_B T} < 1$ the series are immediately summable. The series in the denominator is of the form

$$\sum_{n=0}^{\infty} z^n = \frac{1}{1-z} \tag{16.10}$$

even in the field of complex numbers. The sum is uniformly convergent and possesses derivatives of all orders. The first derivative is

$$\sum_{n=1}^{\infty} n z^{n-1} = \frac{1}{(1-z)^2} \tag{16.11}$$

Multiplication by z gives the series in the numerator of the expression for the average energy of the solid. Then the average energy is

$$\langle E \rangle = 3Nh\nu(1 - e^{-h\nu/k_BT})\frac{e^{-h\nu/k_BT}}{(1 - e^{-h\nu/k_BT})^2} = 3Nh\nu\frac{1}{e^{h\nu/k_BT} - 1} \tag{16.12}$$

The derivative with respect to temperature is the heat capacity

$$C_v = \frac{\partial \langle E \rangle}{\partial T} = 3Nh\nu\frac{e^{h\nu/k_BT}}{(e^{h\nu/k_BT} - 1)^2}\frac{h\nu}{k_BT^2} \tag{16.13}$$

which simplifies to

$$C_v = 3Nk_B \left(\frac{h\nu}{k_BT}\right)^2 \frac{e^{h\nu/k_BT}}{(e^{h\nu/k_BT} - 1)^2} \tag{16.14}$$

Of course, no one knows what led Einstein to this result. The point is that he could have followed the above path of logic. The function is frequently used and as a result is often expressed in either of two alternate forms. One of these is

$$C_v = 3Nk_B \left(\frac{h\nu}{k_BT}\right)^2 \frac{e^{-h\nu/k_BT}}{(1 - e^{-h\nu/k_BT})^2} \tag{16.15}$$

The other expression depends on two of the several definitions of hyperbolic functions

$$\sinh x \equiv \frac{1}{2}(e^x - e^{-x}) \tag{16.16}$$

$$\mathrm{csch\,} x = \frac{1}{\sinh x} \tag{16.17}$$

These allow writing the symmetric form

$$C_v = 3Nk_B \left(\frac{h\nu}{k_BT}\right)^2 \mathrm{csch}^2\left(\frac{h\nu}{k_BT}\right) \tag{16.18}$$

Whichever form is used, the function vanishes at zero temperature and has an upper limit of unity as T grows indefinitely

$$\lim_{T \to \infty} \left[\left(\frac{h\nu}{k_B T} \right)^2 \text{csch}^2 \left(\frac{h\nu}{k_B T} \right) \right] = 1 \qquad (16.19)$$

If N is taken as Avogadro's number, the high temperature limit of the heat capacity is evidently then $3Nk_B = 3R$, which is the rule of Dulong and Petit.

Einstein could not have used the canonical ensemble to derive the heat capacity, for the first exposition of the ensemble was published in the Transactions of the Connecticut Society by J. Willard Gibbs at about the same time Einstein was working, around 1905. In more recent practice, the route to the heat capacity starts with the canonical partition function

$$Q = q^N \qquad (16.20)$$

where

$$q = q_1 q_2 q_3 \qquad (16.21)$$

with

$$q_1 = q_2 = q_3 = \sum_{v=0}^{\infty} e^{-(v+1/2)h\nu/k_B T} = \frac{e^{-h\nu/2k_B T}}{1 - e^{-h\nu/k_B T}} \qquad (16.22)$$

This approach gives the same specific heat, but an average energy that differs by the inclusion of the temperature independent zero point energy. A plot of the Einstein function for a single frequency is shown in Figure 16.1.

The first improvement in the specific heat was provided a few years later by P. Debye. The essence of Einstein's approximation was the use of a single frequency to model whatever complexity the true vibrational structure of a solid presented. The number of frequencies in unit frequency interval is the frequency distribution function, $g(\nu)$. For Einstein, $g(\nu)$ is a Dirac delta function. Debye noted that the solid should sustain a range of low frequency vibrations, much as a string of $N + 1$ identical masses m held together by N springs of

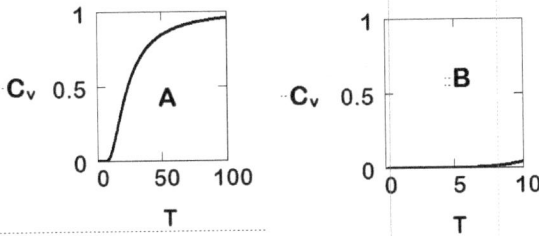

Fig. 16.1. Einstein heat capacity for a single oscillator. The typical values of lattice frequencies lie between a few cm^{-1} and about 100 cm^{-1}. Curve A is for 50 cm^{-1}, and curve B is the same but from 0 to 10 K, illustrating the essentially vanishing heat capacity in the range of temperatures near absolute zero. Experimental curves do not display these long regions of nearly zero specific heat. The Einstein function approaches zero temperature with zero slope.

Hooke's law force constant f will sustain N vibrations of different frequencies lying between 0 and some maximum value. To illustrate this point, the vibrations and heat capacity of the linear chain is treated in the next section.

16.2. The Linear Chain

The problem is simple enough to admit of an exact treatment, rendered more facile by algebraic developments made much later in the mid to late 1930's and not known to Debye. The objective in this treatment is to obtain an improved frequency distribution function, $g(\nu)$.

The essential quantity in this approach is the secular determinant, which in general is

$$|GF - \lambda E| = 0 \qquad (16.23)$$

where $\lambda = \omega^2 = 4\pi^2\nu^2$ and ν is the frequency in inverse seconds. F is the force constant matrix and G is the inverse kinetic energy matrix, the entries for which are described in detail in Wilson, Decius and Cross.[1]

The process is illustrated for a chain of 3 masses. The potential energy for nearest neighbor interactions only may be taken as

$$V = \frac{f}{2}\left[(x_2 - x_1)^2 + (x_3 - x_2)^2\right] \qquad (16.24)$$

where f is the Hooke's law force constant and the masses, numbered sequentially, have instantaneous very small displacements x_i along the line of centers of mass. When these coordinates are used to describe the problem, the non-genuine vibration, a translation of the entire chain along the x axis, is included. To eliminate this motion of zero frequency, internal coordinates S_i are defined in general by the matrix equation

$$S = Bx \qquad (16.25)$$

where the matrix B contains information pertaining to the geometry but not the masses of the molecule. In the three mass case, the internal coordinates are

$$S_1 = x_2 - x_1$$

$$S_2 = x_3 - x_2 \qquad (16.26)$$

If x is defined as a column vector with sequential entries x_1, x_2, x_3 and S is a column vector with the entries S_1, S_2, the B matrix is defined by writing

$$\begin{pmatrix} S_1 \\ S_2 \end{pmatrix} = \begin{pmatrix} B_{11} & B_{12} & B_{13} \\ B_{21} & B_{22} & B_{23} \end{pmatrix} \begin{pmatrix} x_1 \\ x_2 \\ x_3 \end{pmatrix} = \begin{pmatrix} -1 & 1 & 0 \\ 0 & -1 & 1 \end{pmatrix} \begin{pmatrix} x_1 \\ x_2 \\ x_3 \end{pmatrix}$$

$$(16.27)$$

The G matrix elements are given in general by

$$G_{t,t'} = \sum_{i=1}^{3} \frac{1}{m_i} B_{t,i} B_{t',i} \qquad (16.28)$$

The G matrix for the three mass chain is

$$G = \begin{pmatrix} \dfrac{2}{m} & -\dfrac{1}{m} \\ -\dfrac{1}{m} & \dfrac{2}{m} \end{pmatrix} \qquad (16.29)$$

In matrix notation, the potential energy is

$$V = S^T F S \qquad (16.30)$$

where for the three mass case at hand

$$F = \begin{pmatrix} f & 0 \\ 0 & f \end{pmatrix} \tag{16.31}$$

and the GF matrix product is

$$GF = \begin{pmatrix} \dfrac{2}{m} & -\dfrac{1}{m} \\ -\dfrac{1}{m} & \dfrac{2}{m} \end{pmatrix} \begin{pmatrix} f & 0 \\ 0 & f \end{pmatrix} = \begin{pmatrix} \dfrac{2f}{m} & -\dfrac{f}{m} \\ -\dfrac{f}{m} & \dfrac{2f}{m} \end{pmatrix} \tag{16.32}$$

The secular determinant is

$$\begin{vmatrix} \dfrac{2f}{m} - \lambda & -\dfrac{f}{m} \\ -\dfrac{f}{m} & \dfrac{2f}{m} - \lambda \end{vmatrix} = 0 \tag{16.33}$$

Now multiply the determinant by $(\frac{m}{f})^2$, and define $x \equiv 2 - \frac{m}{f}\lambda$. The secular determinant becomes

$$D_2 = \begin{vmatrix} x & -1 \\ -1 & x \end{vmatrix} = 0 \tag{16.34}$$

The roots are obtained by expanding the determinant, $x^2 - 1 = 0$, or $x = \pm 1$. Then $\lambda = \frac{f}{m}, 3\frac{f}{m}$. These are the squared angular frequencies of the genuine vibrations. Incidentally, the chain of 2 masses has an equivalent secular determinant

$$D_1 = |x| = x = 0 \tag{16.35}$$

which yields the diatomic frequency expression $\lambda = 4\pi^2\nu^2 = 2\frac{f}{m}$, or $\nu = \frac{1}{2\pi}\sqrt{\frac{2f}{m}}$, a familiar result. The secular determinant for a chain of 4 masses and 3 bonds is

$$D_3 = \begin{vmatrix} x & -1 & 0 \\ -1 & x & -1 \\ 0 & -1 & x \end{vmatrix} = 0 \tag{16.36}$$

and so on. The general case of N bonds is

$$
D_N = \begin{vmatrix}
x & -1 & & & & \\
-1 & x & -1 & & & \\
& -1 & x & -1 & & \\
& & \ddots & \ddots & \ddots & \\
& & & -1 & x & -1 \\
& & & & -1 & x
\end{vmatrix} = 0 \qquad (16.37)
$$

all entries other than those displayed are zero. The determinant D_N is $N \times N$, and may be expanded in minors

$$
D_N = x D_{N-1} - (-1) \begin{vmatrix}
-1 & -1 & 0 & 0 & 0 & 0 \\
0 & x & -1 & 0 & 0 & 0 \\
0 & -1 & x & -1 & 0 & 0 \\
0 & 0 & -1 & x & -1 & 0 \\
0 & 0 & 0 & -1 & x & -1 \\
\vdots & \vdots & \vdots & \vdots & \vdots & \vdots
\end{vmatrix} \qquad (16.38)
$$

and since further expansion by minors of the large determinant yields only D_{N-2}, the recursion relation ensues

$$
D_N = D_1 D_{N-1} - D_{N-2} \qquad (16.39)
$$

This relation is also obeyed by the Tschebycheff polynomials, defined by

$$
C_n(\varphi) \equiv \frac{\sin((n+1)\varphi)}{\sin(\varphi)} \qquad (16.40)
$$

These polynomials have values determined by the basic trigonometric relation

$$
\sin(a+b) = \sin(a)\cos(b) + \cos(a)\sin(b) \qquad (16.41)
$$

The first few are

$$
C_0 = \frac{\sin(\varphi)}{\sin(\varphi)} = 1 \qquad (16.42)
$$

$$C_1 = \frac{\sin(2\varphi)}{\sin(\varphi)} = 2\cos(\varphi) \tag{16.43}$$

$$C_2 = \frac{\sin(3\varphi)}{\sin(\varphi)} = 4cos^2(\varphi) - 1 \tag{16.44}$$

The general recursion relation obeyed by the Tschebycheff polynomials is

$$C_n = C_1 C_{n-1} - C_{n-2} \tag{16.45}$$

as may be verified for C_2, C_1, C_0 given above. Thus, identifying D_N with C_n, the condition $D_N = 0$ amounts to requiring

$$(N+1)\varphi = l\pi \quad l = 1, 2, 3 \cdots N \tag{16.46}$$

To obtain the roots of the secular determinant, there is no need to expand $D_N = 0$. It suffices to consider D_1. This is

$$D_1 = 2\cos(\varphi) = x = 2\cos\left(\frac{l\pi}{N+1}\right) \tag{16.47}$$

and, as $x = 2 - \frac{m}{f}\lambda$,

$$\lambda_l = 2\frac{f}{m}\left(1 + \cos\left(\frac{l\pi}{N+1}\right)\right) \quad l = 1, 2, 3 \ldots N \tag{16.48}$$

Note that the frequencies are determined by the diatomic molecule expression, modified by a strictly mathematical function with an upper limit ($l = 1$) and a lower limit ($l = N$). The diatomic frequency occurs when $\frac{l\pi}{N+1} = \frac{\pi}{2}$, or $l = \frac{N+1}{2}$. Since this must be an integer, the diatomic frequency occurs in only half of the possible chains, those which have an even number of masses. Although it is used in chains with an odd number of masses, it never appears in the frequencies of the chain. Finally, the frequencies of the chain form a branch called the longitudinal acoustic branch, with frequencies given by

$$\nu(l) = \frac{1}{2\pi}\sqrt{2\frac{f}{m}}\sqrt{1 + \cos\left(\frac{l\pi}{N+1}\right)} \tag{16.49}$$

where N is the number of bonds and $N+1$ is the number of masses. This function is plotted for a chain of 100 masses in Figure 16.2.

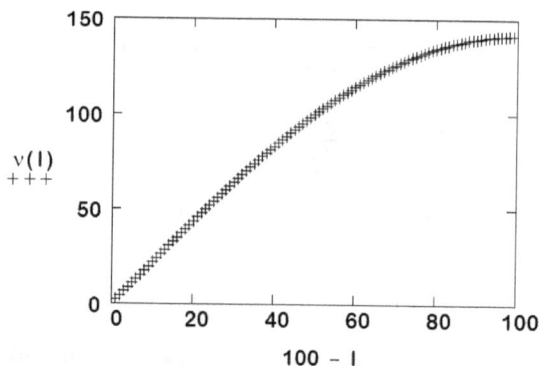

Fig. 16.2.   Longitudinal acoustic frequencies for a chain of 100 identical masses and diatomic frequency taken as 100 cm^{-1}. The plot is of frequency versus $100-l$, in order to turn the plot around to the more customary plot of this branch. The zero frequency is omitted, and individual frequencies are plotted as crosses.

The frequency expression is more conventionally expressed otherwise, but it is relatively easy to convert to the more compact form. First, define a new integer j as $j = N + 1 - l$. Then

$$1 + \cos\left(\frac{l\pi}{N+1}\right) = 1 + \cos\left(\frac{(N+1-j)\pi}{N+1}\right)$$

$$= 1 + \cos\left(\pi - \frac{j\pi}{N+1}\right)$$

$$= 1 - \cos\left(\frac{j\pi}{N+1}\right) \qquad (16.50)$$

since $\cos(a+b) = \cos a \cos b - \sin a \sin b$ and $\cos \pi = -1, \sin \pi = 0$. Further,

$$\cos 2\Theta = \cos^2 \Theta - \sin^2 \Theta = 1 - 2\sin^2 \Theta \qquad (16.51)$$

and therefore

$$1 - \cos 2\Theta = 2\sin^2 \Theta \qquad (16.52)$$

Setting $2\Theta = \frac{j\pi}{N+1}$, the frequency expression becomes

$$\nu(j) = \frac{1}{2\pi}\sqrt{2\frac{f}{m}}\sqrt{2}\sin\left(\frac{j\pi}{2(N+1)}\right) \quad j = 0, 1, 2 \cdots N \qquad (16.53)$$

The equation is written in this way to emphasize the central position of the diatomic frequency. The value of $j = 0$ is included to include the frequency of the translation, a non-genuine vibration but a perfectly good normal coordinate. Also note that the full value of the maximum frequency, $\sqrt{2}\nu_{diatomic}$ is never achieved by any finite chain. Like the well in Zeno's paradox, the frequency may only be approached as a limit as the chain length grows. Finally, if j is allowed to vary unrestrictedly, the frequency becomes multi-valued, as the sine function rises and falls. Negative frequencies are physically impossible, and the expression is therefore usually written with the sine function in absolute value symbols.

$$\nu(j) = \frac{1}{2\pi}\sqrt{2\frac{f}{m}}\sqrt{2}\left|\sin\left(\frac{j\pi}{2(N+1)}\right)\right| \qquad (16.54)$$

When j is unrestricted, a plot of frequency against j has a lumpy appearance, as shown in Figure 16.3. The region from $-(N+1)$ to $+(N+1)$ is called the first Brillouin zone. The zone center is the translation, and the zone edge is the maximum frequency that real

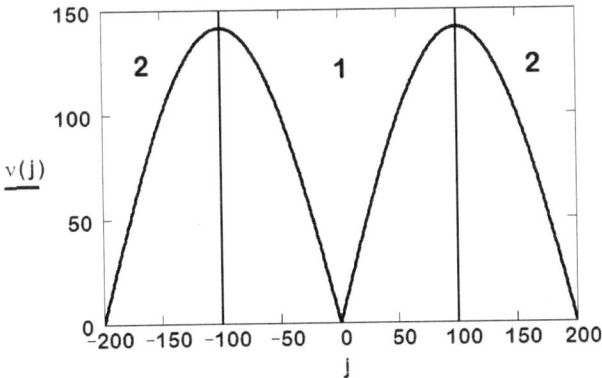

Fig. 16.3.   First two Brillouin zones for the finite chain of 100 identical masses, assuming a diatomic frequency of 100 cm^{-1}. The zones are labelled 1 and 2. The vertical lines mark the inaccessible zone edge points. The translation lies at $j = 0$ and is reproduced as two half points in zone 2 at $j = -200$ and 200. As these are universal curves, individual points are not presented. The pattern iterates indefinitely in either j direction, but physically only exists in the positive direction. The index j varies from 0 to 99 for the 100 mass chain.

chains approach but never attain, $\nu_{max} = \sqrt{2}\nu_{diatomic}$. The second Brillouin zone repeats the frequencies of the first, and runs from $-(N+1)$ to $-(2N+1)$ and $(N+1)$ to $(2N+1)$, and so on.

Returning now to the positive half of the first Brillouin zone, which contains all the physically relevant results, the question that is asked is what is the density of vibrational states, that is, the distribution of states as a function of j? This question is equivalent to defining the number of states in unit frequency interval $G(\nu)$. Given $G(\nu)$, the contribution of the longitudinal lattice modes to the thermodynamic functions can be obtained. Defining

$$\nu_0 \equiv \sqrt{2}\nu_{diatomic} \tag{16.55}$$

rewrite the chain frequencies as

$$\nu(j) = \nu_0 \sin\left(\frac{j\pi}{2(N+1)}\right) \quad j = 0,1,2\cdots N \tag{16.56}$$

and solve this for j

$$j = \frac{2(N+1)}{\pi}\sin^{-1}\left(\frac{\nu(j)}{\nu_0}\right) \tag{16.57}$$

Let $\Delta\nu$ be a frequency difference so small that it encloses a single point on the plot of ν against j, and let $\Delta j = j+1-j = 1$. Then

$$G(\nu)\Delta\nu = \Delta j \tag{16.58}$$

Passing to the limit of vanishing infinitesimals,

$$G(j) = \frac{dj}{d\nu} \tag{16.59}$$

Therefore, the frequency distribution function G is, for the finite chain

$$G(\nu) = \frac{2(N+1)}{\pi}\frac{1}{(\nu_0^2 - \nu^2)^{1/2}} \tag{16.60}$$

This function is plotted in Figure 16.4.

In addition to the stretching vibrations involved in the longitudinal modes, there are two orthogonal bending modes called the transverse acoustic modes. The bends are of lower frequency than

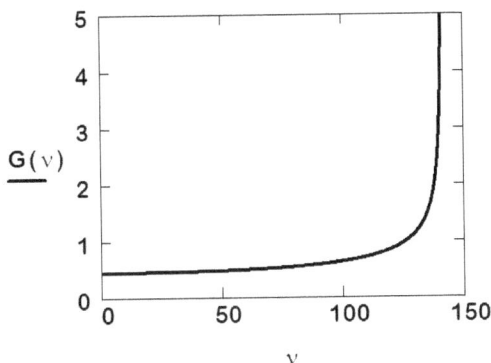

Fig. 16.4. Frequency distribution for the longitudinal vibrations of a finite chain of 100 identical masses with identical nearest neighbor force constants. The diatomic frequency was taken as 100 cm^{-1} and the maximum frequency was 141.42 cm^{-1}.

the stretches, and three masses are required to define a bend. The chain has N bonds, $N + 1$ masses and for the transverse modes in one plane there are two modes of zero frequency, one a translation perpendicular to the chain axis and the other a rotation in the plane of vibration about an axis perpendicular to the chain axis and the plane of bending vibrations and passing through the center of mass of the chain. The identical pattern of modes is repeated in a plane perpendicular to the other plane, these two planes intersecting in the chain axis. Thus, there are in all $3N - 5$ modes of genuine vibration. The transverse modes behave similarly to the longitudinal modes, but have a maximum frequency that is typically about half the maximum longitudinal frequency.

The general vibrational problem involves simultaneous diagonalization of the kinetic energy T and the potential energy V. In Cartesian coordinates, these are

$$2T = \sum_i m_i \dot{x}_i^2 \tag{16.61}$$

$$2V = \sum_{i,j} f_{i,j} x_i x_j \tag{16.62}$$

In general, the kinetic energy can be immediately diagonalized by using Cartesian mass weighted coordinates q_i defined as

$$q_i \equiv \sqrt{m_i} x_i \tag{16.63}$$

In terms of these coordinates the kinetic and potential energy are

$$2T = \sum_i \dot{q}_i^2 \tag{16.64}$$

$$2V = \sum_{i,j} \frac{f_{i,j}}{\sqrt{m_i m_j}} q_i q_j \tag{16.65}$$

The problem is reduced to diagonalizing V while maintaining T in diagonal form. There always exists a set of coordinates Q_k called the normal coordinates that does this. In terms of the normal coordinates, the kinetic and potential energies become

$$2T = \sum_k \dot{Q}_k^2 \tag{16.66}$$

$$2V = \sum_k \omega_k^2 Q_k^2 \tag{16.67}$$

where the normal coordinates are linear combinations of the Cartesian mass weighted coordinates

$$Q_k = \sum_i Q_{ik} q_i \tag{16.68}$$

and the ω_k are the angular frequencies $2\pi \nu_k$, called the characteristic frequencies of the problem, the fundamental frequencies or the normal frequencies. The problem is in general difficult enough to require special computer programming, by methods established in the middle of the twentieth century. In the case of the linear chain, the coefficients Q_{ik} are

$$Q_{ik} = \frac{\sqrt{2 - \delta_{k,0}}}{\sqrt{N+1}} \cos\left(\frac{(2i-1)k\pi}{2(N+1)}\right) \tag{16.69}$$

where N counts the number of bonds or nearest-neighbor force constants and $N+1$ is the number of identical masses. It is rare to obtain such a simple solution to the vibrational problem, made possible by the uniform masses and identical force constants. Numbering the masses and the frequencies is important. In order to use this relation, the counting for the normal coordinates Q_k must start with 0, the normal coordinate of the translation, and proceed to N, rather than from 1 to $N + 1$. Counting for the Cartesian coordinates can begin with either 1 or 0, as convenient. However, the factor $2i - 1$ assumes counting begins with 1. This must be changed to $2i + 1$ if numbering begins with 0.

It is unfortunate that the Cartesian mass weighted coordinates are represented by the letter q while the partition function is also represented by the letter q. In context, there is no confusion, but one must be careful not to confuse the two. The canonical ensemble partition function is

$$Q = \prod_i q_i \tag{16.70}$$

where each partition function q_i is

$$q_i = \frac{e^{-h\nu_i/2k_B T}}{1 - e^{-h\nu_i/k_B T}} \tag{16.71}$$

The energy referred to the bottom of the wells as zero is

$$\langle E \rangle = k_B T^2 \frac{\partial \ln Q}{\partial T} \tag{16.72}$$

$$\langle E \rangle = \sum_{i=1}^{3N} h\nu_i \left(\frac{1}{2} + \frac{1}{e^{h\nu_i/k_B T} - 1} \right) \tag{16.73}$$

The heat capacity is

$$C_v = k_B \sum_{i=1}^{3N} \left(\frac{h\nu_i}{k_B T} \right)^2 \frac{e^{h\nu_i/k_B T}}{(e^{h\nu_i/k_B T} - 1)^2} \tag{16.74}$$

These sums contain 3 times Avogadro's number of terms, and are impractical to perform. At low temperatures, it is practical to perform the sums for chains up to 211 masses (this limit was set by the available computational facility). Figure 16.5 shows the low temperature heat capacity of chains of 100, 150 and 200 masses. For much longer chains and higher temperatures. the alternate form of

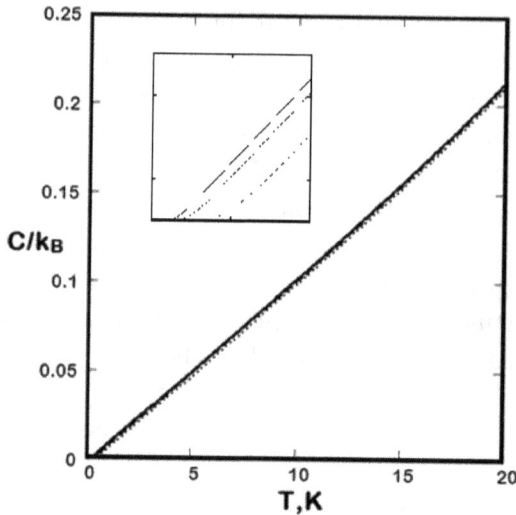

Fig. 16.5.   Low temperature heat capacity for chains of 100 masses (dotted line), 150 masses (dashed line) and 200 masses (solid line). Note the essentially linear behavior, quite different from those of the Einstein or Debye approximations in this region. The heat capacity sum was divided by $N+1$ to produce a universal curve, where N measures the number of bonds, and $N+1$ the number of masses. Insert: T from 0 to 2 K, C/k_B from 0 to 0.02. Note that all curves approach $T = 0$ with zero slope.

the exponentials allows calculation and the heat capacity is shown in Figure 16.6.

For the linear chain, the average frequency, that would be used in an Einstein approximation of the specific heat of the chain is

$$\langle \nu \rangle = \frac{2\nu_{max}}{\pi} \tag{16.75}$$

which requires the integral

$$\int \sin^{-1} x dx = x \sin^{-1} x + (1 - x^2)^{1/2} \tag{16.76}$$

The heat capacity of the linear chain can be approximated by a method analogous to the method used below in the Debye approximation for an isotropic solid. The essence of it is that the chain is divided into equal parts by any mode, just as the harmonics of a stringed musical instrument divide the string into equal parts. Let d

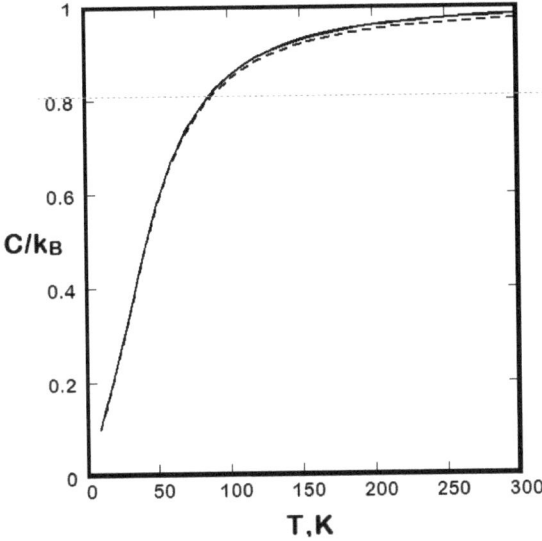

Fig. 16.6. High temperature behavior of the heat capacity of linear chains. Dashed line, 100 masses in chain. Solid lines, 1000 and 100,000 masses in chains. The behavior of chains longer than 1000 masses is identical at higher temperatures.

be the distance between masses in the chain. The length of the wave of minimum wavelength and maximum frequency is $L = (N + 1)d$, as shown in Figure 16.8, while the effect of a change in maximum frequency is demonstrated in Figure 16.7.

Each equal segment measures a half wavelength of the counter-propagating waves that form the standing wave of frequency ν. If j is the number of segments into which the chain is divided, then $\frac{\lambda}{2} = \frac{(N+1)d}{j} = \frac{c}{2\nu}$. When the frequency is the maximum, $j = N + 1$ and $d = \frac{c}{2\nu_0}$. Then $\frac{N+1}{\nu_0 j} = \frac{1}{\nu}$ or $j = \frac{N+1}{\nu_0}\nu$. Then approximately

$$g(\nu) = \frac{\partial j}{\partial \nu} = \frac{N+1}{\nu_0} \qquad (16.77)$$

This is somewhat larger than the zero frequency limit of $G(\nu)$, which is

$$\lim_{\nu \to 0} G(\nu) = \frac{2(N+1)}{\pi \nu_0} \qquad (16.78)$$

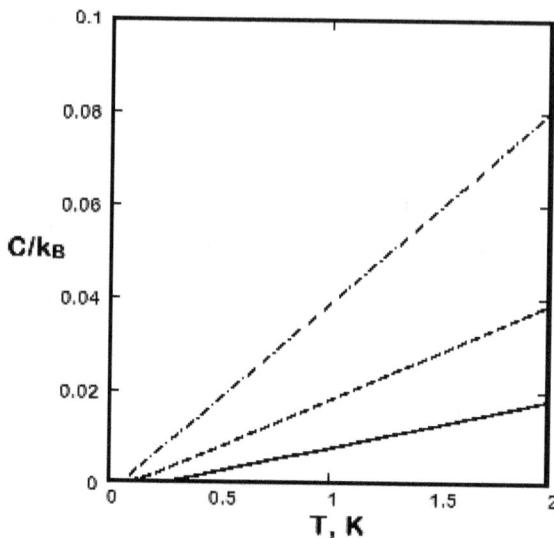

Fig. 16.7.   Heat capacities at very low temperatures for chains of 200 masses. Solid line, maximum frequency 141.4 cm^{-1}. Dashed line, maximum frequency 70.7 cm^{-1}. Dash-dot line, maximum frequency 35.35 cm^{-1}.

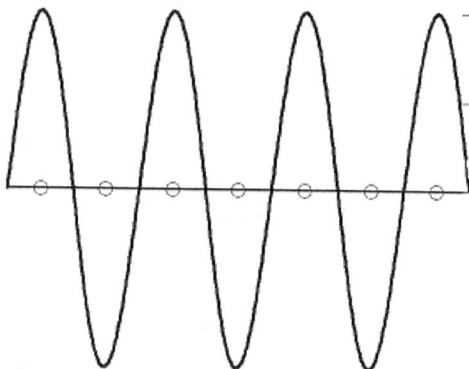

Fig. 16.8.   Minimum wavelength supported by a chain of masses. The length of the wave is $7d$ where d is the distance between nearest neighbor masses.

However, requiring the integral of $g(\nu)$ over the range of allowed frequencies must produce the number of degrees of freedom, including the translation. Clearly, $g(\nu)$ must be $(N+1)/\nu_0$ for

$$\int_0^{\nu_0} \frac{N+1}{\nu_0} d\nu = N + 1 \qquad (16.79)$$

This merits some further discussion. The two distribution functions must converge to the same limit when the frequency tends to zero. This can only mean that the maximum frequency is different, and the Debye maximum must be a factor of $\pi/2 = 1.571$ greater for extremely long chains. Indeed, even for the shortest possible chain of four masses the ratio of 1.57 is closely approached, as the ratio is $\sqrt{1+\sqrt{2}} = 1.554$. Thus

$$\nu_0^D = \frac{\pi}{2}\nu_0 \qquad (16.80)$$

where ν_0^D is the maximum frequency when the end masses are immobile and ν_0 is that when end masses are free to vibrate. Hereafter, ν_0 will be used to designate the maximum frequency, and which is meant will derive from the approximation in use. A further refinement is evident from the 4 mass case. When the ends are free, there are 4 degrees of vibrational freedom, or $N+1$, where N counts the number of bonds. But when the ends are fixed, the number of degrees of freedom drops to two, or $N-1$. Strictly speaking, then, the Debye approximation always has two degrees of freedom less than the freely vibrating chain. Since the approximation is intended to apply to very long chains, there is only a fractional difference of 2 parts in a very large number, which is negligible and the number of masses is close enough.

The energy approximates as

$$E = \int_0^{\nu_0} \left(h\nu \left(\frac{1}{2} + \frac{1}{e^{h\nu/k_BT} - 1} \right) \right) \frac{N+1}{\nu_0} d\nu \qquad (16.81)$$

The zero point energy E_0 is $h\nu_0/4 = h\langle\nu\rangle/2$, as the average frequency is $\nu_0/2$, where

$$\langle\nu\rangle = \frac{\int_0^{\nu_0} \nu g(\nu)d\nu}{\int_0^{\nu_0} g(\nu)d\nu} \qquad (16.82)$$

The functionally operational part is

$$E - E_0 = \frac{(N+1)h}{\nu_0} \int_0^{\nu_0} \frac{\nu d\nu}{e^{h\nu/k_BT} - 1} \qquad (16.83)$$

To investigate the low temperature limiting behavior of the heat capacity of the linear chain, introduce the change in variable $x = h\nu/k_B T$. The energy function becomes

$$E - E_0 = \frac{(N+1)(k_B T)^2}{h\nu_0} \int_0^{x_0} \frac{x e^{-x} dx}{1 - e^{-x}} \qquad (16.84)$$

At low temperatures x becomes large and the exponentials become small. The geometric series is $\frac{1}{1-x} = 1 + x + x^2 + x^3 + \cdots$. Using this, the integral can be rewritten as

$$E - E_0 = \frac{(N+1)(k_B T)^2}{h\nu_0} \int_0^{x_0} x(e^{-x} + e^{-2x} + e^{-3x} + \cdots) dx \qquad (16.85)$$

As the temperature declines towards zero, the upper limit x_0 becomes indefinitely large, and may be replaced by infinity. An integration by parts shows that

$$\int_0^\infty x e^{-nx} dx = \frac{1}{n^2} \qquad (16.86)$$

and the energy becomes

$$E - E_0 = \frac{(N+1)(k_B T)^2}{h\nu_0} \sum_{n=1}^\infty \frac{1}{n^2} \qquad (16.87)$$

Introducing the characteristic temperature $\Theta = h\nu_0/k_B$, this becomes

$$E - E_0 = (N+1)k_B \frac{T^2}{\Theta} \sum_{n=1}^\infty \frac{1}{n^2}$$

The sum converges to the value $\pi^2/6$ according to Jolley.[2] The low temperature limit of the energy function and the specific heat is therefore

$$E - E_0 = (N+1)k_B \frac{T^2}{\Theta} \frac{\pi^2}{6} \qquad (16.88)$$

$$C_L = \frac{(N+1)k_B \pi^2}{3\Theta} T \qquad (16.89)$$

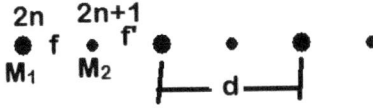

Fig. 16.9. Model for a chain of alternating masses and force constants. The masses M_1 and M_2 are different, and the force constant f is much greater than f'. Each diatomic combination forms a Bravais cell. The centers of the Bravais cells are spaced a distance d apart. If there are $N + 1$ diatomic molecules in the chain, there are $N + 1$ Bravais cells in the chain.

which is very nearly the behavior of the exact sum. C_L is used since the string is composed of mass points and only the length is defined.

A chain of alternating masses and force constants has a multi-variable frequency distribution function. The model to begin treatment of this modification of the one dimensional chain is shown in Figure 16.9.

The analysis is briefly sketched in the following equations. The equations of motion for typical masses are

$$M_1 \ddot{x}_{2n} = -f'(x_{2n} - x_{2n-1}) + f(x_{2n+1} - x_{2n}) \qquad (16.90)$$

$$M_2 \ddot{x}_{2n+1} = -f(x_{2n+1} - x_{2n}) + f'(x_{2n+2} - x_{2n+1}) \qquad (16.91)$$

Displacements are generally represented as

$$x_{2n} = A_1 e^{-2\pi i(\nu t - nkd)} \qquad (16.92)$$

$$x_{2n+1} = A_2 e^{-2\pi i\left(\nu t - \left(n + \frac{1}{2}\right)kd\right)} \qquad (16.93)$$

These lead to the secular determinant

$$\begin{vmatrix} f + f' - 4\pi^2\nu^2 M_1 & -(fe^{\pi ikd} + f'e^{-\pi ikd}) \\ -(fe^{-\pi ikd} + f'e^{\pi ikd}) & f + f' - 4\pi^2\nu^2 M_2 \end{vmatrix} = 0 \qquad (16.94)$$

from which the squared frequencies may be extracted. The result is

$$\nu^2 = \frac{f + f'}{8\pi^2}\left(\frac{1}{M_1} + \frac{1}{M_2}\right)$$

$$\pm \frac{1}{4\pi^2}\left[\left(\frac{f + f'}{2}\right)^2\left(\frac{1}{M_1} + \frac{1}{M_2}\right)^2 - \frac{4ff'}{M_1 M_2}\sin^2 \pi kd\right]^{\frac{1}{2}}$$

$$(16.95)$$

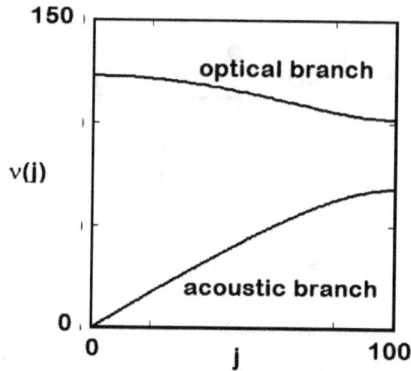

Fig. 16.10.   Acoustic and optical branches for a chain of masses 10 and 15 amu, with force constant f taken from a frequency of 100 cm^{-1} for an isolated diatomic molecule of these two masses and a force constant $f' = 0.5f$.

where $kd = j/2(N+1)$ and $j = 0, 1, 2 \ldots N$. If the minus sign is selected, and $j = 0$ is taken, the frequency is zero. If j is taken as $N+1$, which the chain cannot access, a limiting high frequency is obtained, defining the limits of the acoustic branch. This branch is also called the external branch. If the positive sign is selected, the optical branch is obtained, which is also called the internal branch. These branches are separated by a gap, or a range of frequencies that the chain cannot support. Frequencies above the zone center frequency of the optical branch also cannot propagate along the chain. An exaggerated version of the acoustic and optical branches generated by the frequency calculation above is shown in Figure 16.10. The frequency distribution function is shown in Figure 16.11.

16.3.   The Debye Approximation

A few years after Einstein introduced his heat capacity discussed above, Peter Debye sought an improvement based on the lattice modes of vibration, the longitudinal acoustic and transverse acoustic modes. The modes of vibration of the linear chain serves as a model with an exact solution. Debye did not, however, use this model, but rather one of its properties. The chain is divided into segments of equal length with nodes, or zeros of the motion, at all segment ends.

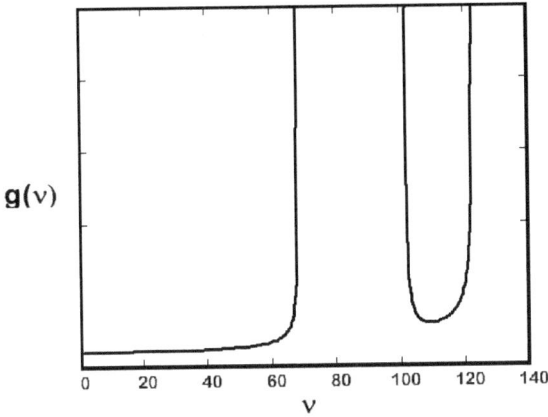

Fig. 16.11. Frequency distribution function for diatomic chain of Figure 16.10. No units are given for $g(\nu)$ since the function plotted is divided by the number of Bravais cells in the chain.

Thus, a chain of length L is divided into two segments when there is one central node, other than the ends which are also nodes in this approximation. In the treatment above, end masses are free to vibrate. For example, in the case of two masses, there are two modes, the translation and the diatomic vibration, and end masses are not fixed for if these were there could be no motion at all. Nonetheless, the translation has no internal nodes, and the vibration has one that lies between the two masses. If the distance between the two masses is d, the end nodes are actually at $-d/2$ and $d+d/2$, taking the origin of coordinates at one of the two masses. The effective length of the wave that divides the chain is then $2d$, not d. The vibration represents two counter propagating waves of the same amplitude, frequency and phase, interfering constructively in a standing wave. This standing wave is the shortest wave this chain will support, and represents the maximum frequency the chain is capable of supporting. For three masses, there are three possibilities, 0, 1 or 2 nodes may be formed, giving the translation (no node), the symmetric stretch (one node at the central mass) and the asymmetric stretch (two nodes between masses). Similar considerations apply to longer chains, and for a chain with N bonds, the maximum number of nodes is N, which defines the shortest wave the chain will support and the maximum frequency. The counter-propagating waves do so at some velocity c, and if the wavelength is λ and the frequency ν, the velocity is

$$c = \lambda\nu \qquad (16.96)$$

Now consider the vibrations of a square or rectangular membrane. First, note that in the Debye approximation the edges of the membrane are fixed in space, rather than free to vibrate as must be the case in any real crystal sheet. This introduces an edge effect that is ignorable in sufficiently large sheets. Second, notice that a row of identical masses with identical force constants lies along each edge, and throughout the interior of the membrane. The edges are expected to be divided into segments of equal length and nodal lines orthogonal to the sides are generated for each node of the chain. In Figure 16.12, a rectangular membrane is shown with maximum deflections out of the plane of the paper. The solid diagonal lines define standing waves composed of waves propagating in directions orthogonal to these lines and represent a maximum deflection out of the plane of the diagram. The dashed lines are similar but represent deflections below the plane of the membrane. One of the many possible normal modes of the membrane is shown. The sides are of length L_1 and L_2, and in general these sides will be divided into k_x and k_y equal parts, where the k_i are positive non-zero integers. A wavelength is the distance between successive crests or between successive troughs, and a half wavelength is the distance between a crest and the nearest trough, as

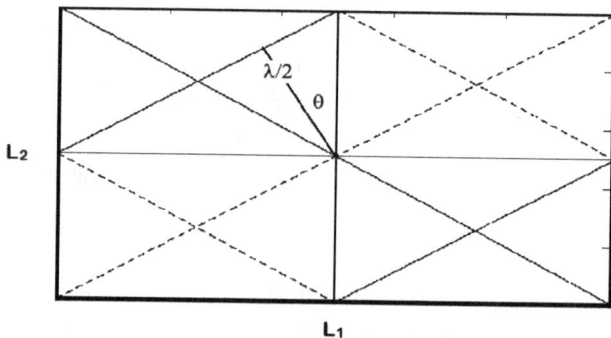

Fig. 16.12.   Schematic diagram of one of the normal modes of vibration of a rectangular membrane. $k_y = 2$ in this example.

shown on the figure. The angle Θ is defined by the distance shown, and the cosine is

$$\cos\Theta = \frac{\lambda/2}{L_2/k_y} \tag{16.97}$$

Using $\cos(\frac{\pi}{2} - \Theta) = \sin\Theta$, it is apparent that the sine is

$$\sin\Theta = \frac{\lambda/2}{L_1/k_x} \tag{16.98}$$

Since $\sin^2\Theta + \cos^2\Theta = 1$, there immediately follows

$$\left(\frac{\lambda/2}{L_1/k_x}\right)^2 + \left(\frac{\lambda/2}{L_2/k_y}\right)^2 = 1 \tag{16.99}$$

This is the equation of a circle of radius $R = 2/\lambda$ and coordinates k_x/L_1 and k_y/L_2. Extension to three dimensions to investigate the normal modes of a rectangular parallelepiped is obvious, and the corresponding relation is the equation of a sphere of radius $R = 2/\lambda$ enclosing all the values of the indices k_i with wavelength λ or greater. To count the number of such points, it is only necessary to measure the volume of the positive octant of the sphere (all the k_i must be positive, by the nature of the problem). However, the unmodified volume contains more than just points, as the dimensions of the rectangular parallelepiped is involved.

Specifically, if the k_i are all set equal to 1, a point at $(\frac{1}{L_1}, \frac{1}{L_2}, \frac{1}{L_3})$ is generated in $(\frac{k_x}{L_1}, \frac{k_y}{L_2}, \frac{k_z}{L_3})$ space, which is one apex of a cell defined by assigning 1 or 0 to each possible value of the k_i. The volume of this cell is $\frac{1}{L_1L_2L_3}$ and it has 8 apices, each of which contributes 1/8 of a point, for a total of 1 point. Then the ratio of a volume in $(\frac{k_x}{L_1}, \frac{k_y}{L_2}, \frac{k_z}{L_3})$ space to the volume of the cell counts the points in the volume.

Each point represents one degree of vibrational freedom. In an isotropic monatomic solid composed of N atoms or molecules there are $3N-6$ longitudinal or transverse lattice modes. For N of the order of Avogadro's number, the 6 translational and rotational degrees of freedom of the whole solid are negligible, and $3N$ will do for the total number of modes. The number of modes with wavelength λ or greater is given by

$$Z = \frac{\frac{1}{8} \; volume \; of \; sphere}{volume \; of \; cell} \tag{16.100}$$

$$Z = \frac{1}{8} \left(\frac{4\pi}{3} \right) \left(\frac{2}{\lambda} \right)^3 L_1 L_2 L_3 \tag{16.101}$$

The volume of the rectangular parallelepiped $V = L_1 L_2 L_3$, and the velocity of propagation of a wave is given by equation (16.96). It is well known that the velocity of propagation for longitudinal waves is larger than that for transverse waves, but this difference is ignored in the Debye theory. Therefore the number of points with frequency equal to or less than the maximum frequency is

$$Z_{max} = 3N = \frac{4\pi}{3} V \frac{\nu_{max}^3}{c^3} \tag{16.102}$$

The number of frequencies with a frequency between ν and $\nu + d\nu$ is dZ.

$$dZ = \frac{4\pi V}{c^3} \nu^2 d\nu \tag{16.103}$$

The fraction of modes in the range ν to $\nu + d\nu$ is

$$\frac{dZ}{Z_{max}} = 3 \frac{\nu^2}{\nu_{max}^3} d\nu \tag{16.104}$$

The distribution function $g(\nu)$ is proportional to this fraction. The integral of $g(\nu)$ is a sum that must add up to the total number of normal modes, including the non-genuine vibrations. In a 3 dimensional crystal, N is the number of atoms (for monatomic solids) or the number of molecules (for molecular crystals), and the appropriate number is $3N$. Thus

$$g(\nu) = 3 \frac{\nu^2}{\nu_{max}^3} \times 3N \tag{16.105}$$

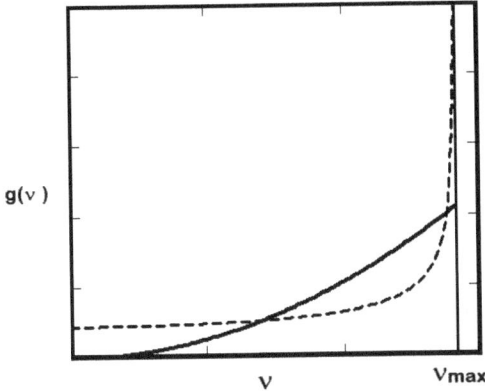

Fig. 16.13. Solid line, Debye's approximate distribution function compared to the linear chain. Dashed line, exact distribution function for the linear chain. Both assume a maximum frequency of the same value. The dashed line terminates slightly above the upper limit of the plot and does not attain ν_{max}, which the Debye function does. Both start from zero frequency. There are important differences in the two functions. The end masses of the linear chain are free to vibrate, whereas the Debye crystal edges are fixed in space. The most important difference is that the Debye function refers to three dimensional space, while the linear chain considers but one.

for the crystal. Integrating this function over the entire range of frequencies gives

$$\int_0^{\nu_{max}} g(\nu)d\nu = 3N \tag{16.106}$$

This function is plotted in Figure 16.13, which is compared with the exact distribution function for the finite chain of uniform masses and force constants. The comparison is apples to oranges, as the Debye function is an approximation in 3 dimensions while the linear chain is an exact solution in one dimension. Indeed, if similar reasoning is applied to a two dimensional sheet, such as graphene, the distribution function will be proportional to the frequency, and if applied to the linear chain, will be given by a constant, reflecting the long region of very slowly varying values of g shown for all but the higher frequencies.

The choice of energy zero is arbitrary, and may be made at pleasure once in any problem. In the crystal, the zero of energy may be chosen as the infinitely separated molecules, and the potential at any molecule in the crystal expanded as a power series in the distance r between centers of mass of the molecules that comprise the crystal. Doing so provides a quadratic term that defines the force constants for intermolecular interaction, a constant term, and higher powers of r. The linear term vanishes from the expansion since the slope at the bottom of a potential energy well is zero. However, the zero of energy may be chosen at the bottom of the potential well in which any molecule is found, leaving the quadratic term as the leading one in the expansion. For sufficiently small displacements, cubic and higher terms may be neglected. For the linear chain, this means only the nearest neighbor force constants remain, and similarly for the 3 dimensional crystal. Then there will be $3N$ frequencies in the crystal, all different.

The frequency distribution function allows approximate evaluation of the energy and heat capacity. Replacing the sums with the integral, the energy E due to the lattice modes is

$$E = \int_0^{\nu_{max}} h\nu \left(\frac{1}{2} + \frac{1}{e^{h\nu/k_B T} - 1} \right) g(\nu) d\nu \qquad (16.107)$$

The heat capacity C_v is the temperature derivative

$$C_v = k_B \int_0^{\nu_{max}} \left[\left(\frac{h\nu}{k_B T} \right)^2 \frac{e^{h\nu/k_B T}}{(e^{h\nu/k_B T} - 1)^2} \right] g(\nu) d\nu \qquad (16.108)$$

The Debye distribution function gives

$$E = \frac{9N}{\nu_{max}^3} \int_0^{\nu_{max}} h\nu \left(\frac{1}{2} + \frac{1}{e^{h\nu/k_B T} - 1} \right) \nu^2 d\nu \qquad (16.109)$$

$$C_v = \frac{9Nk_B}{\nu_{max}^3} \int_0^{\nu_{max}} \left[\left(\frac{h\nu}{k_B T} \right)^2 \frac{e^{h\nu/k_B T}}{(e^{h\nu/k_B T} - 1)^2} \right] \nu^2 d\nu \qquad (16.110)$$

$$C_v = \frac{9Nh^2}{k_B T^2 \nu_{max}^3} \int_0^{\nu_{max}} \nu^4 \frac{e^{h\nu/k_B T}}{(e^{h\nu/k_B T} - 1)^2} d\nu \qquad (16.111)$$

Introducing the classic substitution $x = \frac{h\nu}{k_B T}$, the heat capacity becomes

$$C_v = 9Nk_B \left(\frac{k_B T}{h\nu_{max}} \right)^3 \int_0^{x_{max}} \frac{x^4 e^x}{(e^x - 1)^2} dx \qquad (16.112)$$

The Debye temperature is defined as $\Theta_D = \frac{h\nu_{max}}{k_B}$, and in terms of this temperature the heat capacity becomes

$$C_v = 9R \left(\frac{T}{\Theta_D} \right)^3 \int_0^{\Theta_D/T} \frac{x^4 e^x}{(e^x - 1)^2} dx \qquad (16.113)$$

for one mole. If T is much less than the Debye temperature, the upper limit becomes very large and may be approximated by infinity. The integral becomes a constant, and the heat capacity is seen to be proportional to T^3. A numerical estimate of the proportionality constant may be made by considering the energy at very low temperatures. Apart from the constant contribution of the zero point energy E_0, the functionally important part is, when x is substituted

$$E = \frac{9Nh}{\nu_{max}^3} \left(\frac{k_B T}{h} \right)^4 \int_0^{x_{max}} \frac{x^3 dx}{e^x - 1} \qquad (16.114)$$

At sufficiently low temperatures, x_{max} may be replaced by infinity. The resulting definite integral has the approximate value

$$\int_0^\infty \frac{x^3 dx}{e^x - 1} = \frac{\pi^4}{15} \qquad (16.115)$$

and the low temperature behavior of the energy is

$$E = \frac{3Nk_B}{5} \left(\frac{k_B}{h\nu_{max}} \right)^3 \pi^4 T^4 \qquad (16.116)$$

The low temperature heat capacity is the temperature derivative. For one mole, C_v is

$$C_v = \frac{12\pi^4 R}{5} \left(\frac{T}{\Theta_D} \right)^3 \qquad (16.117)$$

which is Debye's famous result.

At high temperatures, the argument of the exponential in the energy integral becomes small for all frequencies at or below the maximum. Expansion of the exponential is justified, and the energy becomes

$$E = \frac{9Nh}{\nu_{max}^3} \int_0^{\nu_{max}} \frac{\nu^3 d\nu}{1 + h\nu/k_B T + \cdots - 1} \qquad (16.118)$$

or

$$C_v = \frac{9Nk_B}{\nu_{max}^3} \int_0^{\nu_{max}} \nu^2 d\nu \qquad (16.119)$$

The result of the integration is $C_v = 3R$ for one mole. At intermediate temperatures, the function defined by the energy expression is tabulated and series expansions are given in AMS 55.[3]

The average frequency is the one that would be used to approximate the specific heat with an Einstein function. It is

$$\langle \nu \rangle = \frac{\int_0^{\nu_{max}} \nu g(\nu) d\nu}{\int_0^{\nu_{max}} g(\nu) d\nu} \qquad (16.120)$$

The result is $\frac{3\nu_{max}}{4}$.

16.4. The Free Electron

Some metals, like the alkali metals, have a single electron in the valence shell of the isolated atom. When a very large number of such atoms are assembled into a metallic crystal, the quantum mechanics of the resulting periodic structure adds the individual sharply defined energy levels of the isolated atom together into a band of levels so closely spaced as to constitute an energy continuum. Since each level can accommodate two electrons with opposing spins, the band is only half filled at absolute zero. As the temperature of the metal increases, the topmost electrons move into higher energy regions of the band. Doing so absorbs energy from the surroundings, and the heat capacity is not zero. The same thing happens to all the other energy levels of the isolated atom, and in other metals there is always some unfilled energy band that allows electrical conductivity and contributes to the heat capacity as energy is stored in the excitation of electrons in the conduction band. For example, in the transition metals Scandium through Copper, the d shell is being filled and only Zinc has a d^{10} ground state configuration. In this series, in the crystalline metal the $4s$ band overlaps the $3d$ band and all these metals conduct electricity. Zinc is like the alkaline earth metals, in that in these cases the s and p bands overlap, providing space for free electrons. The coinage metals Copper, Silver and Gold, behave at very low temperatures like the alkali metals, and have a composite heat capacity composed of contributions from the lattice vibrations of the nuclei and the electron gas in the conduction band.

To get at the heat capacity of the electrons in a metal, several new concepts are needed. The mean occupation number for fermions was described in the chapter on Fermi-Dirac and Bose-Einstein statistics, and is for one state i of the fermion

$$\langle n_i \rangle = \frac{1}{e^{(\varepsilon_i - \mu)/k_B T} + 1} \tag{16.121}$$

In the metal, the highest occupied level is the Fermi level. At the zero of temperature, all electrons are condensed into the lowest levels the Pauli exclusion principle will allow them to occupy, and the highest possible Fermi level results. Take this level to be $\mu_F(0)$. In a conduction band, as the temperature increases only tiny energy inputs are needed to remove electrons from the highest levels into the higher levels of the band, and the highest filled level decreases. As the temperature increases further, more and more electrons leave the lower levels and the Fermi level declines further. The Fermi level is thus a slowly varying function of temperature.

Basically, $\langle n_i \rangle$ is the number of fermions per state. If the number of electrons in unit energy interval $n(\varepsilon)$ is required, it is the product of the number of electrons per state times the number of states per unit energy interval, $g(\varepsilon)$. That is,

$$n(\varepsilon) = \frac{g(\varepsilon)}{e^{(\varepsilon - \mu_F)/k_B T} + 1} \tag{16.122}$$

Even though electrons in metals are far from a true electron gas, to get an approximation of their behavior assume that these electrons behave as an ideal gas. Then the only energy is kinetic with a mass m

that is different from the rest mass of the electron. It is basic that the number of states accessed by a system is given by the corresponding volume in phase space. For electrons, this is expressed as

$$Nn = \frac{2}{h^3} \int_0^a \int_0^b \int_0^c \iiint dx\,dy\,dz\,dp_x\,dp_y\,dp_z \qquad (16.123)$$

the volume element $dp_x\,dp_y\,dp_z$ can be expressed in spherical coordinates as $4\pi p^2 dp$. For the kinetic energy $\varepsilon = p^2/2m$ and thus $p^2 dp = (2m)^{3/2} \varepsilon \, d\varepsilon^{1/2} = \frac{1}{2}(2m)^{3/2} \varepsilon^{1/2} d\varepsilon$. Then, performing the xyz integration but not the energy integration, there results

$$g(\varepsilon) = \frac{dNn}{d\varepsilon} = \frac{2^{7/2}\pi m^{3/2}}{h^3} V \varepsilon^{1/2} \qquad (16.124)$$

where V is the volume of the metal. This is a parabola symmetric about the horizontal ε axis with vertex at the origin. It is defined only for positive energies, and for negative energies $g(\varepsilon)$ is zero.

With this approximate distribution function in hand, the calculation of the heat capacity of the conduction electrons in a metal can begin. The definition of heat capacity

$$C = \left(\frac{\partial E}{\partial T}\right)_N \qquad (16.125)$$

requires knowing both E and N. For convenience, introduce the Fermi function

$$F_x = \frac{1}{e^x + 1} \qquad (16.126)$$

The number of electrons in the band is

$$N = \int_{-\infty}^{\infty} g(\varepsilon) F_x \left(\frac{\varepsilon - \mu}{k_B T}\right) d\varepsilon \qquad (16.127)$$

and the energy is

$$E = \int_{-\infty}^{\infty} \varepsilon g(\varepsilon) F_x \left(\frac{\varepsilon - \mu}{k_B T}\right) d\varepsilon \qquad (16.128)$$

Thermodynamically, the heat capacity is defined in a constant volume system. Here, the energy is given as a function of T and μ. Then a general variation in E is

$$dE = \left(\frac{\partial E}{\partial T}\right)_\mu dT + \left(\frac{\partial E}{\partial \mu}\right)_T d\mu \qquad (16.129)$$

and as an immediate result

$$\left(\frac{\partial E}{\partial T}\right)_N = \left(\frac{\partial E}{\partial T}\right)_\mu + \left(\frac{\partial E}{\partial \mu}\right)_T \left(\frac{\partial \mu}{\partial T}\right)_N \qquad (16.130)$$

The partial differential coefficient $\left(\frac{\partial \mu}{\partial T}\right)_N$ cannot be immediately evaluated from the equations for E and N. However, the triad relation states

$$\left(\frac{\partial \mu}{\partial T}\right)_N = -\frac{\left(\frac{\partial N}{\partial T}\right)_\mu}{\left(\frac{\partial N}{\partial \mu}\right)_T} \qquad (16.131)$$

and these partial differential coefficients can be evaluated from the equation for N. The heat capacity becomes

$$\left(\frac{\partial E}{\partial T}\right)_N = \left(\frac{\partial E}{\partial T}\right)_\mu - \left(\frac{\partial E}{\partial \mu}\right)_T \frac{\left(\frac{\partial N}{\partial T}\right)_\mu}{\left(\frac{\partial N}{\partial \mu}\right)_T} \qquad (16.132)$$

Evaluating these coefficients from the E and N integrals produces the gargantuan relation

$$C_v = -\frac{1}{k_B^2 T} \int_{-\infty}^{\infty} \varepsilon(\varepsilon - \mu) g(\varepsilon) F_x' \left(\frac{\varepsilon - \mu}{k_B T}\right) d\varepsilon$$

$$+ \frac{1}{k_B^2 T} \frac{\int_{-\infty}^{\infty} \varepsilon g(\varepsilon) F_x' \left(\frac{\varepsilon-\mu}{k_B T}\right) d\varepsilon \int_{-\infty}^{\infty} (\varepsilon - \mu) g(\varepsilon) F_x' \left(\frac{\varepsilon-\mu}{k_B T}\right) d\varepsilon}{\int_{-\infty}^{\infty} g(\varepsilon) F_x' \left(\frac{\varepsilon-\mu}{k_B T}\right) d\varepsilon}$$

$$(16.133)$$

where $F_x' \left(\frac{\varepsilon-\mu}{k_B T}\right)$ is the derivative of the Fermi function. Now by replacing ε in the first integral by $\varepsilon - \mu$ and replacing ε in the first integral in the numerator of the ratio by $\varepsilon - \mu$ makes compensating contributions and leaves the value of the heat capacity unaltered. Introducing the variable $x = (\varepsilon - \mu)/k_B T$ makes the expression considerably more compact. With these changes, the heat capacity is

$$C_v = -k_B^2 T \int_{-\infty}^{\infty} g(\mu + k_B T x) x^2 F_x' dx$$

$$+ k_B^2 T \frac{\left(\int_{-\infty}^{\infty} g(\mu + k_B T x) x F_x' dx \right)^2}{\int_{-\infty}^{\infty} g(\mu + k_B T x) F_x' dx} \qquad (16.134)$$

To get the low temperature behavior of the electronic heat capacity, expand $g(\mu + k_B T x)$ in powers of x. This gives the heat capacity in powers of $k_B T/\mu$, which is a small quantity at low temperatures. The leading term in the expansion is for $x = 0$, and is therefore $g(\mu)$. Dropping all higher powers of x, the low temperature heat capacity becomes

$$C_v \cong g(\mu) k_B^2 T \left(- \int_{-\infty}^{\infty} x^2 F_x' dx \right) \qquad (16.135)$$

The integrand of the integral is a symmetric function and is equal to twice the integral from zero to infinity. An integration by parts shows the value is

$$4 \int_0^{\infty} x F_x dx = 4 \int_0^{\infty} dx (x e^{-x} - x e^{-2x} + x e^{-3x} - x e^{-4x} + \cdots = \frac{\pi^2}{3} \qquad (16.136)$$

The heat capacity becomes

$$C_v \cong \frac{\pi^2}{3} g(\mu) k_B^2 T \qquad (16.137)$$

which establishes the linear dependence on temperature of the low temperature heat capacity. To gain further insight into the low temperature heat capacity, the Fermi level needs to be evaluated. This may be done at absolute zero. The number of free electrons in the conduction band is N, and this is given by

$$N = \int_{-\infty}^{\mu} g(\varepsilon) d\varepsilon \qquad (16.138)$$

The result is immediate

$$N = \frac{2}{3}\mu g(\mu) \tag{16.139}$$

allowing the heat capacity to be written

$$C_v \cong N k_B \frac{\pi^2}{2} \frac{k_B T}{\mu} \tag{16.140}$$

The heat capacity of the electrons in the conduction band is proportional to T, but is too small to make appreciable contribution to the heat capacity of the metal at ambient temperatures. However, at extremely low temperatures the heat capacity of the lattice modes varies as the cube of temperature and eventually the electronic contribution becomes comparable. In this temperature range, the heat capacity of the metal is given by an equation of the form

$$C_v = \gamma T + \alpha T^3 \tag{16.141}$$

Then if the heat capacity data are plotted as C_v/T against T^2, a straight line results, the slope of which is proportional to the Debye temperature and the intercept gives γ directly. Approximate results for Cu, Ag and Au, all d^9 elements, are listed in Table 16.1. Exact results are given in the paper by Corak, *et al.*[4]

The Debye temperature is dominated by the atomic mass while the electronic coefficient remains in a small range, indicating similar effective masses and Fermi levels in these metals. Since the Debye temperature is proportional to the maximum frequency, the ratios of these temperatures should be near the ratios of the square roots of the masses. Assuming the Debye temperature for copper, that for Ag should be near 263 K and that for gold near 195 K. These are a bit higher than the observed, suggesting a loosening of the interatomic

Table 16.1. Debye temperatures and electronic coefficients for the coinage metals.

	Cu	Ag	Au
γ, millijoules mole^{-1}deg^{-2}	0.688	0.610	0.743
Θ_D deg	343	225	164

bonding as the atomic mass increases. The experimental linear plots are impressive, and are best seen by consulting the paper by Corak et al., where slightly more exact values for the entries in Table 16.1 will also be found.

16.5. Compressibility of Crystals

Recently, theoretical interest in compressibility of crystalline solids was awakened in a paper by Astala. Auerbach and Munson.[6] In essence, these authors made the *ansatz* that the energy of a strained isotropic crystal depended on the interaction of the strain l with the normal coordinates Q_k of the characteristic vibrations of the strained crystal. The frequencies of these vibrations appear explicitly nowhere in their theory, though they remark that the frequencies are contained in one of their matrices..

Experimentally, the frequencies of the crystal are measured by Raman spectroscopy at the center of the first Brillouin zone when the crystal symmetry allows spectroscopic activity. Each of these frequencies initiates a branch of frequencies that extends to the edge of the first Brillouin zone, defining a band of frequencies that the crystal will support. In addition, the acoustic branches are present in every crystal, have zero zone center frequencies, and are not spectroscopically active. In principle, it is possible to measure these by neutron diffraction experiments. In a diamond anvil cell, the Raman spectrum of a very small crystal under pressure may be measured, often in back scattered light, and the optical lattice modes with Raman frequencies in the $100 \, \text{cm}^{-1}$ or less range show frequency shifts with increased pressure of the order of about 1 to $10 \, \text{cm}^{-1}$ per GPa. Higher optical frequencies often show somewhat smaller frequency shifts, almost always to higher frequencies.[7] Frequency shifts in soft molecular crystals, such as iodine,[8] show large intensity changes reflecting rearrangement of the molecules in the unit cell as well as changes in electronic structure. The theory here concerns only small pressure increments and associated strains, in what might be described as an initial isothermal compressibility. Such small pressure increments were studied experimentally by x-ray diffraction from single crystals of argon grown carefully from the melt[9] and studied as a function of temperature between a few degrees Kelvin and the triple point near 83 K, and as a function of pressure up to about 20 atmospheres. These studies measured changes in the lattice constant, but provided no information about the vibrational structure of these crystals.

In view of the fact that the experimental evidence concerns frequency changes, the consequences of expanding the frequencies in a power series in the strain and stopping the series at the linear term should be explored. This approach follows to some extent the lead of Astala, Auerbach and Munson in the volume calculations, but otherwise diverges strongly from their approach.

Consider the small strain response of hypothetical monatomic isotropic crystals of cubic symmetry. The vibrations of the crystal may be collected in three branches, each containing Avogadro's number of frequencies for one mole of crystal. A simple model that may be solved exactly is provided by the one dimensional finite chain of $N_b + 1$ identical masses m connected by N_b identical springs each of force constant f. The angular frequencies ω_k of such a chain depend only on the diatomic frequency

$$\omega_d = \sqrt{\frac{2f}{m}} \tag{16.142}$$

and the number of nodes in the standing wave that appears in the chain at each characteristic frequency. The result for the frequencies is

$$\omega_k = \sqrt{2}\omega_d \sin\left(\frac{k\pi}{2(N_b+1)}\right) \quad k = 0, 1, 2, \ldots, N_b \tag{16.143}$$

The chain is not capable of supporting a vibration with $k = N_b + 1$. The amplitudes of the ith mass in the kth mode is given by

$$Q_{ik} = \frac{\sqrt{2 - \delta_{k,0}}}{\sqrt{N_b + 1}} \cos\left(\frac{(2i-1)k\pi}{2(N_b+1)}\right) \tag{16.144}$$

where $\delta_{k,0}$ is the Kronecker delta, equal to 1 when k = 0 but is zero otherwise. The coefficient Q_{ik} refers to the mass weighted Cartesian displacement coordinates, in that the kth normal coordinate Q_k is given by

$$Q_k = \sum_{i=1}^{N_b+1} Q_{ik}q_i \quad k = 0, 1, 2, \ldots, N_b \tag{16.145}$$

and the mass weighted coordinate q_i is, in terms of the longitudinal Cartesian displacement coordinate Δx_i, just $q_i = m^{1/2}\Delta x_i$. When the linear chain is compressed by a small amount, there is nothing in the Q_{ik} that can possibly be a function of this strain. The frequency contains a force constant that could be altered by the strain, as the binding electrons are displaced, thus altering the force constant. Nothing else is capable of varying. Indeed, if the wave corresponding to the maximum frequency is considered as an example, the wavelength is shortened by the compression. Experimentally, the variation in the frequencies is typically expressed as a function of pressure, not strain. However, stress is proportional to strain, and all that is needed to make a conversion is the constant of proportionality..

Written in terms of the normal coordinates, the classical harmonic vibrational Hamiltonian has the form

$$H = \frac{1}{2} \sum_{k=0}^{N} (P_k^2 + \omega_k^2 Q_k^2) \tag{16.146}$$

In the Hamiltonian formalism, the velocities of the coordinates are given by the derivative $\frac{\partial H}{\partial P_k} = \dot{Q}_k$. From the vibrational Hamiltonian, the derivative is $\frac{\partial H}{\partial P_k} = P_k$ and thus for the harmonic vibrational system, $P_k = \dot{Q}_k$. This Hamiltonian applies to the vibrations of a three dimensional crystal. The classical vibrational energy for a three dimensional monatomic isotropic crystal is exactly the same Hamiltonian, but with three times Avogadro's number of non-zero frequencies, omitting the six due to translation and rotation of the crystal as a whole. In general, there are N non-zero acoustic frequencies in all, whatever the size of the crystal.

All the lattice frequencies are small, and the quantized vibrational energy levels are so closely packed that the energy is effectively continuous. Hence a continuous description of the lattice vibrational energy is appropriate. All vibrational energies are non-degenerate. Ignoring the translations and rotations of the crystal as a whole, the partition function in the Gibbs ensemble is

$$\Delta = \frac{1}{h^N} \int_{-\infty}^{\infty} e^{-\sum_k (P_k^2 + \omega_k^2 Q_k^2)/2k_B T} e^{-pV/k_b T} dP dQ dV \qquad (16.147)$$

where $dP = \prod_k dP_k$ and $dQ = \prod_k dQ_k$. When written in this way, the partition function has the dimensions of volume. Partition functions are composed of dimensionless probabilities. The simplest way to convert the partition function to probability factors is to divide the volume element by a standard volume, taking in this way the fractional increment in the volume rather than the volume itself. Take the volume of the unstrained crystal as the relative point, and the volume element as the fractional increment in volume dV/V_0.

The ensemble average volume is

$$\langle V \rangle = -k_B T \left(\frac{\partial \ln \Delta}{\partial p} \right)_{T,N} \qquad (16.148)$$

The cubic unit cell in the unstrained crystal has a lattice constant defining the cell dimension a_0 and a volume $V_0 = a_0^3$, and the volume of the unit cell under initial strain l is $(a_0 - l)^3$. For the crystal under a small strain the volume change $\Delta V = V - V_0$ in each unit cell is

$$\Delta V = \left(a_0^3 - 3a_0^2 l + 3a_0 l^2 - l^3 \right) - a_0^3 \qquad (16.149)$$

The strain is small enough that only linear terms need be retained and therefore the volume change in a unit cell due to initial strain is $\Delta V = -3a_0^2 l$. By the same token, the volume element is $dV = -3a_0^2 dl$. The volume is extensive, and the total volume change in the whole crystal is a number of the order of Avogadro's number N_0 times as much. There is no point to carrying N_0 along when it will cancel in the end, as the isothermal compressibility is intensive. With this omission, the partition function becomes

$$\Delta = -\frac{3a_0^2}{V_0} e^{pV_0/k_B T} \int_{-\infty}^{\infty} \int_{-\infty}^{\infty} \int_0^{a_0} e^{-\sum_k (P_k^2 + \omega_k^2 Q_k^2)/2k_B T}$$

$$\times e^{-3pa_0^2 l/k_B T} \frac{dQ dP}{h^N} dl \qquad (16.150)$$

The pre-integral multiplier e^{pV_0/k_BT} changes e^{-pV/k_BT} into $e^{-p\Delta V/k_BT}$. The limit on the strain integration is set to allow only positive strain. The reason for this is that the pressure is regarded as increasing from zero, and negative pressure is not experimentally available. Negative volume does not exist, and the maximum possible mathematical value for the strain is therefore the unit cell dimension a_0, a physical impossibility since the volume cannot really be reduced to zero. The unstrained crystal is regarded as at equilibrium in a hard vacuum, and from this state only positive pressures are possible, resulting in only positive strains.

The integrations over the momenta are all Gaussian, and produce the result

$$\int_{-\infty}^{\infty} e^{-\sum_k P_k^2/2k_BT} dP = (2\pi k_BT)^{N/2} \tag{16.151}$$

Returning to the frequency–normal coordinate product part of the Hamiltonian, for one value of k the frequency–coordinate product may be written

$$\omega_k^2 Q_k^2 = (\omega_{0k} + \omega'_{0k}l)^2 Q_k^2 \tag{16.152}$$

where $\omega'_{0k} = (\frac{d\omega_{0k}}{dl})_0$, and where the subscript 0 refers to the unstrained crystal. The kth coordinate integral in the partition function is

$$I = \int_0^{a_0} dl \int_{-\infty}^{\infty} dQ_k e^{-(\omega_{0k}+\omega'_{0k}l)^2 Q_k^2/2k_BT} e^{-3pa_0^2l/k_BT}$$

$$= \int_0^{a_0} dl \frac{\sqrt{2\pi k_BT}}{\omega_{0k} + \omega'_{0k}l} e^{-3pa_0^2l/k_BT} \tag{16.153}$$

This integral may be more compactly written

$$I = \frac{\sqrt{2\pi k_BT}}{\omega_{0k}} \int_0^{a_0} \frac{e^{-al}}{1 + r_k l} dl \tag{16.154}$$

where $a = 3a_0^2 p/k_BT$ and $r_k = \omega'_{0k}/\omega_{0k}$ is the ratio of the slope to the frequency. There are N such integrands in the partition function, one for each normal coordinate. The partition function becomes

$$\Delta = -\frac{(2\pi k_B T)^N}{h^N \prod_k \omega_{0k}} \frac{3a_0^2}{V_0} e^{pV_0/k_B T} \int_0^{a_0} \frac{e^{-al}}{\prod_k (1 + r_k l)} dl \qquad (16.155)$$

The pre-integral factor can be written as

$$\frac{(2\pi k_B T)^N}{h^N \prod_k \omega_{0k}} \frac{3a_0^2}{V_0} e^{pV_0/k_B T} = \frac{3a_0^2}{V_0} e^{pV_0/k_B T} \prod_k \left(\frac{k_B T}{h\nu_{0k}} \right) \qquad (16.156)$$

since $\omega_{0k} = 2\pi\nu_{0k}$. This is a product of classical high temperature partition functions for the N vibrational frequencies ν_{0k}, each frequency measured in $\sec^{-1}$ units. This product is independent of the pressure, as are the other factors in equation (16.156).

The product in the denominator of the integrand in equation (16.154) contains all powers of the strain up to and including l^N. Retention of all these powers would be inconsistent with the linear nature of this analysis. Dropping terms of order l^2 or higher, the product may be written

$$1 + \left(\sum_k r_k \right) l = 1 + rl \qquad (16.157)$$

with evident definition of r as the sum of the ratios. Although each ratio of the slope to the frequency is itself a small quantity, it does not follow that a large sum of very small quantities is necessarily a small quantity, and it is therefore necessary to avoid the temptation to expand the denominator as the power series $1 - rl + r^2 l^2 + \cdots$. The partition function can now be written as

$$\Delta = -\frac{3a_0^2}{V_0} e^{pV_0/k_B T} \prod_k \left(\frac{k_B T}{h\nu_{0k}} \right) \int_0^{a_0} \frac{e^{-al}}{1 + rl} dl \qquad (16.158)$$

Changing variables to $y = 1 + rl$, the partition function becomes

$$\Delta = -\frac{3a_0^2}{V_0} \frac{e^{-pV_0/k_B T}}{r} \prod_k \left(\frac{k_B T}{h\nu_{0k}} \right) e^{a/r} \int_1^\infty \frac{e^{-ay/r}}{y} dy \qquad (16.159)$$

The upper limit of integration is really $1 + a_0 r$. However, the quantity r is really large, and the value of the integrand at large values is nearly zero. Little error can accdrue from replacing $1 + a_0 r$ with infinity. Changing variables again, let $t = ay/r$, and the partition function becomes

$$\Delta = -\frac{3a_0^2}{V_0} e^{pV_0/k_B T} \frac{e^{a/r}}{r} \prod_k \left(\frac{k_B T}{h\nu_{0k}} \right) \int_{a/r}^{\infty} \frac{e^{-t}}{t} dt \qquad (16.160)$$

The integral exists, as the integrand diverges only at $t = 0$, and very nearly represents the positive half of a Dirac delta function centered on zero. This deceptively simple integral resists integration in closed algebraic form. The integration can be accomplished in terms of the exponential integral Ei. The result is given by Abramowitz and Stegun[10] as

$$\int_{a/r}^{\infty} \frac{e^{-t}}{t} dt = -Ei(-a/r) \qquad (16.161)$$

Abramowitz and Stegun remark that the function $E_1(x)$ is often used to represent $-Ei(-x)$. Defining x as a/r, equation (16.161) allows evaluation of the partition function as

$$\Delta = \frac{3a_0^2}{V_0} e^{pV_0/k_B T} \frac{e^x}{r} E_1(x) \prod_k \left(\frac{k_B T}{h\nu_{0k}} \right) \qquad (16.162)$$

Multiplying the expression by p/p and $k_B T/k_B T$ does not alter the value but allows it to be expressed as the obviously dimensionless function

$$\Delta = \frac{k_B T}{pV_0} e^{pV_0/k_B T} x e^x E_1(x) \prod_k \left(\frac{k_B T}{h\nu_{0k}} \right) \qquad (16.163)$$

since the quantity x is dimensionless. The pressure dependence is hidden in the parameter x. This partition function is of the form $\Delta = constant \times f_1(p) \times f_2(x)$. Using equation (16.148), the ensemble average volume per unit cell is therefore

$$\langle V \rangle = -k_B T \left(\frac{\partial \ln f_1(p)}{\partial p} \right)_{T,N} - k_B T \left(\frac{\partial \ln f_2(x)}{\partial x} \right)_{T,N} \frac{dx}{dp} \qquad (16.164)$$

Since $x = 3a_0^2 p/rk_B T$ for an isotropic crystal with a constant lattice constant a_0, $\frac{\partial x}{\partial p} = 3a_0^2/rk_B T$. The lattice constant will be held constant in the following, since the effect of compression is measured by the strain. The logarithms are

$$\ln f_1(p) = -\ln p - \frac{pV_0}{k_B T} + \ln \frac{k_B T}{V_0} \tag{16.165}$$

$$\ln f_2(x) = \ln(x) + x + \ln(E_1(x)) + \sum_k \ln\left(\frac{k_B T}{h\nu_{0k}}\right) \tag{16.166}$$

In order to use this relation, we need the derivative of $E_1(x)$ given by Abramowitz and Stegun[10] as

$$\frac{d}{dx} E_1(x) = -\frac{e^{-x}}{x} \tag{16.167}$$

The derivative is

$$\left(\frac{\partial \ln f_2(x)}{\partial x}\right)_{T,N} = \frac{1}{x} + 1 - \frac{e^{-x}}{x E_1(x)} \tag{16.168}$$

and the constant lattice constant contribution to the average volume becomes

$$\langle V \rangle = V_0 + \frac{k_B T}{p} - \frac{3a_0^2}{r}\left(\frac{1}{x} + 1 - \frac{e^{-x}}{x E_1(x)}\right) \tag{16.169}$$

But $\frac{3a_0^2}{r} \times \frac{1}{x} = \frac{k_B T}{p}$, and the terms in $\frac{k_B T}{p}$ exactly cancel, leaving

$$\langle V \rangle = V_0 - \frac{3a_0^2}{r}\left(1 - \frac{e^{-x}}{x E_1(x)}\right) \tag{16.170}$$

The compressibility is essentially given by the derivative

$$\left(\frac{\partial \langle V \rangle}{\partial p}\right)_{T,N} = \left(\frac{\partial \langle V \rangle}{\partial x}\right)_{T,N}\left(\frac{\partial x}{\partial p}\right)_{a_0}$$

$$= -\frac{9a_0^4}{r^2 k_B T}\frac{e^{-x}}{x E_1(x)}\left[1 + \frac{1}{x} - \frac{e^{-x}}{x E_1(x)}\right] \tag{16.171}$$

The strained volume does not vary much from the initial volume and this contribution may be approximately written as

$$\chi = \frac{9a_0^4}{V_0 r^2 k_B T} \frac{e^{-x}}{x E_1(x)} \left[1 + \frac{1}{x} - \frac{e^{-x}}{x E_1(x)}\right] \tag{16.172}$$

The multiplier $\frac{9a_0^4}{V_0 r^2 k_B T} \frac{p}{p} \frac{k_B T}{k_B T} = \frac{3a_0^2 p}{r k_B T} \frac{3a_0^2 k_B T}{V_0 r k_B T p} = x \frac{3a_0^2 p}{r k_B T} \frac{k_B T}{p V_0} \frac{1}{p} = x^2 \frac{k_B T}{p V_0} \frac{1}{p}$. All the other entries in equation (16.172) are dimensionless. Hence the units of χ are those of reciprocal pressure. The final expression for the constant lattice constant contribution to the compressibility due to the acoustic branches is then

$$\chi = \frac{1}{p} \frac{k_B T}{p V_0} \frac{e^{-x}}{E_1(x)} \left[1 + x - \frac{e^{-x}}{E_1(x)}\right] \tag{16.173}$$

This equation displays a behavior analogous to that of an ideal gas as shown in Figure 16.14, and might therefore be called a phonon gas contribution. In this connection, we remark that equation (16.170) may also be expressed as

$$\langle V \rangle = V_0 - \frac{k_B T}{p} \left(x - \frac{e^{-x}}{E_1(x)}\right) \tag{16.174}$$

and this may be regarded as the equation of state of the phonon gas.

The calculation required to plot the function in equation (16.173) in Figure 16.14 was done using the expression given by Abramowitz and Stegun,[10] $E_1(x) = -\gamma - ln(x) - \sum_{n=1}^{\infty} \frac{(-1)^n x^n}{n \times n!}$, where γ is Euler's constant, $0.5772156649\ldots$

Discussion

All crystals have acoustic modes, but none are spectroscopically active in any crystal. There is no information other than the speed of wave propagation, given for argon by Peterson, Batchelder and Simmons[9] as 1640 and 944 ms^{-1} at 0 K for the longitudinal and transverse acoustic branches in this face centered crystal, leading to an estimate of the maximum frequencies of 51.5 and 29.7 cm^{-1}. This result was obtained by assuming the shortest wavelength the lattice will support is twice the lattice constant, corresponding to adjacent argon atoms travelling in opposite directions. These frequencies correspond to characteristic temperatures of 74 and 43 K.

If equation (16.173) has validity, it should calculate the compressibility of crystalline argon at temperatures below the triple point near 83 K. At 50 K, Peterson, Batchelder and Simmons[9] reported that the compressibility was near 4.37×10^{-11} cm^{2}dyne^{-1}, or 4.37×10^{-10} Pa^{-1}. The lattice constant at 50 K is 5.3684 Angstrom. All of the values of compressibility given by Peterson, Batchelder and Simmons would not have a contribution from equation (16.173), for all were calculated from their careful plots of lattice parameter against pressure. The slopes of these plots are all negative and increasingly so with increasing temperature, as the lattice expands, but the variation in the slopes (da_0/dp) with temperature is sigmoidal, not linear. Their tabulated values of the compressibility were calculated from the relation

$$\chi = \frac{3}{a_0} \left| \frac{da_0}{dp} \right| \tag{16.175}$$

None of their compressibilities were obtained by direct measurement. At temperatures near 4 K almost all vibrational contributions are frozen out, except perhaps for a few very low frequencies near the center of the first Brillouin zone. At 50-70 K, the acoustic modes are largely contributing, but equation (16.173) is still not really applicable. Nonetheless the equation makes the point that the compressibility is not given by equation (16.175) alone. There are other contributors concerned with storage of the energy of compression in the vibrational structure of the crystal.

In calculating the parameter x in equation (16.173), the puzzle point is what should be used for the ratio r? Some crystals have Raman active optical acoustic lattice modes, and the pressure dependence of some of these have been reported. It is possible to crystallize both argon and xenon in hexagonal closest packing structures, and this crystal has optical lattice vibrations that use the same masses and force constants the acoustic branches do. The longitudinal optic branch is Raman active at the zone center (k = 0 selection rule) and can be followed as a function of pressure. Both argon and xenon show an increase in Raman frequency with increasing pressure, though with different rates.[11,12] In boron triiodide, Anderson and Lettress[13] report that due to the crystal structure (D_{2h} subgroup of the space group, two molecules per unit primitive cell), the longitudinal optical mode is Raman active and could be followed as a function of pressure up to 10 GPa. The frequency varies strictly linearly over the range from 0.1 GPa to 10 GPa. The ratio[13] r_0 is 5.71/25.9 GPa^{-1}. Frequencies in the branch away from the zone center are silent. In calculating x, the assumptions are that all frequencies in the acoustic branches have the same ratio, and that the pressure dependence of the frequencies are similar to that of the Raman active translational modes of boron triiodice The calculation is not especially sensitive to the value of r, and is more sensitive to the value of the lattice constant. As an example of the sensitivity of the calculation to the value of r, consider the compressibility of a nanocrystal in comparison with that of a macrocrystal. If the macrocrystal has Avogadro's number of molecules, the nanocrystal will have a nanomole of molecules. The number of longitudinal acoustic vibrations for the macrocrystal is of the order of Avogadro's number, the highest frequency lying very nearly on the edge of the first Brillouin zone. The nanocrystal has 10^9 fewer vibrations in this branch, and the highest frequency does not approach the zone boundry as closely. The effect of fewer frequencies diminishes the value of r by a factor of 10^9. This causes the quantity x to increase by this factor, and as a resilt the compressibility increases somewhat. Using twice the value of the lattice constant for argon at 0 K (5.31108 Angstrom) and a temperature of 300 K, and estimating r from the boron triiodide data of Anderson and Lattress,[13] the ratio of compressibility for the nanocrystal to that for the macrocrystal results in about 1.5_2, increasing as the crystal becomes smaller. The physical reason for this behavior is clear. The compression work that appears in the crystal when the pressure is increased is deposited in part in contraction of the lattice constant, and in part in the vibrational structure of the crystal. Nanocrystals have fewer vibrations to absorb the energy change, and the smaller the crystal, the fewer vibrations there are to absorb the compression work. The vibrational work component is partially absorbed by increasing the vibrational frequencies, but some also goes into increasing the excitation of the quantized vibrations. A somewhat related argument is made in a series of papers by Karasevskii and Lubashenko.[5]

In this connection, we note that the increase in frequency with increasing pressure is universal, for the increased pressure must result in a smaller space for vibrational amplitudes to occur. The harmonic oscillator is dominated by a parabolic potential function. If this is crudely modeled by a box potential, the energy levels in the box are given by $E_n = n^2h^2/8mL^2$, where m is the effective mass of the vibration and L the box dimension, while $n = 1, 2, 3 \ldots$ If L is decreased, all the energy levels must increase. This implies that the adiabatic compressibility must always exceed the isothermal, for in the absence of heat flow, each level carries with it the population appropriate to the temperature of the uncompressed crystal. Imposition of isothermal conditions on the crystal allows heat flow across the boundary of the crystal and surroundings, as the populations relax to the new higher energy levels. The Heisenberg uncertainty principle thus requires that there be a vibrational contribution to the isothermal and adiabatic compressibilities, independently of the anharmonicity, which does not enter any of the above arguments.

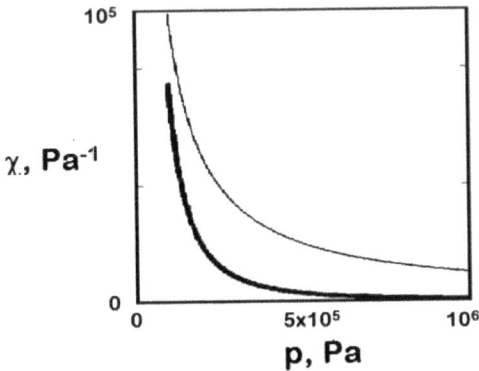

Fig. 16.14.   Comparison of equation (16.173) with the compressibility of an ideal gas. Heavy line, equation (16.173) for an imaginary argon with unrealistically low acoustic frequencies at 50 K. Thin line, ideal gas The similarity in behavior of the two functions suggests that a pressure exerted by a phonon gas contributes to the compressibility of crystals.

The behavior of equation (16.173) as a function of pressure is compared with the behavior of an ideal gas over the same pressure range in Figure 16.14. With the parameters used, equation (16.173) lies well below the ideal gas at all pressures, but unlike the ideal gas, equation (16.173) is strongly dependent on temperature. The trace representing equation (16.173) in Figure 16.14 is for a temperature of 50 K. If the temperature is reduced to 4 K, the resulting trace drops substantially, nearly disappearing into the horizontal axis as most of the vibrational contributions become frozen out.

Problems

1. Give the Einstein heat capacity at constant volume as a function of pressure, assuming that the pressure dependence of the frequency is $\nu = \nu_0 + \frac{d\nu}{dp}p$.
2. Obtain the characteristic frequencies for a linear chain of four identical masses m and potential function $V = \frac{f}{2}[(x_2 - x_1)^2 + (x_3 - x_2)^2 + (x_4 - x_3)^2]$.
3. Show that the low temperature heat capacity of a graphene sheet is proportional to T^2, and obtain an expression for the characteristic temperature Θ_D of this sheet.
4. Evaluate the integral $\int_0^\infty x e^{-nx} dx$ for $n = 1, 2, 3, 4, \ldots$

References

1. E. B. Wilson, Jr., J. C. Decius and P. Cross, "Molecular Vibrations", McGraw-Hill, New York, 1956.
2. L. B. W. Jolley, "Summation of Series", Dover, New York, 1964, formula 305.
3. M. Abramowitz and I. A. Stegun, Ed., Handbook of Mathematical Functions, AMS 55, U. S. Department of Commerce, National; Bureau of Standards, p. 998.
4. W. S. Corak, M. P. Garfunkel, C. B. Satterswaithe and A. Wexler, Phys. Rev. **98**, 1699 (1955).
5. A. I. Karasevskii and V. V. Lubashenko, Low Temp. Phys. **35**, 275 (2009).
6. R. Astala, S. M. Auerbach and P. A. Monson, Phys. Rev. B **71**, 014112 (2005).

7. A. Anderson, Trends in Chem. Phys. **10**, 225 (2002).

8. A. Congeduti, P. Postorino, M. Nardone and U. Buontempo, Phys. Rev. B **65**, 014302 (2001).

9. O. G. Peterson, D. N. Batchelder and R. O. Simmons, Phys. Rev. **150**(2), 703–11 (1966).

10. M. Abramowitz and I. A. Stegun, Nat'l Bur. Stds. AMS 55, 1964, 5th Printing, p. 228.

11. H. Shimizu, N. Wada, T. Kume, S. Sasaki, Y. Yao and J. S. Tse, Phys. Rev. B 77, 052101 (2008).

12. Y. A. Freiman, A. F. Goncharov, S. M. Tretyak, A. Grechnev, J. S. Tse, D. Errandonea, H. K. Mao and R. J. Hemley, Phys. Rev. B 78, 014301 (2008).

13. A. Anderson and L. Lattress, J. Raman Spectrosc. **33**, 173 (2002).

Index

www.ingramcontent.com/pod-product-compliance
Lightning Source LLC
Chambersburg PA
CBHW070743220326
41598CB00026B/3727